Electron transfer reactions

For Elizabeth

Electron transfer reactions

R. D. Cannon
School of Chemical Sciences
University of East Anglia

Butterworths

LONDON BOSTON
Sydney Durban Wellington Toronto

THE BUTTERWORTH GROUP

United Kingdom London	**Butterworth & Co (Publishers) Ltd** 88 Kingsway, WC2B 6AB
Australia Sydney	**Butterworths Pty Ltd** 586 Pacific Highway, Chatswood, NSW 2067 Also at Melbourne, Brisbane, Adelaide and Perth
Canada Toronto	**Butterworth & Co (Canada) Ltd** 2265 Midland Avenue, Scarborough, Ontario, M1P 4S1
New Zealand Wellington	**Butterworths of New Zealand Ltd** T & W Young Building, 77–85 Customhouse Quay, 1, CPO Box 472
South Africa Durban	**Butterworth & Co (South Africa) (Pty) Ltd** 152–154 Gale Street
USA Boston	**Butterworth (Publishers) Inc** 10 Tower Office Park, Woburn, Massachusetts 01801

All rights reserved. No part of this publication may be reproduced or transmitted in any form or by any means, including photocopying and recording, without the written permission of the copyright holder, application for which should be addressed to the Publishers. Such written permission must also be obtained before any part of this publication is stored in a retrieval system of any nature.

This book is sold subject to the Standard Conditions of Sale of Net Books and may not be re-sold in the UK below the net price given by the Publishers in their current price list.

First published 1980

© Butterworth and Co (Publishers) Ltd, 1980

ISBN 0 408 10646 8

British Library Cataloguing in Publication Data

Cannon, R D
 Electron transfer reactions.
 1. Chemical reaction, Conditions and laws of
 2. Electrons
 I. Title
 539.7'2112   QD501   79–41278

ISBN 0–408–10646–8

Typeset by The Macmillan Co. of India Ltd, Bangalore
Printed and bound in England by the Camelot Press, Southampton

Preface

This book describes the mechanisms of electron transfer reactions between metal ions in solution, and the closely-related phenomena of electron exchange between atoms or molecules in the gaseous phase, and in the solid state. The first of these subjects spans the traditional disciplines of inorganic and physical chemistry; the other two are usually considered as aspects of physics. Yet there are basic principles common to all three which justify an attempt at a unified treatment, and in the physical parts of the book, especially, I have tried to draw these out.

With such a wide field, omissions are inevitable, and I am aware that some are more easily regretted than justified. The omission of oxidation–reduction reactions of non-metals is to some extent arbitrary. Electrochemical reaction mechanisms are not treated in detail since they are already well covered elsewhere. Electron transfer in biochemical systems is touched on only briefly. However, studies in the inorganic field have contributed a great deal to the understanding of electron processes in general, and I hope that workers in other disciplines will find the book useful in that respect.

Coverage of the literature, while always selective, has been attempted to the end of 1976; though some later references have also been incorporated. It should be mentioned here that while this book has been in press there have been important advances in the theory of electron transfer reactions, particularly in the study of non-adiabatic reactions by the use of various forms of state-to-state theory. It seems too early to assess the impact these theories will have on the experimental side of the subject. This and other recent work is being regularly reviewed in two of the Chemical Society's Specialist Periodical Reports series: *Inorganic Reaction Mechanisms* and *Electronic Structure and Magnetism of Inorganic Compounds*.

Throughout the book SI units are used, except that the unit of energy is the calorie (1 calorie = 4.184 joule) and the symbol M is used for moles per litre or, more strictly, mol dm^{-3}. Equations involving electrical or magnetic quantities are expressed in the rationalised four-quantity system, some of them apparently for the first time. To facilitate comparison with the older literature, the factors such as $4\pi\varepsilon_0$ or $L/4\pi\varepsilon_0$ are often printed separately in the formulae.

Many colleagues have contributed to this work with discussion and criticism and I am especially grateful to Dr P. Day, Professor W. C. E. Higginson, Dr D. R. Rosseinsky, Dr A. G. Sykes and Dr K. Wieghardt, who all read portions of the manuscript before publication.

The work of writing this book began in 1972 with the award of the first Butterworth Scientific Fellowship. I am grateful to Butterworths for their encouragement to begin a long-meditated project, and also to the University of East Anglia for periods of study leave.

R.D.C.

Contents

Part 1: Electron transfer in the gas phase   1

1 Electron transfer between atoms   3
1.1 Resonance transfer   3
    1.1.1 Qualitative description   3
    1.1.2 Outline of calculation: the two-state approximation   5
    1.1.3 Comparisons with experiment   8
    1.1.4 Rates of reaction   9
    1.1.5 Many-electron atoms   9
    1.1.6 Oscillatory effects due to resonance   10
    1.1.7 Extensions of theory   12
    1.1.8 Multiple electron transfer   14
    1.1.9 Extreme velocity ranges   14
1.2 Electron transfer between unlike atoms   15
    1.2.1 Qualitative description   15
    1.2.2 Semiquantitative models   18
    1.2.3 Quantum-mechanical calculations   22
    1.2.4 Accidental resonance   23
1.3 The Langevin model   24
1.4 Typical reaction rates   25

2 Electron transfer involving molecules   26
2.1 Formal classification of mechanisms   26
2.2 Description of principal reaction mechanisms   28
    2.2.1 Single event: electron transfer without atom transfer   28
    2.2.2 Single event: electron transfer with atom transfer   29
    2.2.3 Double event: electron transfer followed by atom transfer   32
2.3 Factors influencing electron transfer rates   33
    2.3.1 Analogies with atom–atom reactions   33
    2.3.2 The Franck–Condon principle   34
    2.3.3 Reactions with atom transfer: electronic state correlation   40
2.4 Large molecules   44

Part 2: Electron transfer in solution — 47

3 The reaction path (1): oxidation states — 49

- 3.1 Introduction and definitions — 49
  - 3.1.1 Reaction mechanisms — 49
  - 3.1.2 Kinetics — 49
  - 3.1.3 Electron transfer paths — 51
- 3.2 Non-complementary reactions — 52
  - 3.2.1 2:1 reactions — 52
  - 3.2.2 3:1 reactions — 63
- 3.3 Reagents of multiple valence — 68
  - 3.3.1 Reactions of a multi-electron and a one-electron reagent — 68
  - 3.3.2 Reactions of a multi-electron and a two-electron reagent — 72
  - 3.3.3 Reactions of a multi-electron and a three-electron reagent — 76
  - 3.3.4 Reactions between two multi-electron reagents — 76
- 3.4 Free energy profiles — 80
- 3.5 Double electron transfer — 84
  - 3.5.1 General — 84
  - 3.5.2 Examples — 87
- 3.6 Multiple electron transfer — 91
- 3.7 Other reaction pathways — 91
  - 3.7.1 Third-order rate laws — 91
  - 3.7.2 Fractional rate laws — 94
  - 3.7.3 Zero-order rate laws — 95

4 The reaction path (2): binuclear intermediates — 97

- 4.1 Introduction — 97
  - 4.1.1 Definitions — 97
  - 4.1.2 Historical — 97
  - 4.1.3 Outline of mechanisms — 98
- 4.2 The diffusion-controlled mechanism — 100
  - 4.2.1 Theory — 100
  - 4.2.2 Examples — 104
- 4.3 Rate-determining steps — some indirect arguments — 104
  - 4.3.1 Reaction rates — 104
  - 4.3.2 Isotope and non-bridging ligand effects — 106
  - 4.3.3 Activation enthalpies — 106
  - 4.3.4 Cyclic transition states — 108
- 4.4 Labile intermediates: kinetic evidence from catalysed pathways — 108
  - 4.4.1 General — 108
  - 4.4.2 Examples — 115
  - 4.4.3 Free-energy profiles — 121
- 4.5 Labile intermediates: direct observation — 124
  - 4.5.1 Precursor complexes — 125
  - 4.5.2 Successor complexes — 129

5 The mechanism of electron transfer — 132

5.1 Introduction — 132
    5.1.1 Historical — 132
    5.1.2 Summary of mechanisms (inner-sphere) — 134

5.2 Inner- and outer-sphere mechanisms: experimental evidence — 137
    5.2.1 Isolation of binuclear complex — 137
    5.2.2 Transfer of bridging group (inert reactants) — 141
    5.2.3 Transfer of bridging group (labile reactants) — 145
    5.2.4 Isomeric reaction products — 147
    5.2.5 Substitution-controlled electron transfer — 147
    5.2.6 Catalysis — 149
    5.2.7 Linear free energy relationships (LFER) — 149
    5.2.8 Other rate comparisons: bridging groups — 152
    5.2.9 Other criteria — 154

5.3 'Remote attack': experimental evidence — 155
    5.3.1 Inert binuclear complexes — 155
    5.3.2 Transfer of bridging group — 155
    5.3.3 Primary product inferred from decomposition products — 160
    5.3.4 Other evidence — 161

5.4 Outer-sphere mechanisms — 161
    5.4.1 Outer-sphere bridging — 166
    5.4.2 Specific orientation of reactants — 169
    5.4.3 Facial attack — 170

Part 3: Energetics of electron transfer — 173

6 Theory of electron transfer — 175

6.1 Models of the electron transfer process — 175
    6.1.1 Early work — 175
    6.1.2 The two-state description — 178

6.2 The thermal mechanism — 180
    6.2.1 Energy curves — 180
    6.2.2 Transition probability — 182

6.3 The activation process — 186
    6.3.1 Metal–ligand bond stretching — 186
    6.3.2 The solvent as dielectric continuum — 188
    6.3.3 The phonon model — 203

6.4 Inner- and outer-sphere energy terms — 204

6.5 Comparison with experiment (1): rate comparisons and correlations — 205
    6.5.1 The Marcus cross-relation — 205
    6.5.2 Effects of work terms — 207
    6.5.3 Linear free energy relations — 210
    6.5.4 Non-linear free energy relations — 213

		6.5.5 The 'anomalous region'	215
		6.5.6 Activation parameters	215
	6.6	Comparison with experiment (2): absolute rate calculations	217
	6.7	Other theories of electron transfer	218
	6.8	Connection with electrochemical kinetics	220

7 Bridged electron transfer: theory and experiment   223

    7.1 Theoretical models   223
          7.1.1 The polarisation model   223
          7.1.2 The three-state model   224
          7.1.3 Applicability of the models   226

    7.2 Review of theories   227
          7.2.1 Single exchange: the two-state model   227
          7.2.2 Resonance transfer   228
          7.2.3 Superexchange   229
          7.2.4 The chemical mechanism   230

    7.3 Evidence for chemical mechanisms   238
          7.3.1 Detection of intermediate   238
          7.3.2 Free energy relationships   241
          7.3.3 Kinetic isotope effect   247

    7.4 Bridging organic ligands: reducibility of bridging group   247
          7.4.1 Qualitative comparisons   248
          7.4.2 Quantitative correlations   248
          7.4.3 Acid catalysis   255

    7.5 Bridging by halide ions   257

8 Optical electron transfer   267

    8.1 Introduction   267

    8.2 Classification of optical electron transfer systems   268
          8.2.1 Structural classification   268
          8.2.2 Classification by symmetry   270
          8.2.3 The Robin and Day classification   270

    8.3 Theory   271
          8.3.1 Charge localisation: the two-state description   271
          8.3.2 Connection with electron transfer kinetics   272
          8.3.3 The optical transition   274
          8.3.4 The charge transfer absorption band   275
          8.3.5 Thermodynamic considerations   277
          8.3.6 Solvent effects   281

    8.4 Examples of optical electron transfer   282
          8.4.1 'Vertical' ionisation and attachment   282
          8.4.2 Outer-sphere electron transfer   284
          8.4.3 Bridged complexes   285
          8.4.4 Symmetrical inert complexes   287
          8.4.5 Directly bonded systems   291

8.5	Reorganisation energies: comparison with theory	295
	8.5.1 The separate sphere model	295
	8.5.2 The ellipsoid model	296
8.6	Photo-induced electron transfer reactions	296

9 Electron transfer in the solid state — 298

9.1	Outline of theory	298
	9.1.1 Electrical conduction	298
	9.1.2 The activation energy	299
	9.1.3 The hopping mechanism	300
	9.1.4 Connection with reaction kinetics	301
	9.1.5 Connection with optical spectroscopy	301
	9.1.6 The sign of the charge carrier	301
9.2	Examples	304
	9.2.1 Some outer-sphere systems	304
	9.2.2 Halide-bridged systems	305
	9.2.3 Mixed oxide systems	305

References 315
Index 339

Part 1
Electron transfer in the gas phase

Chapter 1
Electron transfer between atoms

1.1 Resonance transfer

1.1.1 Qualitative description

The simplest possible electron transfer reaction is the gas-phase process:

$$H + H^+ \to H^+ + H \tag{1.1}$$

To a chemist, equation (1.1) may not at first sight convey a chemical reaction at all. It is usually said that if in any system the electronic configuration can be written in more than one way, with no difference in energy, the true configuration is neither of the two, but is intermediate to them both and lower in energy than either. The condition of resonance has been loosely described as a situation in which the electron spends a certain fraction of its time in each of its component states; but many authors have warned against this description, sometimes in quite emotive terms[a].

Resonance, however, is a description of the stationary state, in which the nuclei are in fixed positions and the electron distribution is described by the wavefunction of lowest energy. But a system undergoing reaction is by definition a non-stationary state; it is, therefore, perfectly permissible to define a rate constant for the reaction shown in equation (1.1). The rate constant can be calculated from quantum mechanics, and can be determined by experiment. It is of course necessary to remember that the nuclei are indistinguishable, and for that reason even the description of experimental results must be rather carefully phrased[b, c].

(a) Eyring, Walter and Kimball, discussing equation (1.1) conclude that 'from this view-point (*which should not be taken too literally*), the electron oscillates between A and B, . . . ' (reference 397, p. 199). Likewise Coulson, discussing the H_2 molecule: 'People sometimes say that the electrons exchange, or "trade", places with one another. *Such language is full of dangers* . . . ' (reference 256, p. 11); but note that Coulson in particular does make clear the nature of the dangers: resonance is not a 'phenomenon' when it arises in the mathematics of stationary states. See reference 256, pp. 76 and 113

(b) For accounts of the theory of atom—atom collisions and related processes see especially references 87, 92, 154, 220, 243, 525, 676, 737, 738, 769, 792, 898, 1111. For reviews of theoretical and experimental results see references 57, 97, 430, 523, 524, 535, 553, 809, 966, 1079

In a rigorous quantum-mechanical treatment a single particle is regarded as a wave packet, and a beam of particles as a train of waves, each characterised by a certain de Broglie wavelength and energy. The interaction of two beams, or of one beam with an assembly of randomly moving particles, is treated by the application of the Schrödinger wave equation. The same basic theory[769, 792, 989, 1111] is applicable to interactions of radiation with matter, of electrons with atoms and molecules, or, as in the case of interest here, of atoms and molecules with each other. In the case of the reaction shown in equation (1.1), if a beam of hydrogen ions crosses a beam of hydrogen atoms the emergent beams will contain a certain fraction of atoms and ions respectively. Nothing need be (or indeed can be) said about the identity of individual electrons or nuclei before and after the crossing point, and the question of electron transfer between pairs of atoms does not arise.

This is the most general statement of the problem, applicable to the whole range of beam energies. However, as long as the energies are below a certain limit, the nuclei may be treated classically, as particles with definite trajectories, while the wave-mechanical description is retained for the electrons. In this energy range it is permissible to speak of 'electron transfer' and to write equations such as equation (1.1).

The process is sometimes described as follows[d]. Consider the situation in which the nuclei, A and B, are infinitely far apart. The electron, in order to have the lowest possible energy, must be in the vicinity of one of them, say nucleus A, and must occupy a 1s orbital. Suppose that it were then possible to move the nuclei to within a finite distance of each other, and then to remove the restriction on the location of the electron. The electron would oscillate between the nuclei—or to put it more correctly, the wavefunction Ψ, which at any instant in time describes the probability of finding the electron within any specified volume element in space, would change smoothly back and forth, with a definite frequency, from a function with maximum electron density in

---

(c) Experimental methods used in the study of gas-phase electron transfer have been extensively reviewed elsewhere (cf. footnote (b), p. 3), and need not be discussed in detail here. The most informative experiments are those in which the reagents are generated in the form of a parallel beam of atoms or molecules of variable and controlled velocity. If, for example, a beam of ions is led into a gas containing the other reagent as neutral atoms or molecules, it is possible to measure the total extent of reaction either as the attenuation of the ion current, or as the appearance of product species, both in the direction of the original beam; or more generally to monitor the scattering products or unreacted species at various angles to the beam. The reaction cross section is measured as a function of beam energy, which in turn is related to particle velocity

More sophisticated techniques employ two beams, so that the angle of crossing is an additional experimental variable[809]. 'Conventional' methods involving mixing reagent gases under near-equilibrium conditions have not generally been employed (but for these, see Chapter 2, footnote (a), p. 26). Resonance methods such as nmr are applicable but so far there is only one report[1079]

(d) See reference 397, p. 534 and reference 256, p. 76

the vicinity of atom A to one similarly localised at atom B. The frequency would depend on the internuclear distance, and would tend to zero at the limit of infinite separation. Consider now the reacting system with the nuclei in relative motion, and, for simplicity, consider them moving in straight lines. The electron wavefunction oscillates continuously, but the frequency rises from zero at the initial infinite separation of the nuclei to a maximum at the distance of closest approach, then declines to zero at the final infinite separation. There is a certain probability that the whole encounter leads to a net transfer of the electron, and this probability depends on the details of the speeds and trajectories of the nuclei.

In the simplest quantum-mechanical calculations, the nuclei are assumed to move slowly compared with the electrons so that, at any instant, the reacting pair can be considered as a diatomic molecule (Born–Oppenheimer approximation). This molecule has a number of allowed energy states, but as a further approximation, the total wavefunction is expressed in terms of the two lowest states only.

1.1.2 Outline of calculation: the two-state approximation[e]

At any instant in time, the reacting system of equation (1.1) constitutes a molecule-ion, H_2^+, characterised by the internuclear distance R. The total wavefunction ψ and Hamiltonian H obey the Schrödinger wave equation

$$H\psi = E\psi \tag{1.2}$$

which has a set of solutions ψ_m to each of which there corresponds an energy E_m given by

$$E_m = \int \psi_m H \psi_m^* \, d\tau \tag{1.3}$$

where ψ_m^* is the complex conjugate of ψ_m and τ includes all relevant coordinates. The wavefunctions ψ_m can be classed as symmetric or anti-symmetric with respect to interchange of the nuclei. As R increases to infinity, the wavefunctions converge in pairs to those of the isolated hydrogen atom; as R decreases to zero they converge to those of the united atom $^2\text{He}^+$. As an approximation, we consider only the lowest energy states, which have wavefunctions ψ_+ (symmetric, bonding state) and ψ_- (anti-symmetric, anti-bonding state). These may be combined as

$$\psi_+ + \psi_- = \psi_A \tag{1.4}$$

$$\psi_+ - \psi_- = \psi_B \tag{1.5}$$

where ψ_A and ψ_B are functions which, in the limit of $R \to \infty$, converge to the orbitals ϕ_A and ϕ_B of the electron localised in the states $H \ldots H^+$ and

---

(e) For a different method of calculation see reference 589

$H^+ \ldots H$. To these functions there correspond energies E_A and E_B respectively,

$$E_A = \int \psi_A H \psi_A^* \, d\tau \qquad (1.6)$$

$$E_B = \int \psi_B H \psi_B^* \, d\tau \qquad (1.7)$$

which are of course equal, and the energy separation between the bonding and anti-bonding states is

$$\Delta E_\pm = E_- - E_+ = 2(\beta - SE_A)/(1-S^2) \qquad (1.8)$$

where

$$S = \int \psi_A \psi_B^* \, d\tau \quad \text{and} \quad \beta = \int \psi_A H \psi_B^* \, d\tau \qquad (1.9)$$

Still considering the static system, the description can be restated using the time-dependent Schrödinger equation. In place of equation (1.2) we have

$$H\Psi = (ih/2\pi)\partial\Psi/\partial t \qquad (1.10)$$

with solutions

$$\Psi_m = \psi_m \exp(-2\pi i E_m t/h) \qquad (1.11)$$

where the time-independent parts ψ_m, and the energies E_m, are the same as before.

Considering the system now as a reacting pair, with the nuclei in relative motion, the total Hamiltonian H_T varies with time. The time-dependent perturbation theory, however, assumes that H_T can be expressed as

$$H_T = H + H' \qquad (1.12)$$

where H is the same time-independent term as before and H' is a time-dependent term which is assumed to be relatively small. The total wavefunction Ψ_T obeys equation (1.10). In terms of the states previously chosen it may be written as

$$\Psi_T = \alpha_+ \Psi_+ + \alpha_- \Psi_- \qquad (1.13)$$

or

$$\Psi_T = c_A \Psi_A + c_B \Psi_B \qquad (1.14)$$

where $c_A = \frac{1}{2}(\alpha_+ + \alpha_-)$, $c_B = \frac{1}{2}(\alpha_+ - \alpha_-)$. The wavefunctions Ψ are all of the time-dependent form analogous to equation (1.11), but the coefficients α and c are time-dependent only in that they are functions of R, which is varying with the time.

It is then argued that if ϕ_A represents the initial state of the system (i.e. when

$t = -\infty$), the probability P that the final state ($t = +\infty$) will be represented by ϕ_B, is

$$P = c_B(\infty)c_B^*(\infty) \tag{1.15}$$

where $c_B(\infty)$ is the limiting value of c_B at $t = \infty$ and $c_B^*(\infty)$ is its complex conjugate. The detailed calculation, not reproduced here, leads to simultaneous linear equations in $\partial c_A/\partial t$ and $\partial c_B/\partial t$; and when these are integrated to yield c_B the final result is

$$P = \sin^2 Q \tag{1.16}$$

where

$$Q = (\pi/h) \int_{-\infty}^{+\infty} \Delta E_\pm \, dt \tag{1.17}$$

To apply these equations it is necessary to have expressions for $\Delta E_\pm$ as a function of R, and for R as a function of t.

The simplest calculations of $\Delta E_\pm$ are based on the Heitler–London approximation[256, 397, 610]. Other methods have also been used, and, indeed in the case of the H_2^+ system the wave equation can be solved exactly in closed form by the use of a suitable coordinate system[89, 397, 610].

The simplest calculations of $R(t)$ are based on the model shown in *Figure 1.1*[492]. Nuclei A and B move in straight lines and the distance of closest approach is defined as the *impact parameter*, b. The coordinates are chosen so that nucleus A is located at distance b from the origin while nucleus B moves along the x-axis with velocity v from $x = -\infty$ to $x = +\infty$. The probability P of net electron transfer will in general depend on b, and the integral over all possible values of b gives the cross section, σ:

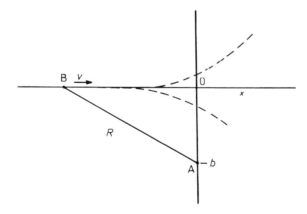

Figure 1.1 Encounter between two atoms. Nucleus A is located at the point $(0, -b)$ where b is the impact parameter; nucleus B approaches along the x axis, with velocity v. The broken curves are typical 'classical' trajectories for repulsive or attractive forces between the atoms

$$\sigma(v) = 2\pi \int_0^\infty P(v, b) b \, db \tag{1.18}$$

This model gives

$$Q = (2\pi/hv) \int_0^\infty \Delta E_\pm (R^2 - b^2)^{-1/2} R \, dR \tag{1.19}$$

More generally, if there are net attractive or repulsive forces, the trajectories will curve near the point of collision, but the impact parameter can still be defined as the minimum separation which would have occurred if the particles had continued to move along their original straight lines.

1.1.3 Comparisons with experiment

For the $H^+ + H$ reaction, Gurnee and Magee[492] performed calculations at two velocities, corresponding to energies of 0.13 and 52 eV. They used first-order perturbation theory, the Heitler–London approximation for $\Delta E_\pm$, and linear trajectories. Dalgarno and Yadav[272] made calculations over the kinetic energy range $E_{kin} = 1 - 10^5$ eV. For this range they could assume classical motion of the nuclei, and separation of the electronic and nuclear motions, but at the higher energies they found it necessary to allow for deflection of the trajectories as a result of the very violent impact. They used first-order perturbation theory, but adopted accurately calculated values of $\Delta E_\pm$ for values of R up to nine times the Bohr radius and Heitler–London values only for greater distances. They also introduced a correction from second-order perturbation theory. The results of these calculations agree well with experimental data[414, 415], as shown in *Figure 1.2*.

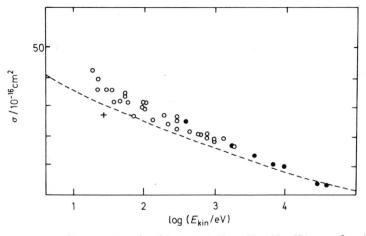

Figure 1.2 Cross sections for the reaction $H^+ + H \to H + H^+$ as a function of kinetic energy E_{kin}. (Based on Figure 12.6 of reference 525). Experimental data points o from reference 414, points ●, reference 415. The broken curve shows the theoretical prediction of Dalgarno and Yadav[272]; the point + shows that of Gurnee and Magee[492]

1.1.4 Rates of reaction

The second-order rate constant for the gas-phase reaction can be related to the cross section provided the distribution of velocities is known. If dN is the number of encounters per unit volume per unit time, with relative velocities in the range v to $(v+dv)$, we have

$$k = L^{-1}[A]^{-1}[B]^{-1} \int dN \qquad (1.20)$$

where $[A]$ and $[B]$ are molar concentrations and L is Avogadro's constant. The rate constant k is that normally used in chemistry, and has the dimensions (concentration)$^{-1}$ × (time)$^{-1}$, with the conventional units litre mol^{-1}s^{-1} which we shall write as M^{-1}s^{-1}. Rate constants quoted in the physics literature are usually the quantity $L^{-1}k$, with dimensions (volume) × (time)$^{-1}$, and units cm^3 s^{-1} (or more explicitly cm^3 molecule^{-1} s^{-1}). Again assuming straight-line trajectories and constant velocities, the kinetic theory of gases gives an expression for dN as a function of temperature[858] which leads to

$$k = L(\mu_{AB}/k_B T)^{3/2}(2/\pi)^{1/2} \int_0^\infty \sigma(v)\exp(-\mu_{AB}v^2/k_B T)v^3\, dv \qquad (1.21)$$

where μ_{AB} is the reduced mass of the reacting pair and k_B is Boltzmann's constant. The problem of low-energy collisions with curved trajectories has also been discussed[91].

For the special case of collisions between hard spheres, where the cross section becomes independent of velocity, equation (1.21) reduces to

$$k = L(8k_B T/\pi\mu_{AB})^{1/2}\sigma \qquad (1.22)$$

This expression can be used to obtain a rough estimate of the rate constant for the reaction of equation (1.1) at room temperature. If we consider the H$^+$ + H system as an ideal gas, the mean speed of the particles is $\bar{v}' = (2k_B T/\mu_{AB})^{1/2}$ and the mean relative velocity [858] is $\bar{v} = \bar{v}'/\sqrt{2} = 1.6 \times 10^3$ m s^{-1}. From Gurnee and Magee's calculations[492] the cross section at this velocity is of the order of 10^{-18} m^2. In other words, electron transfer takes place on average over distances of about 10 Å. From equation (1.22), the rate constant is $k \simeq 10^{13}$ M^{-1}s^{-1}.

1.1.5 Many-electron atoms

A number of other symmetrical resonant processes have been observed, with rare gas or metal atoms

$$A^+ + A \to A + A^+ \qquad (1.23)$$

where A = He, Ar etc., Li, Na, etc., or Hg, etc. The simplest theoretical treatment[492, 874, 969, 970] assumes that the transferring electron moves independently of all others and, as before, has only two states, symmetric and anti-symmetric. The treatment is then exactly the same as for the reaction $H^+ + H$ and leads to the same equations (equations 1.16–1.19), so that the problem of calculating cross sections reduces again to the problem of calculating $E_\pm$ as a function of R. Various wavefunctions have been used for this purpose including approximately hydrogenic functions, various empirically modified functions[969] and a model which considers in some detail the screening effect of the inner electrons[970].

Accurate solutions have also been obtained for the three-electron case ($He^+ + He$)[759]. A different procedure applicable to heavy atom systems has been developed by Smirnov[953, 954], using a surface integration in place of equation (1.9). Smirnov's method allows for some mutual distortion of the atomic wavefunctions which the LCAO approximation naturally excludes, but a recent analysis[548] has shown the two methods to be closely related. It is interesting to note that in the energy regions most studied by experiment, the bulk of the electron transfer occurs over relatively long distances (up to ten times the Bohr radius); thus, for a reacting system the LCAO is a much better approximation to the actual wavefunction than it is in the case of the equilibrium configuration of a normal diatomic molecule.

Agreement with experiment is generally good. The data have been compiled and reviewed several times[524, 535, 548, 874]. A general trend is that for different reacting systems, at a given velocity, reaction cross sections increase with increasing ease of ionisation of the atom A. This intuitively reasonable result has been confirmed for a large number of cases (*Figure 1.3*) and is in accord with detailed theoretical predictions. Indeed, in most treatments, the ionisation energies of the atoms are built into the empirical equations used for the atomic orbitals[874].

1.1.6 Oscillatory effects due to resonance

It will be noted that the probability P is an oscillating function of the impact parameter (equation 1.16). This stems directly from the oscillating, time-dependent wavefunction. To put it loosely, if the electron is continually jumping from one atom to the other, its position depends on whether the length of time spent in the encounter corresponds to an odd or even number of jumps. This effect has actually been observed directly, in two classes of experiment: low-angle scattering measurements and total cross-section measurements.

Low-angle scattering measurements

If the reacting particles impinge at sufficiently high energy, and if the products are collected at one specific angle of scatter, then by varying the impact energy the distance of closest approach can be varied. (Alternatively, it may be said that if the nuclei pass sufficiently closely, the impact parameter approximates to zero, hence varying the energy amounts to varying the duration of the collision.) Lockwood and Everhart[687] studied the $H^+ + H$ charge transfer

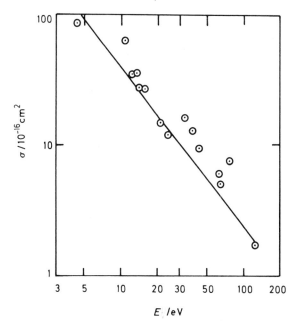

Figure 1.3 Correlation between cross section σ for symmetrical exchange reactions $A^+ + A \to A + A^+$, and ionisation energy E_i for the process $A \to A^+ + e^-$, in the gas phase[523]

process under these conditions, by passing a beam of H^+ ions into a gas containing H atoms, and measuring the flux of H^+ emitted at a small fixed angle from the original beam. They found that the probability varied with energy in a periodic manner as expected (*Figure 1.4*). The energies involved were of such magnitude as to produce rather violent collisions, with impact parameters as small as $0.026\,a_0$ (where a_0 is the Bohr radius).

For a more detailed discussion of these and related experiments, see references 675, 738, and for the theory see references 738 and 1097.

Total cross section measurements

Since the total cross section σ is an average over all impact parameters, it might be expected that periodicity would be lost as a result of the integration (equation 1.17) and the dependence of the cross section on v would be monotonic. Earlier experiments appeared to show this, but since 1965 a number of cases have been discussed in which the total cross section showed a periodicity with increasing velocity[738, 842-844]. It has been pointed out by Smith[957] that the failure of earlier theories to predict this effect was merely due to too gross an approximation in the method of calculation. The usual integration method involved replacing the oscillating part of $P(b)$ by its

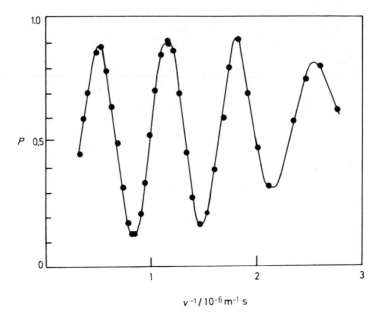

Figure 1.4 Probability P of the reaction $H^+ + H \rightarrow H + H^+$, at a scattering angle of 3°, showing periodic dependence on velocity v (here plotted as the reciprocal)[687]

average value $P = \frac{1}{2}$, and this is valid only when the period of oscillation itself varies monotonically. This in turn will be true only if the energy term $E_\pm$ varies monotonically with R. For the H_2^+ case, $\Delta E_\pm$ does in fact vary monotonically[675], but calculations on the heavier atoms have since shown that $E_\pm$ goes through a maximum at a value of R which is within the range of interest for electron transfer processes. In the case of the Cs_2^+ molecule for example, the maximum is calculated to occur at $R \simeq 7\ a_0$. Smith showed that the effect of this is that the overall value of σ does indeed vary with v in a periodic fashion.

Actual calculations of $\Delta E_\pm(R)$ are now available for several alkali metal systems M_2^+ and have been used to predict reaction cross sections. Agreement between theory and experiment, while not perfect, is convincing evidence that the physical basis of the phenomenon is correctly understood[146, 212, 271, 750].

The reaction $He^+ + He$ has been discussed theoretically. Here too there is a maximum in $\Delta E_\pm(R)$, but it occurs at rather a small distance ($R \simeq 1.4\ a_0$) and the effect of this is that the oscillations in $\sigma(v)$ are too small in amplitude to be observed[60, 119, 242, 395].

1.1.7 Extensions of theory

(a) Other hydrogen-like systems

Zhdanov[1121] has calculated cross sections for some one-electron systems with higher nuclear charges ($He^{2+} + He^+$; $O^{8+} + O^{7+}$).

(b) Different ground state orbitals

In the perturbed stationary state treatment, the transferring electron is assumed to have zero angular momentum — i.e. the orbitals ϕ_A and ϕ_B are s orbitals. Hodgkinson and Briggs[548] have extended the treatment to cover orbitals of higher angular momentum. They note that in such cases, there are various possible different electronic states for the reacting pairs, with total angular momentum quantum number $m = 0, 1, 2 \, (\Sigma, \pi, \Delta$ etc. states). In general, the transfer probability is lower for interactions via the π, Δ etc. states than for the Σ state; and hence the overall transfer cross section, which must include contributions from all these, is lower than would otherwise be predicted.

(c) Breakdown of the two-state approximation

Participation of higher excited electronic states of the pseudo-molecule leads[34, 35, 97, 675] to the possibility of the crossing of energy levels, and the distinction between adiabatic and non-adiabatic descriptions. Although some examples of symmetrical resonant transfer have been considered from this point of view it will be more convenient to discuss these in the section 1.2.

(d) Highly excited reactants

When the electron donor atom is excited close to its ionisation limit, the outermost electron moves almost independently of the other electrons. It has been pointed out that capture of the electron from such an atom is analogous to the capture of a free electron — a much studied process. The examples discussed have, however, all been asymmetrical charge transfer, including atom–molecule reactions[741].

(e) Translational electronic energy

The first-order perturbation method introduces approximations additional to the ones already mentioned. Bates and MacCarrol[92], in particular, have pointed out that it does not take proper account of the translational motion of the electron as it is carried along by the moving nuclei. In spite of the defects inherent in the method, however, the final result is found to be satisfactory and this is attributed to the fact that in slow collisions most of the electron transfer takes place over relatively large distances. Momentum transfer is more important in high-energy collisions[85, 405, 695].

(f) Alternate reaction pathways

Another important possibility, not considered here, is that, given sufficient energy, the reactants may emerge in electronically excited states. Methods of calculation of cross sections for the various possible product states, have been reviewed[822].

1.1.8 Multiple electron transfer

Calculations have been made[410, 492] for reactions of the type

$$A^{2+} + A \rightarrow A + A^{2+} \tag{1.24}$$

It is assumed that the wavefunctions for $A^{2+} \ldots A$ and $A \ldots A^{2+}$ change smoothly into one another, so that in effect the two electrons are transferred simultaneously. Using the same two-state perturbation method as for one-electron transfer, the theoretical expressions are precisely analogous to equations (1.8), (1.15) etc. above and the functions Ψ_A and Ψ_B can be approximated by the two-electron wavefunctions for the He atom[410]. On comparing different two-electron transfer systems it is found experimentally[416] that cross sections increase in the order Ne < Ar < Kr ⩽ Xe. The controlling factor is the total ionisation energy of the two transferring electrons[410], just like the one-electron ionisation energy in one-electron transfer.

Gurnee and Magee[492] pointed out that double electron transfer could also occur by a stepwise process,

$$A^{2+} + A \rightarrow A^{+} + A^{+} \rightarrow A + A^{2+} \tag{1.25}$$

but they conjectured that this would not compete effectively with the one-electron process since the individual steps would be non-resonant.

The phenomenon of an oscillatory dependence of the cross section on energy, described above for one-electron transfer, has also been observed in two-electron transfer[223, 645].

Very recently, transfers of up to four electrons have been observed in single collisions, i.e. in the reaction $Ar^{5+} + Ar \rightarrow Ar^{+} + Ar^{4+}$, but these are at high energies, 10^5–10^6 eV[626, 771a].

1.1.9 Extreme velocity ranges

At very high impact velocities, the Born–Oppenheimer approximation becomes invalid and entirely different treatments are required. These have been extensively reviewed[222, 525, 737, 738, 769], but will not be described here.

At low impact velocities, the assumption that the atoms pass each other in straight lines must clearly break down. Owing to the attractive force between the atoms, lowering the velocity leads to curved trajectories until, below a certain critical velocity, a molecule is formed which can rotate for a finite period of time before separating into products. Data on the reaction $Kr^{+} + Kr$ at low energy appear to support this[956], the cross sections at velocities $\simeq 10^5$ cm^{-1} s being some 50% above the theoretical prediction[548]. A general classical model for low-energy collisions is described below (p. 24). In addition calculations on the $H^{+} + H$ system have indicated the possibility of a quantum mechanical resonance effect. At energies below 0.5 eV, the cross section is predicted to be a periodic function of energy, the maxima corresponding to allowed values of the rotational angular momentum of the H_2^{+} molecule[745]. At still lower energies (10^{-5} eV), allowance has to be made for the fact that the anti-bonding state of the molecule has a stable configuration at an internuclear distance of 12.5 times the Bohr radius[744]. This is almost certainly a general phenomenon.

1.2 Electron transfer between unlike atoms (see footnotes b, c, pp. 3, 4)

1.2.1 Qualitative description

The equation

$$A^+ + B \rightarrow A + B^+ \tag{1.26}$$

denotes a one-electron transfer reaction between two different monatomic species in the gas phase. The actual charges need not be as shown: the positive sign is used merely to denote the direction of transfer. This reaction is a fundamentally different process from resonance transfer. In the low-energy range where the motion of the nuclei can be treated as classical, it is possible to construct curves showing potential energy as a function of internuclear distance R, for the two zero-order electronic states represented by $A^+ \ldots B$ and $A \ldots B^+$. The curves will differ according to whether either the reactants or the products pair has an energy minimum corresponding to formation of a stable molecule. *Figure 1.5* illustrates the possible situations when neither has such a minimum. Energies of the initial and final states are $U_i^0(R)$ and $U_f^0(R)$ respectively, both being functions of the internuclear distance R. The superscript zeros refer to zero-order states. The energy difference is

$$\Delta U^0(R) = U_f^0(R) - U_i^0(R) \tag{1.27}$$

which at infinite separation becomes the difference in the ionisation energies of nuclei A and B:

$$\Delta U(\infty) = U_f(\infty) - U_i(\infty) = I_B - I_A \tag{1.28}$$

(The superscripts are omitted here since there is no interaction).

Taking the example of an endothermic reaction, $(\Delta U(\infty) > 0)$ *Figure 1.5(a)* illustrates the case where the curves do not intersect, and $\Delta U^0(R)$ is large for all values of R. In the course of a collision, the point P representing the progress of the reaction will move in from infinity along the U_i^0 curve until the relative kinetic energy of the reacting pair is entirely converted into potential energy, and then recede again to infinity. The extent of this movement will depend on the initial kinetic energy and on the impact parameter. It is intuitively evident that no reaction will occur, but actually this statement needs to be qualified, as follows.

If the initial kinetic energy is sufficient to carry the reacting pair to a point such as P', where $U_i^0(R) > U_f(\infty)$ then it is possible for the reactants to recede along the U_f^0 curve with correspondingly less kinetic energy, provided that a mechanism exists for effecting the transition. The required mechanism is coupling of the nuclear and electronic motions and is significant only when the velocities of nuclei and electron are comparable. In such a case, the potential

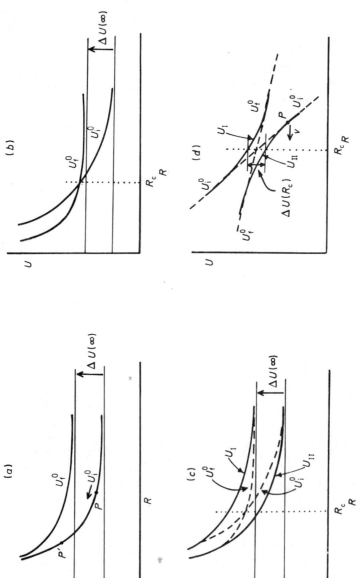

Figure 1.5 Energy curves for an electron transfer reaction between unlike atoms (equation 1.26). $U_i^0(R)$ denotes the energy of the state $(A^+ + B)$ and $U_f^0(R)$ the energy of the state $(A + B^+)$ as functions of the interatomic distance R. (a) without crossing; (b) with crossing, zero interaction case; (c) with crossing and strong interaction; (d) with crossing and weak interaction, idealised model with linear first-order curves (see p. 20). In Zener's equation, equation (1.45), p is the probability that a system represented by a point P moving with horizontal velocity v will first follow the lower full curve and then return along the broken $U_f^0(R)$ curve

energy is no longer a function solely of the positions of the nuclei, and diagrams of the type of *Figure·1.5* are not applicable. When nuclear motion is slow relative to electronic motion, the Born–Oppenheimer approximation is valid. The same argument holds for an exothermic reaction in which $\Delta U^0(R) < 0$ for all values of R. Although this seems surprising at first sight, it is required by the principle of microscopic reversibility. The rate constants k_{for}, k_{rev}, and cross sections σ_{for}, σ_{rev} of forward and reverse reactions are related by

$$k_{for}/k_{rev} = \sigma_{for}/\sigma_{rev} = \exp[-\Delta U(\infty)/kT] \tag{1.29}$$

There is still the possibility of a vertical electronic transition, at any particular internuclear distance, with absorption or emission of a photon, but this is a three-body collision and highly improbable — unless the atoms are in a stable or metastable configuration represented by a minimum in the energy curve[1109].

Figures 1.5(b)–(d) illustrate cases in which the energy curves intersect at some critical distance, $R = R_c$. In general the two zero-order electronic states corresponding to pure configurations $A^+ \ldots B$ and $A \ldots B^+$, with energies U_i^0 and U_f^0, will interact and the energies are represented by the continuous curve $U_I(R)$ and $U_{II}(R)$. On traversing the lower curve from right to left, the character of the wavefunction changes from one describing the configuration $A^+ \ldots B$ to one describing $A \ldots B^+$.

In the limit of zero interaction, we have a true crossing of the curves, and (again provided that nuclear motion is slow compared with electronic) the reactants approach and recede along the same curve, with no net electron transfer. However, a more typical case is that shown qualitatively in *Figure 1.5(d)*, in which the energy curves are well approximated by U_i^0 or U_f^0, except over a small range of R values near the crossing point where the curves separate by an amount $\Delta U(R_c)$, equal to twice the resonance energy

$$\Delta U(R_c) = U_I(R_c) - U_{II}(R_c) = 2\beta(R_c) \tag{1.30}$$

This model leads to a general prediction of the form of the dependence of reaction rate, or cross section, on the mutual kinetic energy, or velocity, of the reactants. In the extreme of low velocity (depending on the precise shapes of the energy curves), reaction may be very slow because the system fails to reach the crossing point at all — though even so, the probability will rise to a maximum as the internuclear distance passes through its minimum. However, given enough energy to reach the crossing point, and if the nuclear motion is so slow that the electronic configuration can always adjust to a minimum potential energy, the reactants will follow the lower potential energy curve as they pass through the crossing and no reaction will follow. This is the limit of *adiabatic* behaviour. The cross section will increase with increasing velocity, to a maximum, then tail off again at higher velocity. This is because at low velocities, the atoms fail to reach the crossing-point, while at high velocities they pass through the point too quickly for reaction to occur — this latter is the *non-adiabatic* condition. This predicted behaviour of the cross section has been verified for a large number of electron transfer reactions[521, 522]. (See also section 1.2.3).

1.2.2 Semi-quantitative models

Although quantum-mechanical calculations, of varying degrees of sophistication, have been used to predict cross sections of different reactions as functions of impact energy, several relatively simple arguments have also been used. They combine features of the quantum-mechanical or classical models and, although only semi-quantitative, they give useful insights into the mechanisms of these reactions.

(a) Adiabatic and near-adiabatic reactions

An approximate criterion for an adiabatic collision was proposed[736] by Massey in 1949, namely that the effective duration Δt of the collision should be large compared with the time constant τ of the relevant electronic motion. The effective duration Δt can be expressed as

$$\Delta t = a/v \tag{1.31}$$

where v is the relative velocity of the colliding pair and a is the effective distance of motion over which electron transfer can be described as 'likely to occur'. The electronic time constant τ is estimated with the aid of the Uncertainty Principle as

$$\tau = h/2\pi\Delta U \tag{1.32}$$

where ΔU is the energy change associated with the electron transfer process. This in turn varies with the internuclear distance. In his original discussion, Massey[736] took $\Delta U = \Delta U(\infty)^{(f)}$; but a more appropriate choice is $\Delta U(R_c)$. Thus Massey's criterion for an adiabatic collision is the condition

$$v \ll (2\pi a/h)\Delta U(R_c) \tag{1.33}$$

where a is a distance comparable with atomic dimensions. Expressing the relative velocity v in terms of the reduced mass μ_{AB} of the colliding pair and the kinetic energy E_{kin} of relative motion we have

$$v = (E_{kin}/\mu_{AB})^{1/2} \tag{1.34}$$

and equation (1.33) becomes

$$\sqrt{E_{kin}} \ll (2\pi/h)\Delta U(R_c)\sqrt{\mu_{AB}} \tag{1.35}$$

The maxima in the plots of σ against v (*Figure 1.6*) may then be interpreted as points where $\Delta t \simeq \tau$, thus the velocity at the maximum is

$$v_{max} \simeq (2\pi/h)\Delta U(R_c)a \tag{1.36}$$

Using the approximation $\Delta U(R_c) = \Delta U(\infty)$, it has been shown that v_{max} does indeed correlate with $\Delta U(\infty)$ for a wide range of reactions, with a constant

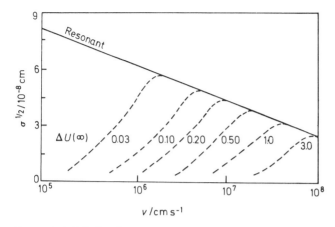

Figure 1.6 Calculated cross sections, as functions of velocity v, for a family of exchange reactions $A^+ + B \rightarrow A + B^+$ *in which the mean ionisation energies of A and B are kept constant (and in this case equal to the ionisation energy of the H atom). The broken curves refer to different energies* $\Delta U(\infty)$*, as indicated. The full line refers to the resonant case* $A = B = H$*,* $\Delta U(\infty) \equiv 0$*. (After Rapp and Francis*[874]*)*

value of a, approximately 7 Å[520]. This may be regarded as the average distance travelled by one atom, in its trajectory past the other atom, during which it is likely to suffer electron transfer.

For the variation of cross section with impact velocity in the adiabatic range, Massey and Burhop[737] proposed the empirical equation

$$\sigma = A \exp(-\gamma a |\Delta U(\infty)|/hv) \tag{1.37}$$

where a is the adiabatic parameter, and A, γ are constants. This equation has been successfully used to correlate a wide range of experimental data[521, 522].

(b) Critical reaction distance

In certain cases, the internuclear distance R_c at which the energy curves cross can be estimated simply. Consider, for example, an endothermic reaction

(f) What Massey actually wrote (reference 736, p. 261) was (apart from changes of notation): 'We denote the initial state by (i), the final state, the energy of which differs from state (i) by ΔE, by (f). When [the reactants] are at a given finite distance apart the state of the system will fluctuate between being mainly like (i) and mainly like (f) with a frequency of order $\Delta E/\hbar$ [where $\hbar = h/2\pi$]. If the number of such fluctuations during the collision is high, the conditions are nearly adiabatic and the chance of finding [the reactants] in state (f) after the collision is very small. The time of collision is of order a/v where v is the relative velocity and [a is] a length of the order of atomic dimensions . . . . For a considerable chance of a collision we must have $\Delta E a/\hbar v \gg 1$.' It is clear that the quantity ΔE mentioned here is what we have termed $\Delta U(R)$; but in the subsequent analysis of data, Massey quotes values of $\Delta U(\infty)$

($\Delta U(\infty) > 0$) which involves the net separation of charge:

$$A + B \rightarrow A^+ + B^- \qquad (1.38)$$

At distances such that the main force between the ions is Coulombic, we have

$$U_f^0(R) = U_f(\infty) - (1/4\pi\varepsilon_0)(e^2/R) \qquad (1.39)$$

(where ε_0 is the permittivity of free-space) while U_i is effectively constant

$$U_i^0(R) = U_i(\infty) \qquad (1.40)$$

The condition at the crossing point is that

$$\Delta U^0(R_c) = U_f^0(R_c) - U_i^0(R_c) = 0 \qquad (1.41)$$

whence

$$e^2/4\pi\varepsilon_0 R_c = U_f(\infty) - U_i(\infty) = \Delta U(\infty) \qquad (1.42)$$

$$R_c = (4\pi\varepsilon_0)(1/e^2)\Delta U(\infty) \qquad (1.43)$$

If R_c proves to be sufficiently large compared with the atomic diameters, it can be assumed that Coulomb's law is indeed observed, and the calculation is valid. Typical values are indeed large: for example, the reaction $Na + Cl \rightarrow Na^+ + Cl^-$, has $R_c = 20a_0$, i.e. 10.6 Å; and the same can be true for exothermic reactions, when the products are of like charge, as for example, $Ar^{2+} + He \rightarrow Ar^+ + He^+$, $R_c = 9a_0$[93].

(c) Transition probability

For a net reaction to occur, it is necessary for the electron to transfer as the reactants approach, but not as they recede again, or vice versa. Thus if the probability of the transition is p, the probability of reaction, P, is given by

$$P = 2p(1-p) \qquad (1.44)$$

The simplest calculation of transition probability is based on arguments put forward by Landau[646], Zener[1120] and Stueckelberg[975]. The most important assumption is that electron transfer, while not restricted to specific internuclear distances is nevertheless confined to a small range of distances on either side of the crossing point. The electronic states corresponding to the energy curves $U_i^0(R)$ and $U_f^0(R)$ (*Figure 1.5*) are then regarded as the basis of the wavefunction. The transition probability is calculated by time-dependent perturbation theory. The usual assumption is made, that the kinetic energy of nuclear motion is small compared with electronic energy and a further approximation is that the energy curves are linear over the relevant range of distances. The final result is

$$p = 1 - \exp(-\delta) \qquad (1.45)$$

$$\delta = \left(\frac{\pi^2}{h}\right)[\Delta U(R_c)]^2 \bigg/ \left[\left(\frac{d}{dt}\right)(U_i^0 - U_f^0)\right]_{R=R_c} \qquad (1.46)$$

where the derivative is evaluated at the crossing point, $R = R_c$. Introducing the velocity $v = dR/dt$, which is either assumed to be constant or else assigned its value at the crossing point, this gives

$$\delta = \left(\frac{\pi^2}{hv}\right)[\Delta U(R_c)]^2 \bigg/ \left[\left|\left(\frac{d}{dR}\right)(U_i^0 - U_f^0)\right|\right]_{R=R_c} \tag{1.47}$$

The validity of equation (1.47) in the light of the assumptions made, has been discussed by Bates[86], and more recently by Baede[57]. The extension to systems with more than two states, involving a succession of curve crossings has also been discussed[713, 738]. An entirely different approach, advocated by Smith[958] and discussed by several other authors[60, 119, 242, 395, 676] uses the 'diabatic curves' U_I and U_{II} as the basis set.

(d) Cross sections of non-adiabatic reactions

Equation (1.47) can be combined with the previous estimates of R_c to predict approximately the magnitude of the cross section for non-adiabatic reactions[88, 90, 93, 94, 151, 269]. Consider again the endothermic ionisation reaction, equation (1.38). Assuming again that the crossing point occurs at distances where the prevailing interatomic forces are electrostatic we have equations (1.39) and (1.40) which give

$$\left(\frac{d}{dR}\right)(U_i^0 - U_f^0) = -\left(\frac{1}{4\pi\varepsilon_0}\right)\left(\frac{e^2}{R^2}\right) \tag{1.48}$$

and, taking the value of R_c from equation (1.43),

$$\left|\left(\frac{d}{dR}\right)(U_i^0 - U_f^0)\right|_{R=R_c} = (4\pi\varepsilon_0)e^{-2}[\Delta U(\infty)]^2 \tag{1.49}$$

Expressing the velocity v in terms of the relative kinetic energy E_{kin} of the reacting pairs, we obtain from equation (1.46):

$$\delta = \left(\frac{\pi^2}{h}\right)\left(\frac{\mu_{AB}}{2E_{kin}}\right)^{1/2}\left(\frac{1}{4\pi\varepsilon_0}\right)e^2\frac{[\Delta U(R_c)]^2}{[\Delta U(\infty)]^2} \tag{1.50}$$

Since we are considering only non-adiabatic reactions we may put $\delta \ll 1$, so that $p \simeq (1+\delta)$ and $P \simeq 2\delta$. Then taking equation (1.18) in the approximate form

$$\sigma(E) = P\pi R_c^2 \tag{1.51}$$

we obtain

$$\frac{\sigma}{\pi a_0^2} = 2\pi\left(\frac{\pi^2 a_0 m_H^{1/2}}{h\sqrt{2}}\right)\left(\frac{1}{4\pi\varepsilon_0}\right)^3\left(\frac{e^2}{a_0}\right)^3\left(\frac{M}{E_{kin}}\right)^{1/2}\frac{[\Delta U(R_c)]^2}{[\Delta U(\infty)]^4} \tag{1.52}$$

where a_0 is the Bohr radius, m_H is the mass of the hydrogen atom, and M is the reduced mass in atomic units. In this way cross sections have been calculated for various systems. For more exact treatments of atom–ion collisions equation (1.39) is modified to include point-charge—induced-dipole terms[94].

More generally, the expression for δ can be substituted into equation (1.45) to predict the variation of $\sigma(E)$ as the kinetic energy E_{kin} is varied from the adiabatic to the non-adiabatic range[94, 526, (g)].

(e) Resonance energies

Similar calculations can also be performed in reverse, to obtain the resonance energy $\beta = \frac{1}{2}\Delta U(R_c)$ from measured cross sections. For this purpose, Bates and Moiseiwitsch[94] used an integrated expression for the cross section, analogous to equation (1.8), namely

$$\sigma = 4\pi R_c^2 I(\delta) \qquad (1.53)$$

where

$$I(\delta) = \int_1^\infty \exp(-\delta x)[1 - \exp(-\delta x)] x^{-3} \, dx \qquad (1.54)$$

and where δ has been defined already. Moiseiwitsch[760] verified that the integral $I(\delta)$ passes through a maximum with increasing δ, namely $I = 0.113$ when $\delta = 0.424$. (This corresponds to a transition probability, $p = \exp(-\delta) = 0.65$.) Hence, from the measured maximum cross section, R_c is obtained by equation (1.53), and the energy $\Delta U(R_c)$ by equation (1.53).

1.2.3 Quantum-mechanical calculations

Several authors[91, 874] have carried out calculations for transfer between unlike atoms by a general method, similar to that for resonance transfer. Again, only the motion of a single transferring electron is considered, and the wavefunction Ψ for this electron is written as the sum of two terms as in equations (1.4) and (1.5), except that Ψ_A and Ψ_B denote orbitals which reduce to the atomic orbitals ϕ_A and ϕ_B as R tends to infinity, and the energies E_A and E_B are different. The final result, which involves approximations and mathematical details not reproduced here, is complex, but Rapp and Francis[874] showed that it correctly reduces to the resonant case on setting $\Delta U(R) = 0$.

The dependence of cross section on impact energy has the expected form. With increasing energy, the cross section rises to a maximum and then falls. The initial rise had also been deduced much earlier in a detailed calculation on the H^+ + He reaction[739]. In the high-energy range, the cross section for a given energy is the same as that calculated for a resonant process of the type $X^+ + X \rightarrow X + X^+$, where the ionisation potential of atom X is the mean of the ionisation energies of atoms A and B. For a series of reactions $A^+ + B$, all with

(g) *Reference* 769, p. 351 f

the same mean ionisation potential, plots of $\sigma^{1/2}$ against v form a family of curves, all converging to the same curve for the resonant case, $\Delta U(\infty) = 0$, and at velocities below the maxima, the cross sections for a series of reactions increase as $\Delta U(\infty)$ decreases to zero (*Figure 1.6*).

All the limitations inherent in the perturbed two-state method, which were briefly reviewed in section 1.1.7, naturally apply also to the mixed-atom system. Some recent theoretical studies of reactions at low energy may be consulted for detailed discussions[97, 582, 592, 751].

The periodic dependence of cross section on energy, discussed in section 1.1.6 in connection with resonant systems can also be observed in reactions between unlike atoms. A recently reported example[1063] is the reaction $Li^+ + Na \rightarrow Li + Na^+$.

1.2.4 Accidental resonance

If the energy change associated with the reaction happens to be close to zero we have what is sometimes called 'accidental resonance'. Although some authors have assumed that accidental resonance necessarily leads to very fast rates of electron transfer, comparable with the rates for true resonance, Bates and Lynn[91] have argued that this is not necessarily the case. When the nuclei of the reacting atoms are far apart, there are two possible eigenfunctions describing the system, which happen fortuitously to be degenerate or nearly so; one corresponding to the state $A^+ \ldots B$, the other to $A \ldots B^+$. The proposed reaction is a transition between these states; but this transition is highly improbable owing to the small overlap. As the particles approach, the overlap increases, but so also in general does the electronic interaction, hence the states cease to be degenerate, and if the energy curves never cross, there may be no condition favourable to electron transfer and the reaction will be much slower than in the resonant case. If there is a crossing point at some critical value of R, the mechanism is the same as the non-resonant cases discussed above. Even if the two energy curves are close together for a range of distances from infinity down to the typical impact parameter, it still does not follow that the process approaches the resonance transfer. The necessary condition is that the integrals in equations (1.6) and (1.7) should be the same, and this is not in general true for different atoms.

It would not be surprising, however, if, for many reactions between different but similar atoms, such as $He^+ + Ar$, or more especially between neighbouring transition metal atoms, like $Cu^+ + Ni$, the above integrals were nearly the same and the reaction rates thus comparable with true resonant processes. In that case, in spite of the fundamental difference between these and the true resonant process, reactions between atoms which happen to have $\Delta U(\infty) = 0$ will, as a group, tend to exhibit higher cross sections than those which are energetically unfavourable. This has indeed been observed and commented upon[736, 769, 874]; see especially reference 87.

1.3 The Langevin model

As already mentioned, theories of reaction rate based on linear trajectories break down when the impact energy is very low since interatomic forces cause deflections in the trajectories on collision. Also, the lower the energy the greater the possibility of forming collision complexes, and if the lifetime of such a complex is long compared with the period of oscillation of the wavefunction, calculations of the type described above become irrelevant.

Under these conditions, a purely classical model originally devised by Langevin[649] in 1905, and elaborated by Gioumousis and Stevenson[463] has frequently been used[404, 411, 1076]. Consider a reaction between a charged and an uncharged particle. The charged particle induces a dipole moment on the uncharged particle, so that there is a mutual electrostatic attraction. Langevin showed that if the impact parameter b is greater than a certain value b_0, the charged particle will merely be deflected; for $b = b_0$, it will enter a closed orbit; and for $b < b_0$ it will be captured by the uncharged particle and will spiral toward the centre. The critical value is

$$b_0 = (4\pi\varepsilon_0)^{-1/2} (4e^2\alpha/\mu_{AB} v^2)^{1/4} \tag{1.55}$$

and the cross section is

$$\sigma = \pi b_0^2 = (e/2\varepsilon_0 v)(\alpha/\mu_{AB})^{1/2} \tag{1.56}$$

where α is the polarisability of the uncharged particle. The radius of the closed orbit is $b_0/\sqrt{2}$, so the model is applicable only when the sum of radii is less than this. With a Maxwell distribution of velocities, the rate constant becomes

$$k = (e/2\varepsilon_0)(\alpha/\mu_{AB})^{1/2} \tag{1.57}$$

The same result has been obtained from transition-state theory, by calculating the energy of the activated complex on the ion-dipole model[396].

Rate constants calculated in this way play a role in gas kinetics somewhat analogous to the diffusion rate constant in solution kinetics. For the interaction between particles at low energies, they give the upper limit estimate for the rates of all possible reactions—for it should be noted that in most systems, electron transfer is only one of several possible outcomes of a collision (others include energy transfer and ionisation to give a free electron).

Bohme, Hasted and Ong[134, 135] have proposed combining the ion-dipole model with the Massey adiabatic model, by assuming that the maximum value of the cross section may be identified with the Gioumousis–Stevenson value.

The ion-dipole potential energy function has also been used[745, 1060] as the basis of some quantum-mechanical treatments of reaction rate. Its validity has also been critically examined and, on the basis of a model calculation using the $H^+ + H$ system, it has been argued[459] that quantum-mechanical resonance effects are significant for collisions occurring, if not at thermal energies, then at least in the energy range 0.1–0.2 eV.

1.4 Typical reaction rates

Rate constants for resonant transfer estimated as above (pp. 9ff) do not differ greatly from the $H^+ + H$ value, $k \simeq 10^{13}$ $M^{-1}s^{-1}$, and it is probably safe to guess that none will be lower than $\simeq 10^{12}$ $M^{-1}s^{-1}$ at room temperature. Langevin-type calculations for heavier atoms lead to a similar conclusion, provided the total energy change associated with the reaction is close to zero. Results quoted by Watson[1076] for reactions of H^+ with various atoms are all of the order of 10^{-9} $cm^3 s^{-1}$, that is, 6×10^{11} $M^{-1}s^{-1}$. On the other hand, experimental measurements of the reaction $H^+ + O \rightarrow O^+ + H$ have yielded[404] a rate constant appreciably lower than these calculations would suggest, i.e. 3.75×10^{-10} $cm^3 s^{-1}$ (2.3×10^{11} $M^{-1}s^{-1}$) at 300 K. Another obvious possibility is that reactions involving like charged ions (either as reactants or as products) will prove to be slower than those involving uncharged species. Hasted and Chong[526] have reported data for a number of reactions of the type $A^{n+} + B \rightarrow A^{(n-1)+} + B^+$. The reaction

$$Kr^{2+} + He \rightarrow Kr^+ + He^+ \tag{1.58}$$

is an example of such a reaction with a nearly zero change in energy ($\Delta U(\infty) \simeq +0.02$ eV). Cross sections for this reaction are nearly constant, over the energy range 400 to 3000 eV, $\sigma \simeq 0.2 \times 10^{-20}$ m^2. For lower energies the cross section is presumably less than or equal to this, and for a temperature of 300 K, we may estimate, as before, a rate constant $k \simeq 10^{10}$ $M^{-1}s^{-1}$.

Chapter 2
Electron transfer involving molecules

2.1 Formal classification of mechanisms[a]

When one or both of the reactants is a molecule, new factors enter into the description of the electron transfer process and it could be argued that it is only at this point, where three or more atoms are involved, that we can begin to speak of 'mechanisms' as chemists understand the term. These mechanisms are distinguished according to the nature and sequence of relative motions of the atoms. The number of mechanisms which have been defined for electron transfer reactions in the gas phase is already considerable, and will clearly increase as new systems are studied; however, already it is possible to attempt a classification, as shown in *Table 2.1*.

The most obvious subdivision is between reactions which lead only to the exchange of electrons and those which lead to structural change. This corresponds to some extent with the solution chemists' division between outer-sphere and inner-sphere reactions (*see* chapter 5). The former class have commonly been called 'charge exchange reactions', actually a rather confusing term since atom transfer reactions can also lead to net charge exchange.

A second subdivision is according to the number of discrete steps involved in the overall reaction event; this must necessarily depend on the interpretation of various lines of evidence and on the degree of sophistication in the definition of 'steps'. Even with an atom transfer reaction it is possible to distinguish various stages in the progress of a reaction, for example complex formation at low energies. The most useful division is into two types: concerted 'single event' processes, and double events in which electron transfer occurs first and is followed by atom transfer.

Yet another distinction can be made between processes in which the reactant atoms move along straight, or slightly curved trajectories, and those in which they bond together for a sufficient length of time to undergo some other

(a) For general introductions, and basic theory of molecular collision processes, see especially references 419, 665 (and see also footnote (b) in Chapter 1, p. 3). Further theoretical treatments are given in references 225, 270. Experimental techniques are reviewed in references 409, 1004, 1005, and of the many general review articles surveying both theoretical and experimental results, the ones I have chiefly consulted are references 149, 232, 406, 450, 622, 809, 833

Table 2.1 Classification of gas-phase ion–molecule electron transfer reactions

	Without transition state	With transition state
Reaction without atom-transfer 'charge exchange'	$A^+ + B \rightarrow A + B^+$ 'Resonant transfer' (when $A = B$); 'Quasi-resonant transfer' ($A \neq B$)	$A^+ + B \rightarrow [AB^+]^\ddagger \rightarrow A + B^+$
Reaction with atom transfer — Single event	$AX + B \rightarrow AX + B$ 'Stripping' (A and B scattered forwards) F & L Class I (A and B scattered backwards)	$AX + B \rightarrow [AXB]^\ddagger \rightarrow A + XB$ F & L Class III
Reaction with atom transfer — Double event 'Harpoon mechanism'	$AX + B \rightarrow AX^- + B^+ \rightarrow A + XB$ 'Spectator stripping' F & L Class II	$AX + B \rightarrow AX^- + B^+ \rightarrow [AXB]^\ddagger \rightarrow A + XB$ F & L Class III

process. If the process is merely rotation or exchange of energy between a few of the possible vibrational modes we speak of a 'transition state' in the sense of Eyring and Polanyi, but if there is the possibility of more extensive rearrangement, or of reaction with another molecule, we speak of a discrete chemical entity, i.e. an 'intermediate'.

Some of the better-known reaction mechanisms have been given picturesque names. The names most generally used are indicated in *Table 2.1*. Also included is a sub-classification of atom transfer reactions, due to Fluendy and Lawley[419] which is described more fully in section 2.2.

An important contrast between the discussions of reaction mechanisms in the gas phase and in solution is the fact that the energy of interaction of the reactants can be controlled and varied over a much wider range in the gas phase than in solution, and that in general different reaction mechanisms apply in different energy ranges[1102, 1103]. Broadly speaking, increasing the energy produces two effects: It allows more violent collisions (smaller impact parameters) and it reduces the collision time. In the limit of very low mutual energies, collision complexes can be formed and the transition state theory provides a valid model. Equilibration of energy can occur between different vibrational modes, and the more complex the reacting molecules, the higher the energy range over which this will be true (or so we may expect). At short reaction times, there may be no collision complex, and the direction of scatter of the product particles will be related to the direction of approach. Equilibration of energy is no longer possible, and 'stripping' mechanisms become relevant. At still shorter times, no atomic motions within molecules are possible, and the electron transfer process is non-adiabatic. Finally, at very high energies, the relative velocity of the two reactants is no longer slow compared with the motion of electrons; the Born–Oppenheimer approximation breaks down, and so does the validity of traditional chemical symbolism.

2.2 Description of the principal reaction mechanisms

2.2.1 Single event: electron transfer without atom transfer

Reactions which involve no net chemical change other than electron transfer, and which appear (at least on present evidence) to proceed in a single step, are typified by equations (2.1–2.4).

$$He^+ + N_2 \rightarrow He + N_2^+ \quad (2.1)$$

$$NO_2 + I^- \rightarrow NO_2^- + I \quad (2.2)$$

$$H_2^+ + H_2 \rightarrow H_2 + H_2^+ \quad (2.3)$$

$$Ar^+ + CH_4 \rightarrow Ar + CH_4^+ \quad (2.4)$$

The mechanisms can be rationalised by the fact that the molecules and molecule-ions on either side of these equations are 'stable'—i.e. strongly internally bonded species, while the interaction between them is known from ordinary chemical evidence to be weak. When the reactants and products are identical, as in equation (2.3), the situation is similar to resonant transfer between atoms. The reaction can only be detected by the observation that the molecules in a particular beam emerge with increased or decreased charge. Reactions of this type are indeed sometimes called 'resonant'—see p. 33.

In cases where molecular beam methods have been used to vary the energy of the reactants and to monitor the energies of the products, it has usually been found that, within certain limits of reactant energy, very little kinetic energy is transferred. In other words, the molecules pass each other at relatively high speed and/or they exchange the electron over relatively long distances without much deflection from their trajectories. At low kinetic energies however, definite interactions may occur. Another variant on this mechanism is temporary partial electron transfer. An example is the non-reactive collision process $H^+ + H_2 \to H^+ + H_2$. Energy is transferred from the vibrational levels of H_2 to the translational levels of H^+, and it is postulated that the passing H^+ ion abstracts electron density from the H–H bond, weakening the bond and lowering the vibrational frequency[460].

2.2.2 Single event electron transfer with atom transfer

Many reactions which involve the transfer of an atom can be called electron transfer reactions in the sense that the atoms or molecules suffer a change in oxidation state. This is fairly obvious in the reaction

$$Cs + SF_6 \to CsF + SF_5 \tag{2.5}$$

and in other reactions described below, such as equation (2.8), in which the metal atom changes from a zerovalent to a univalent state. On the other hand, reactions such as

$$H_2 + CO^+ \to H + HCO^+ \tag{2.6}$$

can be classed as electron transfer only in a formal sense. According to conventional definitions the carbon atom has undergone 'reduction'. Its oxidation number has changed from III to II while that of the hydrogen has changed from 0 to I. It could equally be argued, however, that the carbon has been oxidised from state III to state IV, while the transferred hydrogen had changed from 0 to $-I$. The theoretical model of such a reaction that is most generally useful takes no note of either of these definitions.

A genuine ambiguity arises with the reaction

$$O^- + O_2 \to O_2^- + O \tag{2.7}$$

which could proceed by a bridged mechanism with transfer of the oxygen atom, or by direct electron transfer. Studies by isotopic labelling have shown that at the lowest energies (the reaction is endothermic with a threshold energy $\Delta U(\infty) \simeq 1.0 \text{ eV}$) both processes occur[833], but at higher energies electron transfer predominates over atom transfer[1059]. The reaction (2.10) below is ambiguous in a different way. It could proceed by transfer of a hydrogen ion from H_2^+ to H_2, or of a hydrogen atom from H_2 to H_2^+. Presumably this could be decided by studying related reactions such as $D_2^+ + H_2$.

Evidence that some atom transfer reactions proceed by way of relatively long-lived transition states or collision complexes, while others may be termed 'direct', comes mainly from molecular beam experiments. Techniques and methods of interpretation of data are reviewed elsewhere[419, 450, 622, 665, 883]. The relevant information consists of the distribution of energies and scatter directions of the reaction products, relative to the energies and directions of the beams of reagents. For analysis the results are presented in polar diagrams of the type shown in *Figure 2.1*. The origin is the centre of mass of the reacting particles, the angular coordinate is the scattering angle, and in *Figure 2.1* the

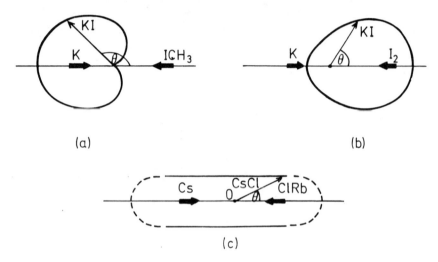

Figure 2.1 *Idealised polar diagrams showing the direction of scattering of reagents in reactive collisions. In each case the reagents are shown approaching along the axis; O is the centre of mass, and the curve shows intensity of scattering of one of the reagents as function of angle θ. (a) Class I behaviour, scattering of KI in the reaction K + ICH₃; (b) Class II behaviour, scattering of KI in the reaction K + I₂; (c) Class III behaviour, scattering of CsCl in the reaction Cs + ClRb. (After Fluendy and Lawley, reference 419, figures 8.6 and 8.7)*

radial coordinate is the intensity of flux of one of the product species in each particular direction. In *Figure 2.2*, the radial coordinate is the velocity of the scattered particles, and contours have been drawn for different values of the intensity[462]. The diagram contains the essential qualitative characteristics of the reaction, from which in turn mechanistic conclusions can be drawn.

The reaction

$$\overrightarrow{CH_3I} + \overleftarrow{K} \rightarrow \overleftarrow{CH_3} + \overrightarrow{IK} \tag{2.8}$$

is an example of Fluendy and Lawley's[419] Class I. The products are scattered mainly backwards, that is to say the CH_3 and K atoms suffer reversal of

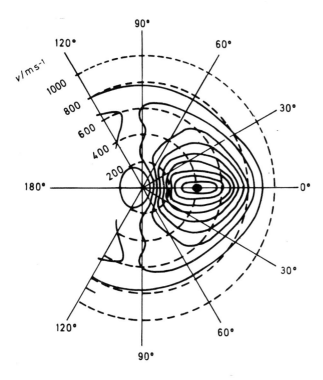

Figure 2.2 Experimental polar diagram for scattering of KI in the reaction $K + I_2$ (compare with Figure 2.1(b)). The radial coordinate for each contour is the velocity of scattering, and the contours denote equal increments of flux intensity, on an arbitrary scale. (After Gillen, Rulis and Bernstein[462])

direction while the transferring I atom continues in its original direction, as indicated by the arrows. The cross section is small ($\simeq 30$ Å^2), comparable with the cross section of a non-reactive collision. The process can be visualised as a violent collision in which energy is efficiently transferred from the vibrational modes of the C–I bond to those of the K–I bond. Momentarily, the atoms assume a configuration K ... I ... CH$_3$. This formally resembles the transition state of an electrophilic substitution or bridged-electron transfer reaction in solution (such as the reactions $Ag^+ + ICH_3$ or $Cr^{2+} + ICH_3$), but it must be assumed to be of much shorter duration.

The reaction

$$Cs + ClRb \rightarrow CsCl + Rb \qquad (2.9)$$

is an example of Class III. The products are scattered, equally forwards and backwards and this is taken to imply that the intermediate configuration CsClRb has a lifetime long enough to permit rotation about the central Cl atom before dissociation into products.

It seems clear that in general a reaction may proceed with or without a transition state, depending on the initial kinetic energy of the reactants. An example is

$$H_2^+ + H_2 \rightarrow H_3^+ + H \tag{2.10}$$

Cross sections for this reaction have been studied as a function of the relative kinetic energy of the reactants, and of the vibrational energy level of the H_2^+ ion. Two kinds of behaviour have been observed[233]: at low impact energies, the cross section *decreases* with increasing vibrational energy, but at higher impact energies, the reverse is true. It has been suggested that at low energy, the reaction path is by way of a collision complex, but at high impact energy there is a changeover to a stripping mechanism[232, 350, 539, 709], probably by hydrogen ion transfer. The structure of the low-energy collision complex has been discussed in the light of data on kinetic isotope effects[1005].

2.2.3 Double event: electron transfer followed by atom transfer

The reaction

$$\vec{K} + \overleftarrow{Br_2} \rightarrow \overrightarrow{KBr} + \overleftarrow{Br} \tag{2.11}$$

illustrates a well-characterised class of reactions in which the products are scattered mainly forwards. The potassium atom and the liberated bromine atom continue in their original directions, while the transferred bromine atom is reversed. The cross sections are large, comparable with those of simple charge transfer. On this basis, Magee[709] proposed the two-step 'harpoon' mechanism. When the approaching reactants reach some critical distance, electron transfer occurs rapidly. If the original reactants were uncharged they become oppositely charged. They are then drawn together by the coulombic force and the reaction is completed:

$$K + Br_2 \rightarrow K^+ + Br_2^- \rightarrow KBr + Br \tag{2.12}$$

The potassium atom has in effect 'harpooned' the bromine molecule in order to capture the bromine atom.

When the reactants are not both neutral molecules the term 'harpooning' becomes less apt and is not in fact used, but two-step mechanisms may still occur, as in equation (2.26) below.

The Magee mechanism, equation (2.12), does not distinguish reactions which have transition states (or intermediates) from those which have not. As before, a transition state or long-lived intermediate leads to equal scattering of the reactants forward and backwards. The scattering pattern observed in the reaction of equation (2.11) implies a concerted mechanism, since the products retain a 'memory' of their initial energy and direction. This is most simply explained by the 'spectator stripping' mechanism[419, 1073b]. Referring again to equation (2.12), it is argued that when electron transfer occurs the bond between the two Br atoms is weakened to such an extent that it is practically broken, giving a Br atom and Br^- ion, but the Br atom is relatively unaffected. The K^+ and Br^- ions converge into the KBr molecule and move away from the Br atom. Relative to the centre of mass, the KBr molecule moves in one direction, and the Br atom, to conserve momentum, must move in the other. Relative to the non-transferring Br atom, however, the effect is that the K atom strips off the transferring Br atom, leaving the other Br atom standing as 'spectator'. Reactions of this type form Fluendy and Lawley's Class II. It is recognised that in terms of relative forward and backward scattering, and of cross section, there is a range of observable behaviour varying from Class II to Class I.

In general, transition states will be favoured at low kinetic energy and spectator stripping mechanisms at higher energies. In the case of reactions of the type

$$K + Br_2 \rightarrow K^+ + Br^- + Br \qquad (2.13)$$

it has been shown both experimentally[1045] and theoretically[527] that at still higher energies the stripping mechanism is superseded by a third possibility—total breakdown into three fragments.

2.3 Factors influencing electron transfer rates

2.3.1 Analogies with atom–atom reactions

Many discussions of the rates of reaction involving molecules have been based on the same principles as have been outlined in chapter 1 for atom–atom reactions. Thus rates for the $H_2^+ + H_2$ exchange and other symmetrical processes have been estimated by adaptations of the two-state impact parameter method[492, 874]. Giese[459] has estimated theoretically the resonance energy of the $H_2^+ + H_2$ system. Arguments based on the Langevin model (p. 24) have been used extensively to estimate approximate rate constants[413]. Some unsymmetrical reactions with very small energy changes can be considered as quasi-resonant. For example, the rate of the reaction $CO^+ + Kr \rightarrow CO + Kr^+$ (with $\Delta U(\infty) \leqslant 0.01$ eV) is similar to the rates of the symmetrical exchange reactions $Kr^+ + Kr$ and $CO^+ + CO$[956]. Exothermic charge transfer seems as a general rule to be very rapid ($k \geqslant 6 \times 10^{12}$ M^{-1} s^{-1} at 300 K)[409]. The Langevin model is also extensively used[1004] with appropriate modifications for dipolar or quadrupolar reactants[148, 346].

For reactions between unlike molecules, the Massey adiabatic criterion may be applied (*see* Chapter 1, equation (1.33)). It has been confirmed for example that for the reaction

$$N_2^+ + Ar \rightarrow Ar^+ + N_2 \qquad (2.14)$$

the plot of cross section against reactants' kinetic energy shows a maximum consistent with the changeover from the adiabatic to the non-adiabatic condition[554], whereas for the quasi-resonant process

$$N_2^+ + N_2 \rightarrow N_2 + N_2^+ \qquad (2.15)$$

the cross section decreases with increasing energy as expected (*Figure 1.6*). More sophisticated calculations make allowance for covalent and ion-dipole forces in the states before and after electron transfer, but the results are similar[1117]. Maxima also occur in the plots of $\sigma(E)$ for some rearrangement reactions, e.g. for the reaction of equation (2.8)[643, 686].

The initial step of a harpoon mechanism has been discussed in a similar way. The critical distance R_c for electron transfer between neutral species is calculated approximately as given in chapter 1 (equation 1.43) and the cross section is obtained as πR_c^2. The results are generally of the right order of magnitude. For several such reactions the plots of cross section against kinetic energy show maxima[59, 345, 612, 612a] from which the resonance energy can be estimated. For the reaction $Br_2 + K$, the estimated reaction distance is 4.5 Å and the resonance energy is 0.047 eV[59].

With molecules as reactants it becomes important to distinguish different vibrational energy states: these are in effect different reactants and may have quite different reactivities. For the reaction

$$HD^+ + Ar \rightarrow Ar^+ + HD \tag{2.16}$$

the plot of cross section against reactants' kinetic energy shows several maxima, the spacing of which agrees with the spacing of vibrational levels of the ion HD^+. The maxima have been explained by applying the Massey criterion separately to each of the vibrational states[116].

2.3.2 The Franck–Condon principle

For charge exchange reactions—unaccompanied by atom transfer or molecular rearrangements—an important limiting case is that in which the transfer event is rapid compared with internal molecular motions. Whether or not this condition holds will depend on the speed of the collision and on the nature of the molecules concerned, but when it does hold the transfer process must be governed by the Franck–Condon principle. Consider, for example, the reaction

$$A^+ + XY \rightarrow A + XY^+ \tag{2.17}$$

where XY is a diatomic molecule and A^+ is monatomic. If the electron transfer act is instantaneous, the overall reaction consists of two steps

$$A^+ + XY \rightarrow A + (XY^+)^* \rightarrow A + XY^+ \tag{2.18}$$

the intermediate $(XY^+)^*$ having the same bond distance as XY. Thus the first step is analogous to the 'vertical' photo-ionisation reaction process

$$XY \rightarrow (XY^+)^* + e^- \tag{2.19}$$

An energy state diagram for equation (2.19) is shown in *Figure 2.3*. The probability of a transition from one electronic state (i) to another (j) is

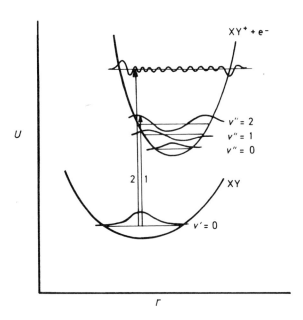

Figure 2.3 Energy curves for a diatomic molecule XY and the ionised form XY^+, showing the operation of the Franck–Condon principle

proportional to the overlap integral between the two vibrational wavefunctions, known as the Franck–Condon factor:

$$FC - \int \psi_i \psi_j^* \, dr \qquad (2.20)$$

For the lowest vibrational level v_0 of any electronic state, the wavefunction has a single maximum; for the higher states it has two maxima at the extreme values of r, and smaller maxima in between. In *Figure 2.3*, transition 1 is allowed because the maxima of the vibrational wavefunctions largely coincide, but transition 2 is forbidden.

Applying this principle to the electron transfer reaction in equation (2.17) the argument is that if the ionisation energy of A exactly matches the energy of an allowed ionisation of XY, the reactants and products have a common energy level with a large vibrational overlap, but if the ionisation energy of A coincides with a forbidden ionisation of XY the vibrational overlap is small. Therefore, the reaction is slower in the second case than in the first, in spite of the fact that the overall energy change ($\Delta U(\infty)$ for equation (2.17)) is more favourable.

The essential conditions for this argument are that the duration of the collision is negligible and that all relevant parameters other than the FC factor are equal. In the photo-ionisation process we are comparing two sets of transitions between the same two electronic states (XY and $(XY^+)^*$ in equations (2.18) and (2.19)), hence the electronic transition moments are equal. The Franck–Condon factor merely determines the shape of the resulting vibrationally broadened absorption band, not its overall intensity. In the electron transfer process, however, we are comparing reactions of one molecule XY with two different reductants. If the interactions of the electronic wavefunctions of the molecule XY with those of the two ions are significantly different, these differences may outweigh the effects of the Franck–Condon factors.

One type of reaction, not strictly an electron transfer, which undoubtedly is controlled by Franck–Condon factors is collision between two neutral species with ejection of an electron (Penning ionisation), for example, the reaction

$$NO + Hg^* \rightarrow NO^+ + Hg + e^- \qquad (2.21)$$

where Hg^* denotes an excited state of the Hg atom[162]. The photoelectron spectrum of NO shows a fine structure with peaks at constant intervals, interpreted as transitions from the lowest vibrational level of NO to the first seven levels of the electronic ground state of NO^+ ($v'' = 0, 1, \ldots, 6$). The reaction in equation (2.21) produces electrons with a spectrum of energies with similar fine structure, corresponding to the production of NO^+ in states $v'' = 0$ to 5 inclusive; a correlation of the two spectra shows close agreement in the relative intensities[162].

Turning to electron transfer reactions proper, the examples which illustrate Franck–Condon effects may be classed according to the methods used to vary the relative energies of reactants and products.

(a) Variation of the kinetic energy of the reactants

Reactions of the type

$$M + B \rightarrow M^- + B^+ \qquad (2.22)$$

have been widely used for the determination of electron affinities of both atoms and molecules[77, 349, 407, 562]. In equation (2.22) B is a halide ion or other species of known ionisation energy and M is the atom or molecule of unknown electron affinity. In general it is found that if $\Delta U(\infty) > 0$ no reaction will occur, but as the collision energy is increased a threshold energy is reached after which the reaction cross section begins to rise with increasing energy, though in practice there is always a 'tail' at low energy due to the distribution of thermal energies of the reacting molecules (*Figure 2.4(a)*). Even when the contribution from the spread of thermal energies is subtracted there is still a tail at low energies which is due to Franck–Condon factors: the curve of cross section against energy can then be analysed to yield both the adiabatic and the vertical electron affinity of the oxidant M. (*Figure 2.4(b)*)[114, 659]. A much-studied example is the molecule NO_2. Reductants used include I^-[114, 659], and the atoms Li, Na, K[58] and Cs[785]. The Franck–Condon effect in this case is mainly attributed to difference in bond angle[(b)] between NO_2, $134°$[763], and NO_2^- ion, $115°$[210].

(b) For a calculation of the Franck–Condon parameters for the photo-detachment process $NO_2^- \rightarrow NO_2 + e^-$, see reference 538; and see also reference 889

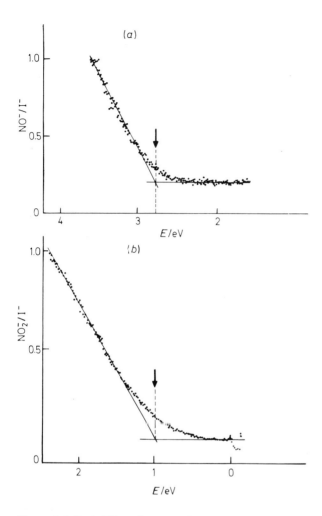

Figure 2.4 Probability of reaction $I^- + X \rightarrow I + X^-$, *(expressed as a ratio of* X^- *produced to* I^- *supplied), as function of the relative kinetic energy E of reactants:* (a) $X = NO$; (b) $X = NO_2$. *In each case the arrow denotes the threshold energy, and the longer 'tail' in (b) is attributed to the Franck–Condon effect, due to the difference in geometry between* NO_2 *and* NO_2^-. *(After Berkowitz et al.[114])*

(b) Variation of the internal energy of the reactants

A number of studies[22, 253, 654] have been made in which one of the reactants is a diatomic molecule and the rates of reaction of particular vibrationally excited states have been determined. For example, for the reaction

$$He^+ + N_2 \rightarrow He + N_2^+ \tag{2.23}$$

the product N_2^+ has been shown, by means of its emission spectrum, to be in the electronic state $C^2\Sigma_u^+$, with vibrational quantum number $v'' = 4$ [654].

This is an almost exactly resonant charge transfer, having $\Delta U(\infty)$ between 0 and -0.01 eV. The reaction is fast and the photoelectron spectrum of N_2 shows that the ($v' = 0$ to $v'' = 4$) transition has the required large Franck–Condon factor. In contrast, the corresponding reaction with the Ne^+ ion is too slow to be detected, in spite of being highly exothermic ($\Delta U(\infty) = -6.0$ eV):

$$Ne^+ + N_2 \rightarrow Ne + N_2^+ \tag{2.24}$$

The rate constant is less than 10^{-14} cm^3 s^{-1} at 300 K ($k < 6 \times 10^6$ M^{-1} s^{-1}), or less than one reaction event for every 10^5 Langevin collisions, but it increases rapidly with increasing population of the higher vibrational levels of N_2[22]. Among the vibrational levels of N_2^+, the one nearest to resonance (i.e. giving $\Delta U(\infty) \simeq 0$ for equation (2.24) with ground state N_2) is the level $v'' = 10$; and the Franck–Condon factor is extremely small. From these and similar observations, it has been concluded that in general 'the overriding consideration necessary for a fast thermal-energy charge transfer reaction of an atomic ion with a molecule is that the molecular ion have an energy level which is resonant with recombination energy of the approaching ion, and that this resonant energy possesses a favourable FC factor with the ground state of the neutral molecule'[654(c)].

The reaction

$$N_2O + Ba \rightarrow N_2 + BaO \tag{2.25}$$

has been studied[1108] under orthodox chemical conditions, i.e. with the reactants in random thermal motion. It is considered to proceed by the two-step process

$$N_2O + Ba \rightarrow N_2O^- + Ba^+ \rightarrow N_2 + BaO \tag{2.26}$$

The electron transfer step is subject to Franck–Condon restrictions and it has been argued that the N_2O molecule reacts preferentially via non-linear vibrationally excited states. The measured increase of population of these states with temperature accounts correctly for the observed temperature dependence of the reaction rate. The importance of this mode is explained by the fact that the electron-affinity of N_2O increases with the bending of the molecule[408]. (The electronic ground state of N_2O^- may be compared with those of the isoelectronic NO_2 molecule and CO_2^- ion.)

When both reactants are molecules, the theory becomes correspondingly more complex. The reactions $H_2^+ + H_2$, $D_2^+ + D_2$ have been studied in some detail[528, 913, 968] and Flannery and co-workers[252, 694, 765] have published theoretical analyses of several reactions of the type

$$AB^+ + AB \rightarrow AB + AB^+ \tag{2.27}$$

where AB are diatomic molecules such as H_2, O_2, CO. For different initial vibrational states of the reactants, and for a range of ion kinetic energies the

---

(c) For clarity the wording of this quotation has been slightly amended

authors calculate total charge transfer cross sections and the distributions of the two product molecules over their respective vibrational states. Not surprisingly, the higher the initial kinetic energy, the higher are the product states which have to be included[252, 694, 765].

The reaction

$$He_2^+ + N_2 \rightarrow N_2^+ + 2He \qquad (2.28)$$

is unusual in that when an electron is added to He_2^+, the immediate product He_2 is in a strongly repulsive state. Hence the electron affinity of He_2^+ depends greatly on the He–He bond distance, and within the range of zero point vibrations of He_2^+, the electron affinities vary continuously from 18.2 to 20.3 eV. The product N_2^+, identified by its emission spectrum as being in the state $B^2\Sigma_g^+$, also has an electron affinity within this range, consistent with the resonance condition; and the population of vibrational states is consistent with the Franck–Condon factors for the 'vertical' process $N_2 \rightarrow N_2^+ + e^-$ [100, 661].

(c) Use of different reactants

There have been a number of studies of series of closely related reactions; for example, of the rare gas ions with a common molecule. Bowers and Elleman[147] studied the reactions

$$A^+ + CH_4 \rightarrow A + CH_4^+ \qquad (2.29)$$

with A = He, Ne, Ar, Kr, Xe. The photoelectron spectrum of CH_4 consists of broad bands, due to the close spacing of the vibrational levels[170], but the relative intensities of the spectrum at the five energies corresponding to the ionisation energies of the rare gas atoms were taken as measuring the relevant Franck–Condon factors. The rate constants for the five reactions correlate well with these intensities, in the order $He \gg Ne \ll Ar < Kr \gg Xe$. They do not correlate with the ionisation energies themselves, nor with rate constants calculated from a Langevin-type model, both of which vary monotonically along the rare gas series.

Similar results have been reported with a number of other polyatomic molecules, such as C_2H_6, C_3H_8, SiH_4 [147], NH_3 and PH_3 [219]. The conclusion that such rate comparisons are governed mainly by Franck–Condon factors is not, however, universally accepted, and some of the data on which it is based (though not the examples cited above) have been challenged. In particular it appears that reactions with Kr^+ and Xe^+ as oxidants may, in some cases at least, be more rapid in relation to other rare gas ions than is predicted on the basis of Franck–Condon factors alone[518]. This should not cause any surprise: it is far from obvious that the electronic interactions between Xe^+ and N_2 will be identical with those between He^+ and N_2.

As pointed out on p. 35, the Franck–Condon principle applies only when the electron transfer process is effectively instantaneous, that is, when the electronic wavefunctions of the reacting pair change rapidly compared with motions of the atoms. The conditions under which this ceases to be the case have been discussed by a number of authors[219, 406, 682]. For example, in the reaction $NO_2 + Cs \rightarrow NO_2^- + Cs^+$ near the threshold energy, the relative velocity of the reactants is 3.3×10^5 cm s^{-1}. Nalley et al.[784] argue that the effective orbital radius of the most loosely bound electron of Cs is about 10 Å, and the time taken for a caesium atom to cross a circle of this diameter is therefore 3×10^{-13} s. During this time, the NO_2 molecule can execute its bending vibration about 10 times and its stretching vibration even more quickly; hence, it has ample time to adjust to the configuration appropriate to the NO_2^- product. More generally, if the relative velocity of the reactants is low enough, and the interaction between them is strong, one or both molecules may be distorted gradually as the 'collision' progresses[406]. This is analogous to the reorganisation effects which govern electron transfer in solution (see Chapter 6). In the limiting case that a collision complex is formed, charge transfer may be rapid in spite of an unfavourable Franck–Condon factor[406].

2.3.3 Reactions with atom transfer: electronic state correlation

Reactions of the type

$$H_2^+ + He \rightarrow H + HHe^+ \qquad (2.30)$$

$$H_2 + He^+ \rightarrow H + HHe^+ \qquad (2.31)$$

$$H_2^+ + Ar \rightarrow H + HAr^+ \qquad (2.32)$$

$$H_2 + Ar^+ \rightarrow H + HAr^+ \qquad (2.33)$$

with atom transfer, must involve close approach of the reactants, and the electronic energy levels must change substantially as a reaction proceeds. The reaction rates and pathways of several such reactions have been successfully rationalised by means of molecular orbital correlations. For example, it is observed that reaction (2.30) proceeds readily but reaction (2.31) does not, although the latter is thermodynamically the more favoured ($\Delta U(\infty) = 0.8$ eV and -8.3 eV respectively), and the same is true of the analogous reaction of Ne; for Ar, however, both reactions (2.32) and (2.33) proceed readily. Mahan[710, 711] has discussed these systems using two different but equivalent formalisms: orbital correlation and state correlation.

The orbital correlation diagram[710] is shown in *Figure 2.5*. In *Figure 2.5(a)* the reactants are written $(HH + He)^+$, which may denote either of the pairs of reactants in equations (2.30) and (2.31); and the products as $(H + HHe)^+$ which may denote the observed products of the two reactions, or other possibilities such as $H^- + HHe^{2+}$. The orbitals shown on the reactants side are the lowest energy (1s) orbital of He, and the two lowest energy orbitals of H_2, bonding σ_g and anti-bonding σ_u^*. On the products side they are the 1s orbital of H, and the two lowest orbitals of HHe$^+$, bonding 1σ and excited 2σ. The simplest possible collision geometry is selected: the He atom is assumed to approach along the axis of the H_2^+ molecule, and the H$^+$ ion is assumed to depart in the same direction. The two lowest orbitals are then found to correlate directly via the

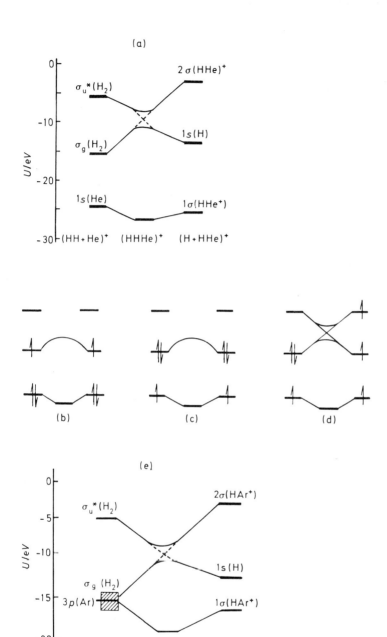

Figure 2.5(a) Orbital correlation diagram for the system $(H-H-He)^+$. *Orbital occupancy for the reactions (b)* $H_2^+ + He \rightarrow H + HHe^+$; *(c)* $H_2 + He^+ \rightarrow H^- + HHe^{2+}$; *(d)* $H_2 + He^+ \rightarrow H + (HHe^+)^*$. *(e) Orbital correlation diagram for the reacting systems* $H_2^+ + Ar \rightarrow H + HAr^+$ *and* $H_2 + Ar^+ \rightarrow H + HAr^+$. *The shaded area on the left hand side indicates doubt as to the relative energies of the* $3p(Ar)$ *and* $\sigma_g(H_2)$ *orbitals. (After Mahan[710])*

lower lying molecular orbital of the intermediate (HHHe)$^+$ configuration. The σ_g orbital of H_2 on the other hand correlates with 2σ of HHe$^+$, and the σ_u^* of H_2 with the 1s orbital of H. This leads to a crossing of the zero-order states as shown by the broken lines in *Figure 2.5(a)*; but the crossing is avoided by resonance as shown by the full lines. The reactants of equation (2.30) are represented by entering three electrons in the lowest possible orbitals on the left and the products by three electrons in the lowest orbitals on the right as shown in *Figure 2.5(b)*. It is evident that a smooth transition is possible from one to the other, as is observed. The reactants of equation (2.31) are represented as shown in *Figure 2.5(c)*, from which it appears that they cannot lead to the same products. They correlate adiabatically with H$^-$ + HHe^{2+} as shown in *Figure 2.5(c)*, or alternatively, if the two electrons of H_2 proceed to different orbitals, with the still higher state consisting of the H atom and the excited molecule HHe$^+$ $(1\sigma^1)(2\sigma^1)$, as shown in *Figure 2.5(d)*. Since this has one electron in the bonding and one in the anti-bonding orbital, it would presumably dissociate into He$^+$ and H.

The similar behaviour of the $(H_2 + Ne)^+$ system can be explained in a similar way. For the $(H_2 + Ar)^+$ system on the other hand there is a significant difference (*Figure 2.5(e)*). The ionisation energies of an argon atom and of H_2 are nearly the same, hence the reactant orbitals $3p(Ar)$ and $\sigma_g(H_2)$ must be placed at nearly the same energy level in the diagram. As before, both these orbitals have the appropriate symmetry to correlate with 1σ of (HAr$^+$) and with 1s of H, but since these energies are similar it is not possible uniquely to associate the $3p(Ar)$ with 1σ (HAr$^+$), and $\sigma_g(H_2)$ with 1s(H) as before. Thus both sets of reactants have an adiabatic minimum energy pathway to the products as shown, and both equations (2.32) and (2.33) can be explained.

Mahan[711] has since restated these arguments in terms of the total energies of the electronic states involved. In *Figure 2.6(a)*, curve A is the ground state potential energy curve for the H_2^+ molecule as a function of intermediate distance and curve B is the ground-state curve for the HeH$^+$ molecule. Taking the energy of the fully separated atomic species (H + H$^+$ + He) as the zero, the two curves may be superimposed and considered as sections through the three-dimensional energy surface of the ground-state of the system (H ... H ... He)$^+$. Since both diatomic species (represented by the minimum points a, b) correlate adiabatically to the separated atomic state, they may be expected to correlate adiabatically with each other via some more direct route, symbolised by the dotted arrow. This corresponds to the reaction shown in *Figure 2.5(b)* and equation (2.30). Curve C is the energy curve of the ground state of H_2, leading to the separated-atom state H + H + He$^+$, and the reactants of equation (2.31) are represented by point c. Curve C does not correlate adiabatically with curve B, so no smooth transition corresponding to equation (2.31) is possible. The most feasible reaction path suggested by the diagram is from curve C to curve A', the curve of the anti-bonding state of H_2^+ in H_2^+ + He, via the crossing point. This leads only to the separated atoms H + H$^+$ + He.

The analogous diagram for the argon system is shown in *Figure 2.6(b)*. The

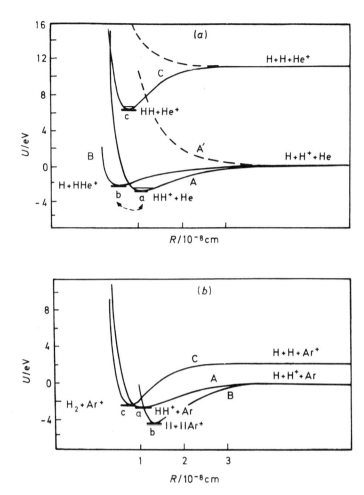

Figure 2.6 Potential energy curve correlation diagrams: (a) for the system $H_2^+ + He \rightleftarrows H + HHe^+$; (b) for the systems $H_2^+ + Ar \rightleftarrows H + HAr^+ \rightleftarrows H_2 + Ar^+$ (After Mahan[711])

reactants and products of equation (2.32), shown at points a and b, again correlate adiabatically with the separated-atom state $(H + H^+ + Ar)$, and presumably they are connected to each other by a low-energy pathway. However, the alternative reactants $(H_2 + Ar^+)$ of equation (2.33) represented by point c, are now at much lower energy owing to the lower ionisation energy of Ar. Thus the two curves A and C cross each other when the H–H distance is close to the equilibrium distance of the H_2 molecule. When the H_2–Ar distance is small (not shown in the figure) resonance is possible and there is a low-energy adiabatic pathway from c to a, and thence to b as observed.

The alternative reaction pathway, c to a without further progress to b, is also known. It is one of the examples of near-resonant transfer discussed above (equation 2.16). Which of the pathways is actually followed will depend on details of the collision. If the energy and impact parameter are such that the Ar^+ atom passes rapidly by the H_2 molecule (a 'grazing' collision), simple electron transfer may be expected; but at lower energy and/or with a more direct collision, atom transfer will be more probable. Detailed calculations of the reaction probability, for the two reaction paths, as functions of the impact parameters, are consistent with this qualitative statement[218].

Orbital correlation arguments have been successfully applied to many other reactions including those which could be classed as purely atom transfer or substitution processes, for example

$$D_2 + CO^+ \rightarrow D + DCO^+ \tag{2.34}$$

and the success of this method in predicting the feasibility and steric course of 'electrocyclic' and other reactions between organic molecules is well known[1107].

2.4 Large molecules

There are very few kinetic data for electron transfer involving large molecules, and it is not yet possible to make any but the most obvious generalisations.

Ast et al.[44] studied the reaction

$$Ar^{2+} + C_6H_6 \rightarrow Ar^+ + C_6H_6^{+*} \tag{2.35}$$

by a mass-spectrophotometric method, the benzene molecule being supplied in a high-energy beam, the Ar^{2+} ions as randomly moving 'targets'. The bulk of the reaction took place with no loss of kinetic energy, implying that the $C_6H_6^+$ product was of an excited state such that $\Delta U(\infty) \simeq 0$. It is expected[536] that this will normally be possible since the vibrational levels of a large molecule are so close together as to form almost a continuum. So long as reactants and products are reasonably well matched in energy, vibrational adjustment can occur in such a way as to produce an exact resonance. In another series of reactions of the type

$$C_6H_5X^+ + C_6H_5Y \rightarrow C_6H_5X + C_6H_5Y^+ \tag{2.36}$$

where X, Y are various functional groups, the dependence of cross section upon beam energy has the form characteristic of resonance transfer[159].

Henglein and Muccini[536] have drawn attention to a number of instances at which cross sections have been measured for two symmetrical processes and also for the unsymmetrical process involving the same reactants (in solution chemistry parlance, the cross reaction):

$$A^+ + A \rightarrow A + A^+ \tag{2.37}$$

$$B^+ + B \rightarrow B + B^+ \tag{2.38}$$

$$A^+ + B \rightarrow A + B^+ \tag{2.39}$$

In some cases where both A and B were molecules, the cross reaction was faster than either of the two symmetrical processes. Examples include $H_2^+ + H_2O$, $H_2O^+ + C_6H_6$. This phenomenon is well known for solution reactions, and provided that A and B do not differ too widely in structure, the rate of the cross reaction is related to the corresponding energy change (see p. 205). No such correlation was noted however for the gas phase systems.

Part 2

Electron transfer in solution

Chapter 3
The reaction path (1): oxidation states

3.1 Introduction and definitions

3.1.1 Reaction mechanisms

The aspects of a reaction mechanism in solution which can be studied experimentally may conveniently be grouped into four classes which may be regarded as separate research problems: (a) the reaction path, (b) the composition of the transition states, (c) the structures of the transition states, and (d) the specific motions within the transition state which lead to the overall reaction. By *reaction path* we understand the complete network of elementary reactions connecting reactants to products via any intermediates. The connecting reactions are the *elementary steps*. The concept of a *transition state* has been much discussed but in the remainder of this book we shall use, without further analysis, a simple definition: the transition state of an elementary reaction is a high-energy intermediate which undergoes unimolecular breakdown into products with a characteristic rate constant $k^{\ddagger} = RT/Lh$. The four problems listed above could be regarded as a programme for determining the mechanism of any given reaction, though not necessarily to be undertaken in that order.

In this chapter we discuss together (a) and (b), reaction pathways and kinetics, as far as changes in oxidation state are concerned; in chapter 4 we consider the same two problems, further subdividing the reaction steps where there is evidence of association between oxidant and reductant and in chapter 5 we review problem (c). In the present state of the subject we shall have relatively little to say about problem (d).

3.1.2 Kinetics

The *stoicheiometry* of a chemical reaction is represented by the balanced chemical equation

$$a\text{A} + b\text{B} + \ldots = p\text{P} + q\text{Q} + \ldots \tag{3.1}$$

where A, B, ... and P, Q, ... denote reactants and products, and the stoicheiometric coefficients, $a, b \ldots, p, q \ldots$ are integers with no common factor. With this convention, the rate of reaction is

$$\text{Rate} = -\frac{1}{a}\left(\frac{d[A]}{dt}\right) = -\frac{1}{b}\left(\frac{d[B]}{dt}\right) = \ldots = +\frac{1}{p}\left(\frac{d[P]}{dt}\right) = \text{etc.} \quad (3.2)$$

The rate of an elementary reaction is in general, given by an expression of the form

$$\text{Rate} = k[A]^\alpha[B]^\beta \ldots \quad (3.3)$$

where A, B, etc. are the reagents which have been found by experiment to influence the rate, and the exponents α, β, ... are not in general equal to a, b, ... According to the transition state theory, this rate law implies the mechanism

$$\alpha A + \beta B + \ldots \underset{}{\overset{K^\ddagger}{\rightleftharpoons}} T^\ddagger \xrightarrow{k^\ddagger} \text{products} \quad (3.4)$$

where the transition state $T^\ddagger$ has the chemical composition $A_\alpha B_\beta \ldots$, and $k^\ddagger K^\ddagger = k$. We shall abbreviate this to

$$\alpha A + \beta B + \ldots \xrightarrow{k} \text{products} \quad (3.5)$$

using the arrow notation to distinguish a mechanistic from a stoicheiometric equation. It is clear from these equations that in general the empirical composition of the transition state as deduced from the form of the rate law is subject to the same ambiguities as the composition of any other complex deduced from equilibrium studies: the rate law does not indicate the number of solvent molecules, nor does it show any other reagent which may be present at constant concentration throughout a series of experiments.

When a slow step is preceded by rapid equilibria, as in a reaction of the type

$$A + B \underset{}{\overset{K_1}{\rightleftharpoons}} C \xrightarrow{k_2} \text{products} \quad (3.6)$$

the rate law may be written in alternative ways which are entirely equivalent, e.g.

$$\text{Rate} = k_2[C] \quad (3.7)$$

$$\text{Rate} = k_2 K_1[A][B] \quad (3.8)$$

The two steps of (3.6) and (3.7) may also be aggregated into one, thus

$$A + B \underset{}{\overset{K_2^\ddagger}{\rightleftharpoons}} T^\ddagger \xrightarrow{k^\ddagger} \text{products} \quad (3.9)$$

with $k^\ddagger K_2^\ddagger = k_2' = k_2 K_1$. Equation (3.9) specifies the *net activation process*[808] without specifying the route by which the transition state is formed. Parameters $\Delta G_2'^\ddagger$, $\Delta H_2'^\ddagger$, $\Delta S_2'^\ddagger$, defined by $\Delta G_2'^\ddagger = \Delta G_1 + \Delta G_2^\ddagger$ etc., are called net activation parameters. A similar procedure may be used to describe consecutive steps even when these are kinetically distinguishable. For example, the equations

$$A + B \underset{k_{-1}}{\overset{k_1}{\rightleftharpoons}} X \tag{3.10}$$

$$X + B \xrightarrow{k_2} \text{products} \tag{3.11}$$

represent a bimolecular step leading to a high-energy intermediate X, followed by a second bimolecular step leading to the products. Application of the steady-state criterion $d[X]/dt = 0$ leads to the rate law

$$\text{Rate} = k_1 k_2 [A][B]^2/(k_{-1} + k_2[B]) \tag{3.12}$$

Although the mechanism contains three rate constants, the rate law contains only two independent experimental parameters. These may be expressed in various ways (e.g. k_1 and k_2/k_{-1}), but for many purposes a good choice is k_1 and $k_1 k_2/k_{-1} = K_1 k_2$. These refer, respectively, to the activation process for step 1, $A + B \rightleftharpoons T_1^{\ddagger}$, and to the net activation process for steps 1 and 2 together, i.e. $A + 2B \rightleftharpoons T_2^{\ddagger}$ where $T_2^{\ddagger}$ is the transition state for step 2. The rate constant $K_1 k_2$ is that which would have been measured if the equilibrium in equation (3.10) had been rapid in comparison with the step in equation (3.11).

For the sake of brevity sequences such as equations (3.10) and (3.11) will often be written in a single line as

$$A + 2B \underset{-1}{\overset{1}{\rightleftharpoons}} X + B \xrightarrow{2} \text{products} \tag{3.13}$$

it being understood that where a reagent appears on both sides of the same step (as here one molecule of B appears on both sides of step 1) that reagent does not occur in the transition state. Also as here, we shall generally omit the k's and K's when writing mechanistic equations, and insert only the subscript numbers or letters.

3.1.3 Electron transfer paths

The typical electron transfer reaction in solution is between reactants each having only two stable oxidation states, differing by one unit:

$$A^+ + B = A + B^+ \tag{3.14}$$

Here A and B denote metal ions with appropriate co-ordinated ligands and, as in the previous chapters, the + sign is used to denote the higher of the two valency states involved, regardless of the actual ionic changes. Usually in such cases, the mechanism is bimolecular, i.e. of the form

$$A^+ + B \to A + B^+ \tag{3.15}$$

and the rate law is second order, i.e. of the form

$$\text{Rate} = k[A^+][B] \tag{3.16}$$

The principal variants of this mechanism are prior addition of ligands to one or both of the reactants, and stepwise processes involving binuclear intermediates, with or without additional ligands. The latter are considered in detail in chapter 4.

Overall multiple electron transfer must occur when the two reagents each have only two valencies differing by n units.

$$A^{n+} + B = A + B^{n+} \tag{3.17}$$

In almost all cases, however, the intermediate valencies, though unstable, are also known, and the possibility that a two or three-electron transfer reaction proceeds in successive steps with high-energy intermediates must always be considered.

When the valence changes of oxidant and reductant differ, for example

$$mA^{n+} + nB = mA + nB^{m+} \tag{3.18}$$

the mechanistic possibilities are more varied. If we assume that all the elementary steps are bimolecular, such reactions must involve unstable oxidation states as intermediates. Some of the intermediates which have been postulated have been studied independently, but others are known only from the kinetic studies. The individual steps may be one- or two-electron transfer reactions, and they too may require addition or loss of other ligands. Finally, if either or both reagents have more than two accessible valence states all of the above possibilities may be open.

Following the definitions of Halpern[506], reactions of one-to-one stoicheiometry in which the valence changes of oxidant and reductant correspond, are termed *complementary*; the others *non-complementary*.

3.2 Non-complementary reactions[a]

In this section we briefly review reactions in which the oxidant and reductant have only two normal valencies, and the valence changes do not correspond. Complete lists of these and other non-complementary reactions are given in *Tables 3.1–3.3*.

3.2.1 2:1 reactions

(a) General

The principal two-electron reagents are the metals of the B groups of the Periodic Table, and some of the neighbouring transition metals, notably Pt^{IV}/Pt^{II} and Au^{III}/Au^{I}. The middle valencies such as tin(III), platinum(III),

---

[a] For reviews see especially references 505, 506

etc., if known at all, are characterised only as short-lived intermediates[b]. Various non-metals also come into this category but are not considered here. Some other metals could be classed as two-electron reagents in the sense that the intermediate valency, though well-characterised, is unstable with respect to disproportionation: for example U^V is unstable with respect to U^{IV} and U^{VI} and Pu^V is unstable with respect to Pu^{IV} and Pu^{VI}. In practice, however, these metals generally act as one-electron reagents. One-electron reagents are numerous and vary from those such as Ce^{IV}/Ce^{III}, for which no other valency is generally considered possible, to others which have additional valencies, unstable under ordinary conditions, but which must be considered kinetically. Thus chromium(II) is generally oxidised only to chromium(III), but in some cases it yields chromium(IV) as a short-lived intermediate. Titanium(IV) is normally reduced only to titanium(III), and although titanium(II) is also known it has not yet been detected in a metal–metal electron transfer reaction.

(b) Classification of mechanisms[b]

When the two-electron reagent is the oxidant, the stoicheiometric equation is

$$A^{2+} + 2B = A + 2B^+ \tag{3.19}$$

The alternative case with the two-electron reagent as reductant can be regarded as the reverse of equation (3.19) and need not be discussed separately. The principal alternative mechanisms are mechanism a—initial one-electron transfer with intermediate A^+, and mechanism b—initial two-electron transfer with intermediate B^{2+}. The general kinetic equations are derived by applying the steady-state criterion to the unstable intermediate.

Mechanism a

$$A^{2+} + B \underset{-a1}{\overset{a1}{\rightleftharpoons}} A^+ + B^+ \tag{3.20}$$

$$A^+ + B \xrightarrow{a2} A + B^+ \tag{3.21}$$

$$\text{Rate} = \frac{k_{a1}k_{a2}[A^{2+}][B]^2}{k_{-a1}[B^+] + k_{a2}[B]} \tag{3.22}$$

Mechanism b

$$A^{2+} + B \underset{-b1}{\overset{b1}{\rightleftharpoons}} A + B^{2+} \tag{3.23}$$

$$B^{2+} + B \xrightarrow{b2} B^+ + B^+ \tag{3.24}$$

(b) For the detection of lead(III) and tin(III), see e.g. references 140, 141. For indium(II) see references 163, 1019. For platinum(III) and thallium(II), see references cited in text

$$\text{Rate} = \frac{k_{b1}k_{b2}[A^{2+}][B]^2}{k_{-b1}[A]+k_{b2}[B]} \tag{3.25}$$

If the two intermediates are of a similar stability the possibility of an interconversion reaction may be considered:

$$A^+ + B^+ \underset{-i}{\overset{i}{\rightleftharpoons}} A + B^{2+}. \tag{3.26}$$

This in effect means that both mechanisms operate, simultaneously and with cross coupling, as may be seen by writing the complete network:

$$\begin{array}{c} A^+ + B^+ + B \\ {}_{a1}\nearrow \quad {}_{-a1}\swarrow \quad \quad \searrow {}_{a2} \\ A^{2+} + B + B \quad \quad -i \updownarrow i \quad \quad A + B^+ + B^+ \\ {}_{-b1}\nwarrow \quad {}_{b1}\searrow \quad \quad \nearrow {}_{b2} \\ A + B^{2+} + B \end{array} \tag{3.27}$$

Other pathways can be obtained by including reactions between two unstable species:

$$A^+ + A^+ \rightarrow A^{2+} + A \tag{3.28}$$

$$A^+ + B^{2+} \rightarrow A^{2+} + B^+ \tag{3.39}$$

but as yet there is no compelling evidence for any of these.

The range of mechanisms widens still further when other intermediate oxidation states are included, though there is still only a finite number of possibilities as long as the steps are restricted to bimolecular one- or two-electron transfer. Of these, the only one considered in the literature[506], and not as yet verified with an actual example, is the state B^- formed by an initial disproportionation:

Mechanism c

$$B + B \underset{-c1}{\overset{c1}{\rightleftharpoons}} B^+ + B^- \tag{3.30}$$

$$A^{2+} + B^- \xrightarrow{c2} A + B^+ \tag{3.31}$$

$$\text{Rate} = \frac{k_{c1}k_{c2}[A^{2+}][B]^2}{k_{-c1}[B^+]+k_{c2}[A^{2+}]} \tag{3.32}$$

The distinctive feature of these rate laws is the occurrence of product terms in the denominator. This has the effect that the reaction is slowed down by the accumulation of products (sometimes called the *mass-law retardation effect*). In some of the earlier work this led to confusion and the true form of the rate law was not appreciated until further experiments were carried out with extra quantities of reaction products added at the start of the reaction. Testing for retardation by products is now standard practice and, when observed, is strong evidence for a stepwise mechanism.

Depending on the rate constants and reagent concentrations the three rate laws may assume limiting forms according to whether step 1 or 2 is rate-determining:

Mechanism a
$$\begin{cases} \text{Rate} = k_{a1}[A^{2+}][B] & \text{when } k_{-a1}[B^+] \ll k_{a2}[B] \quad (3.33) \\ \text{Rate} = K_{a1}k_{a2}[A^{2+}][B]^2/[B^+] & \text{when } k_{-a1}[B^+] \gg k_{a2}[B] \quad (3.34) \end{cases}$$

Mechanism b
$$\begin{cases} \text{Rate} = k_{b1}[A^{2+}][B] & \text{when } k_{-b1}[A] \ll k_{b2}[B] \quad (3.35) \\ \text{Rate} = K_{b1}k_{b2}[A^{2+}][B]^2/[A] & \text{when } k_{-b1}[A] \gg k_{b2}[B] \quad (3.36) \end{cases}$$

Mechanism c
$$\begin{cases} \text{Rate} = k_{c1}[B]^2 & \text{when } k_{-c1}[B^+] \ll k_{c2}[A^{2+}] \quad (3.37) \\ \text{Rate} = K_{c1}k_{c2}[A^{2+}][B]^2/[B^+] & \text{when } k_{-c1}[B^+] \gg k_{c2}[A^{2+}] \quad (3.38) \end{cases}$$

Two of the possible limiting laws occur twice, hence if either of these is found experimentally, it is not possible to distinguish the mechanisms kinetically, but other lines of evidence may still be open. In some instances, the rates of individual steps may be known from other studies, or the equilibrium constant K_1 may be known, and it may be possible to show whether the above inequalities hold or not. More definitely, it may be possible to obtain evidence of the existence and nature of the reactive intermediate required by the mechanism. Ideally this would be done by applying a physical test sensitive enough to detect the intermediate at low concentration: for example, the intermediate chromium(V) has been detected by e.s.r. in some reactions of chromium(VI) with organic reductants[494, 757, 964]. More usually indirect chemical tests are employed. These are of two sorts: a reagent may be added which is expected to generate additional concentrations of the intermediate and so catalyse the reaction; alternatively, the added reagent may be such as to react with the intermediate and suffer oxidation or reduction in the process. *Induced* reactions, so called, have been extensively documented.

(c) Examples (see Table 3.1)

Thallium(III)–Thallium(I). In the group of reactions with thallium(III) as oxidant, the most powerful reductant so far used is chromium(II). There are no kinetic data other than the observation that the reaction is too fast to follow, but there is evidence that the primary step is a two-electron transfer, mechanism b. The chromium(III) product is the binuclear species $Cr_2(OH)_2^{4+}$ which would be expected to result from the reproportionation step b2:

$$Tl^{III} + 2Cr^{II} \xrightarrow{b1} Tl^{I} + Cr^{IV} + Cr^{II} \xrightarrow{b2} Tl^{I} + (Cr^{III})_2 \quad (3.39)$$

Table 3.1 Rate laws and mechanisms of 2:1 non-complementary reactions (stoicheiometry $A^{2+} + 2B = A + 2B^+$)

Reagents[a] A^+	B^+	Mechanism[b]	Rate-determining step(s)[b]	References
Tl^{III}	Cr^{II}	b	?	37
Tl^{III}	V^{III}	a	1	276
Tl^{III}	Fe^{II}	a	1, 2	41, 398–9, 424, 461, 544, 598, 930
Tl^{III}	$Os(bipy)_3^{2+}$	a	1	588
Tl^{III}	V^{IV}	a	1	545, 994
Tl^{III}	$Fe(bipy)_3^{2+}$ [c]	a		
Tl^{III}	$Fe(phen)_3^{2+}$	a	1	456
Tl^{III}	$Ru(phen)_3^{2+}$	a	1	756
Tl^{III}	Ce^{III} [d]	a	1, 2	339
			2	487, 1069
Tl^{III}	Mn^{II} [d]	a	2	399, 907
Tl^{III}	Co^{II} [d]	a	1, 2	42, 398, 399
Tl^{III}	Np^{VI} [d]	a or b	2	1032
Pu^{VI}	Ti^{III}	a	1	869
Mn^{VII}	V^{IV}	a	1	762
Pu^{VI}	U^{IV}	a	1	793
U^{VI}	Fe^{II} [d]	a	1, 2	118
Sn^{IV}	Fe^{II} [d]	a	1, 2	1089
Sn^{IV}	Fe^{II} [d]	a	2	934[e]
Sn^{IV}	V^{IV} [d]	a	2	279
Sn^{IV}	Ce^{III} [d]	a	2	168, 370[e]
Sn^{IV}	Mn^{II} [d]	a	2	712
Sn^{IV}	Co^{II} [d]	a or b	2	277, 1089
In^{III}	Fe^{II} [d]	a	2	1018
$AuCl_4^-$	Fe^{II}	a	2	888
$Pt(NH_3)_5Cl^{3+}$	V^{II}	a	2	63
$PtCl_6^{2-}$	V^{II}	a	2	63
$Pt(NH_3)_5Cl^{3+}$	$Cr(bipy)_3^{2+}$	a	2	101
trans-$PtL_2X_4^-$	Fe^{2+} [f][g]	a	1, 2	838, 839
trans-$PtL_2X_4^-$	$Fe(CN)_6^{4-}$ [g]	a[h]	1	78
trans-$PtCl_4(OH_2)_2$	Ce^{III} [d]	a[h]	2	1119, 865[k]
$PtCl_6^{2-}$	$IrCl_6^{3-}$ [d]	a	1, 2	510
trans-$PtL_2X_4^-$	$Fe(C_5H_5)_2$ [g]	a[h]	1	840
$Pt(NH_3)_5Cl^{3+}$	Cr^{II}	b	1	102
trans-$PtCl_4(OH_2)_2$	V^{IV} [d]	a	2	865[k]

(a) Aqueous medium, generally with high salt concentration, but free of strongly complexing anions except where otherwise stated. When the reagents are specified only by oxidation state it is to be understood that they occur in whatever complex forms are predominant under these conditions stated, e.g. thallium(I) is Tl^+, but thallium(III) is partially hydrolysed to $TlOH^{2+}$ or $Tl(OH)_2^+$. In most cases the rate is sensitive to hydrogen ion concentration, and generally this has been studied in detail, but the information is omitted here

The reaction with vanadium(IV) is first order in each reagent. This rules out mechanism c which otherwise might be considered plausible; and mechanism b, involving vanadium(VI) is clearly also ruled out. This leaves only mechanism a, with step a1 rate-determining.

$$Tl^{III} + 2V^{IV} \xrightarrow{a1} Tl^{II} + V^V + V^{IV} \xrightarrow{a2} Tl^I + 2V^V \qquad (3.40)$$

The reaction with iron(II), when first investigated[424, 598] appeared also to be of first order in each reagent. However, both original groups of workers found slight decreases in specific rate as the reaction progressed, and Ashurst and Higginson[41], in a definitive study, showed that the reaction was retarded by iron(III). This established that the reaction proceeds by mechanism a, taking the form

$$Tl^{III} + 2Fe^{II} \underset{-a1}{\overset{a1}{\rightleftarrows}} Tl^{II} + Fe^{III} + Fe^{II} \xrightarrow{a2} Tl^I + 2Fe^{III} \qquad (3.41)$$

It is interesting to note that in the presence of a platinum metal surface, the inhibitory effects disappear, and the rate law reverts to[461]

$$\text{Rate} = k_{a1}[Tl^{III}][Fe^{II}] \qquad (3.42)$$

with the same rate constant, k_{a1}[544]. Evidently the metal surface catalyses the forward reaction a2 by two simultaneous electrode reactions, such as

$$Tl^{II} + e^- \rightarrow Tl^I \qquad (3.43)$$

$$Fe^{III} \rightarrow Fe^{II} + e^- \qquad (3.44)$$

The reaction with cobalt(II) shown in *Table 3.1* was actually studied in reverse, as the oxidation of thallium(I) by cobalt(III). The rate law observed was close to the form

$$\text{Rate} = k_{-2}[Tl^I][Co^{III}] \qquad (3.45)$$

but a slight retardation was observed when cobalt(II) was added. There being no retardation by the other product thallium(III), the mechanism would

---

(b) See text, p. 53
(c) Similar rate laws are found with a number of other reductants of the general type ML_3^{2+}, where M is Fe, Ru and L is aromatic dietertiaryamine
(d) Reaction actually studied in reverse, $A + 2B^+ = A^{2+} + 2B$
(e) Halide ion catalysis
(f) Reaction studied in both directions
(g) L is $MeNH_2$, $EtNH_2$, $PrNH_2$, $\frac{1}{2}$en; where X is Cl, Br
(h) Assumed
(i) L is Et_3As, $n-Pr_3P$, Et_3P, Et_2S, piperidine; X is Cl, Br. Reaction in non-aqueous solvents
(j) No detectable reaction $Mn^{III} + Tl^I$, but reaction $Mn^{II} + Tl^{II}$ has been measured
(k) Various other platinum complexes also studied

appear to be the reverse of mechanism a:

$$2Co^{III} + Tl^{I} \underset{a2}{\overset{-a2}{\rightleftharpoons}} Co^{III} + Co^{II} + Tl^{II} \overset{-a1}{\longrightarrow} 2Co^{II} + Tl^{III} \qquad (3.46)$$

The form of the rate law shown in *Table 3.1* has been derived by combining the experimental rate law with the equilibrium expression

$$k_{a1}k_{a2}/k_{-a1}k_{-a2} = [Tl^{III}][Co^{II}]^2/[Tl^{I}][Co^{III}]^2 \qquad (3.47)$$

The reaction with cerium, given as $Tl^{III} + 2Ce^{III}$, was also studied in reverse, as $Tl^{I} + 2Ce^{IV}$. The extraordinary slowness of the latter reaction, in spite of the favourable redox potentials, was remarked on as early as 1933[940]. In the earliest experiments, no reaction could be detected at all, and a solid compound has been prepared, $Tl^{I}Tl^{III}Ce_2^{IV}(SO_4)_4$, which shows no tendency to decompose internally. The reaction is catalysed by platinum black [940], and by added manganese(II), and in separate experiments, Shaffer showed that cerium(IV) oxidises manganese(II) successively to manganese(III) and manganese(IV), while manganese (IV) (in the form of hydrous MnO_2) readily oxidises thallium(I). A possible mechanism for the catalysed reaction is therefore

$$2Ce^{IV} + Tl^{I} + Mn^{II} \rightarrow Ce^{III} + Ce^{IV} + Tl^{I} + Mn^{III}$$

$$\rightarrow 2Ce^{III} + Tl^{I} + Mn^{IV} \rightarrow 2Ce^{III} + Tl^{III} + Mn^{II} \qquad (3.48)$$

Later, the reaction $Tl^{III} + 2Ce^{III}$ was studied in nitric acid media[487]. The rate law contained a term first order in each reagent, but also a term zero order in thallium, suggestive of either H_2O or NO_3^- as a reductant.

Observations on Thallium(II). Besides the metal–metal electron transfer processes described here, thallium(II) has been detected as the product of free radical reactions such as

$$Tl^{I} + OH \rightarrow Tl^{II} + OH^{-} \qquad (3.49)$$

where the hydroxy radical is generated from water by pulse radiolysis[213], and in photolytically induced reactions such as

$$[Tl^{III} \ldots OH]^{2+} \overset{h\nu}{\longrightarrow} [Tl^{II} \ldots OH]^{2+} \longrightarrow Tl^{II} + OH \qquad (3.50)$$

Recently two groups of workers[398, 930] have succeeded in generating Tl^{2+} in high enough concentrations for direct observation, using a combination of equations (3.49) and (3.50). These workers have recorded the absorption spectrum of Tl^{2+} and have measured rates of several reactions, including equation (3.41), step a2, and equation (3.46), step a2. These data have yielded the equilibrium constants of the elementary steps and thence the Tl^{+}/Tl^{2+} and Tl^{2+}/Tl^{3+} redox potential[398, 399].

Tin(IV) – Tin(II). All 2:1 reactions involving this couple have been studied as oxidations of tin(II).

The reaction with cerium(IV) is first order in each reagent, and since cerium(II) can safely be excluded, only mechanism (a) (reverse) need be considered:

$$2Ce^{IV} + Sn^{II} \xrightarrow{-a2} Ce^{IV} + Ce^{III} + Sn^{III} \xrightarrow{-a1} 2Ce^{III} + Sn^{IV} \quad (3.51)$$

The reaction with cobalt(III) has the analogous rate law and induction experiments confirm the existence of an unstable intermediate[1089], but they do not distinguish between the alternatives tin(III) (mechanism a) and cobalt(I) (mechanism b).

The reaction

$$2Fe^{III} + Sn^{II} = 2Fe^{II} + Sn^{IV} \quad (3.52)$$

is the subject of a considerable literature dating back as far as 1895[815]. Bray and Gorin[158] in 1932 proposed mechanism b (reverse) with iron(IV) as intermediate; but Weiss[1081] in 1944, reworking some of the older data, established the rate law (equation 3.54) consistent with mechanism a (reverse):

$$2Fe^{III} + Sn^{II} \underset{a2}{\overset{-a2}{\rightleftharpoons}} Fe^{II} + Fe^{III} + Sn^{III} \xrightarrow{-a1} 2Fe^{II} + Sn^{IV} \quad (3.53)$$

$$\text{Rate} = \frac{k_{-a2}k_{-a1}[Fe^{III}]^2[Sn^{II}]}{k_{-a1}[Fe^{III}] + k_{a2}[Fe^{II}]} \quad (3.54)$$

Some earlier workers had overlooked the retardation by iron(II) since, as the equation shows, the extent of retardation diminishes when iron(III) is present in excess. Retardation is also very slight in the presence of chloride ions, presumably because complexation of the iron and tin species changes the ratio k_{-a1}/k_{a2}. Wetton and Higginson[1089] found evidence of the reactive intermediate, again presumably tin(III), by the induced reduction of cobalt(III) complexes.

Mercury(II) – Mercury(0) Table 3.2. The stable reduced form of mercury is the binuclear ion Hg_2^{2+}, of formal oxidation state $+1$, hence reactions of mercury(II) with one-electron reductants have the stoicheiometry

$$2Hg^{II} + 2B = (Hg^{I})_2 + 2B^+ \quad (3.55)$$

which does not strictly conform to the definition of a non-complementary reaction (though the reverse reaction might be said to conform). Mercury(I) however occurs in rapid equilibrium with the oxidation states II and 0, and the relevant equilibrium constant is known

$$(Hg^{I})_2 \rightleftharpoons Hg^{II} + Hg^{0} \quad (3.56)$$

(where Hg^0 denotes the solvated atom, not the bulk metal). It is convenient to

Table 3.2 Summary of 2:1 non-complementary reactions of the $Hg^{II}/Hg^{I}/Hg^{0}$ system

Reductant (B)	Stoicheiometry $Hg^{II} + 2B = (Hg^{I})_2 + 2B^{+}$	
	Rate expression	References
Cr^{II}	$k[Hg^{2+}][B]$	341
V^{II} (a)	$k[Hg^{2+}][B]$	483
Fe^{II}	$k[Hg^{2+}][B]$	2
V^{III} (b)	see text	481

Oxidant (B^{+})	Stoicheiometry $(Hg^{I})_2 + 2B^{+} = Hg^{II} + 2B$	
	Rate expression	References
Ce^{IV}	$k[(Hg^{I})_2][B^{+}]$	702
Co^{II}	$k[(Hg^{I})_2][B^{+}]$	906
Ag^{II} (c)	$k[(Hg^{I})_2][B^{+}]$	545
Np^{VII}	$k[(Hg^{I})_2][B^{+}]$	1029
$Fe(phen)_3^{3+}$	$k[(Hg^{I})_2][B^{+}]$	291
$Ru(bipy)_3^{3+}$	$k[(Hg^{I})_2][B^{+}]$	291
$IrCl_6^{2-}$ (c)	$k[(Hg^{I})_2][B^{+}]$	1116
Mn^{III}	see text	902

(a) There is also a complementary two-electron reaction (see text, p. 74)
(b) For an additional reaction path, see text, p. 75
(c) Reaction studied by catalysis of the overall reaction $Ce^{IV} + Hg_2^{2+}$.

discuss first the reactions of Hg_2^{2+} with one-electron oxidants. With one possible exception, these all follow the rate law

$$\text{Rate} = k[(Hg^{I})_2][B^{+}] \tag{3.57}$$

which requires a three-atom transition state. The most generally suggested mechanism is as follows, where 'Hg_2^{III}' denotes a mixed- or average-valency intermediate (see p. 270).

$$B^{+} + (Hg^{I})_2 \longrightarrow B + Hg_2^{III} \tag{3.58}$$

$$Hg_2^{III} \longrightarrow Hg^{II} + Hg^{I} \tag{3.59}$$

$$B^{+} + Hg^{I} \longrightarrow B + Hg^{II} \tag{3.60}$$

Since there is no evidence that equation (3.59) is reversible, the first two steps may be summed to give

$$B^{+} + (Hg^{I})_2 \rightarrow B + Hg^{II} + Hg^{I} \tag{3.61}$$

Higginson et al.[545] have written the two steps consecutively as shown.

McCurdy and Guilbault[702] on the other hand, considered equation (3.61) as 'concomitant breaking of the $(Hg-Hg)^{2+}$ bond and electron transfer'. However, unless the reverse of equation (3.59) is kinetically significant, the two alternatives are indistinguishable.

Reactions of mercury(II) with one-electron reductants, again with one possible exception, obey the rate law

$$\text{Rate} = k[\text{Hg}^{II}][\text{B}] \tag{3.62}$$

which requires only a two-atom transition state. Many reaction sequences are consistent with this, including the following analogues of mechanism a and b discussed above, i.e.

Mechanism a':

$$\text{Hg}^{II} + \text{B} \underset{-a1}{\overset{a1}{\rightleftharpoons}} \text{Hg}^{I} + \text{B}^{+} \tag{3.63}$$

$$\text{Hg}^{I} + \text{B} \xrightarrow{a2} \text{Hg}^{0} + \text{B}^{+} \tag{3.64}$$

$$\text{Hg}^{0} + \text{Hg}^{II} \xrightarrow{3} (\text{Hg}^{I})_2 \tag{3.65}$$

Mechanism b':

$$\text{Hg}^{II} + \text{B} \underset{-b1}{\overset{b1}{\rightleftharpoons}} \text{Hg}^{0} + \text{B}^{2+} \tag{3.66}$$

$$\text{B}^{2+} + \text{B} \xrightarrow{b2} \text{B}^{+} + \text{B}^{+} \tag{3.67}$$

$$\text{Hg}^{0} + \text{Hg}^{II} \xrightarrow{3} (\text{Hg}^{I})_2 \tag{3.65}$$

But a more economical third possibility is the reverse of equations (3.58)–(3.60) above, i.e. the following, with step 1 rate-determining.

Mechanism a'':

$$\text{Hg}^{II} + \text{B} \underset{-a1}{\overset{a1}{\rightleftharpoons}} \text{Hg}^{I} + \text{B}^{+} \tag{3.68}$$

$$\text{Hg}^{II} + \text{Hg}^{+} \underset{-2}{\overset{2}{\rightleftharpoons}} \text{Hg}_2^{III} \tag{3.69}$$

$$\text{Hg}_2^{III} + \text{B} \xrightarrow{3} (\text{Hg}^{I})_2 + \text{B}^{+} \tag{3.70}$$

The kinetic consequences of this mechanism can be worked out by applying the steady–state condition to both intermediates Hg^{I} and Hg_2^{III}. It seems likely however that step 2 will be rapid in both directions so that the two intermediates will always be in mutual equilibrium, in which case the full rate laws for the forward and backward reactions may be written as

$$\text{Rate}_+ = \frac{k_1 k_3 [\text{Hg}_2^{\text{II}}][\text{B}^+]^2}{k_3[\text{B}^+] + k_{-1} K_2^{-1}[\text{Hg}^{\text{II}}][\text{B}]} \tag{3.71}$$

$$\text{Rate}_- = \frac{k_{-1} k_{-3} K_2^{-1}[\text{Hg}^{\text{II}}]^2[\text{B}]^2}{k_3[\text{B}^+] + k_{-1} K_2^{-1}[\text{Hg}^{\text{II}}][\text{B}]} \tag{3.72}$$

but neither of these composite forms has yet been observed.

Of the exceptional rate laws mentioned above, that of the reaction $\text{Hg}^{\text{II}} + \text{V}^{\text{III}}$ has two terms corresponding to parallel paths, one of them being mechanism a with both steps 1 and 2 kinetically significant (see p. 75).

The reaction $\text{Mn}^{\text{III}} + (\text{Hg}^{\text{I}})_2$ has the stoicheiometry

$$2\text{Mn}^{\text{III}} + (\text{Hg}^{\text{I}})_2 = 2\text{Mn}^{\text{II}} + 2\text{Hg}^{\text{II}} \tag{3.73}$$

and the rate law

$$\text{Rate} = A\frac{[\text{Mn}^{\text{III}}][(\text{Hg}^{\text{I}})_2]}{[\text{Hg}^{\text{II}}]} + B\frac{[\text{Mn}^{\text{III}}]^2[(\text{Hg}^{\text{I}})_2]}{[\text{Mn}^{\text{II}}]} \tag{3.74}$$

where A and B are empirical rate constants. The first term corresponds to the reverse of mechanism a', with step $-a2$ rate-determining

$$(\text{Hg}^{\text{I}})_2 \underset{}{\overset{-3}{\rightleftharpoons}} \text{Hg}^{\text{II}} + \text{Hg}^0 \tag{3.75}$$

$$\text{Hg}^0 + \text{Mn}^{\text{III}} \xrightarrow{-a2} \text{Hg}^{\text{I}} + \text{Mn}^{\text{II}} \tag{3.76}$$

$$\text{Hg}^{\text{I}} + \text{Mn}^{\text{III}} \longrightarrow \text{Hg}^{\text{II}} + \text{Mn}^{\text{II}} \tag{3.77}$$

so that $a = K_3^{-1} k_{-a2}$. The second term corresponds to a mechanism which so far appears to be unique:

$$\text{Mn}^{\text{III}} + \text{Mn}^{\text{III}} \underset{}{\overset{-b2}{\rightleftharpoons}} \text{Mn}^{\text{II}} + \text{Mn}^{\text{IV}} \tag{3.78}$$

$$\text{Mn}^{\text{IV}} + (\text{Hg}^{\text{I}})_2 \xrightarrow{k'} \text{Mn}^{\text{II}} + 2\text{Hg}^{\text{II}} \tag{3.79}$$

with $b = K_{b2}^{-1} k'$. The k' step is rate-determining, with two mercury atoms in the transition state, hence in the reverse direction it would seem to require a three-body collision. Presumably this step must be broken down into a sequence of two-body collisions, at least one of which involves some binuclear intermediate other than $(\text{Hg}^{\text{I}})_2$. Several such sequences can be written down, but the evidence so far available does not distinguish between them. No example of mechanism b' is known. One of two pathways in the reaction $\text{Hg}^{\text{II}} + \text{V}^{\text{II}}$ could be classified under this heading (i.e. reaction (3.145) below), but under the conditions of the experiments, step 2 does not occur and what is observed is the complementary two-electron reaction, step 1. In the reverse direction, as already mentioned, one of two pathways in the reaction $(\text{Hg}^{\text{I}})_2 + \text{Mn}^{\text{III}}$

involves the disproportionation equilibrium, K_{b2}^{-1}, but the rate-determining step equation (3.79) does not correspond to step $-b1$.

For any mechanism involving mononuclear mercury(I) as intermediate, there is an alternative route to the Hg_2^{2+} product i.e.,

$$Hg^I + Hg^I \xrightarrow{4} (Hg^I)_2 \quad (3.80)$$

but in no case has this been found to be kinetically significant. (For example, in mechanism a', if step (3.65) is replaced by a rate-determining step (3.80) we have

$$\text{Rate} = k_4[Hg^I]^2 = k_4 K_{a1}^2 [Hg^{II}]^2 [B]^2 / [B^+]^2 \quad (3.81)$$

but this has never been observed.) Equation (3.80) has often been discussed, but is generally discounted on the grounds that although it must be rapid, the concentration of Hg^I is too low.

Other binuclear complexes Wharton, Ojo and Sykes[1091] have reported a case in which a binuclear complex containing two one-equivalent reducing centres is oxidised in successive one-equivalent steps (equation (3.82), where $(Mo^V)_2 = Mo_2O_4(edta)^{2-}$, $Ir^{IV} = IrCl_6^{2-}$).

$$(Mo^V)_2 + 2Ir^{IV} = 2Mo^{VI} + 2Ir^{III} \quad (3.82)$$

The rate law shows retardation by iridium(III), implying transition states of the composition $Mo^V Mo^V Ir^{IV}$ and $Mo^{VI} Mo^V$:

$$\text{Rate} = a\lfloor (Mo^V)_2 \rfloor \lfloor Ir^{IV} \rfloor / (1 + b\lfloor Ir^{III} \rfloor) \quad (3.83)$$

The proposed reaction sequence is an example of mechanism a"(reverse), except that step -2 is slow:

$$(Mo^V)_2 + Ir^{IV} \underset{3}{\overset{-3}{\rightleftarrows}} Mo^{VI}Mo^V + Ir^{III} \quad (3.84)$$

$$Mo^{VI}Mo^V \xrightarrow{-2} Mo^{VI} + Mo^V \quad (3.85)$$

$$Mo^V + Ir^{IV} \xrightarrow{-a1} Mo^{VI} + Ir^{III} \quad (3.86)$$

In these equations the steps are numbered to conform with equations (3.68)–(3.70) above. Thus the empirical parameters in equation (3.83) are $a = k_{-3}$, $a/b = K_{-3} k_{-2}$.

3.2.2 3:1 reactions

The only examples in this category are reactions of the chromium (VI)/chromium(III) couple with various one-electron reagents. These are

Table 3.3 Rate laws and mechanisms of reactions of chromium(VI) with one-electron reductants

$$Cr^{VI} + 3B \underset{-1}{\overset{1}{\rightleftharpoons}} Cr^{V} + B^{+} + 2B \underset{-2}{\overset{2}{\rightleftharpoons}} Cr^{IV} + 2B^{+} + B \underset{-3}{\overset{3}{\longrightarrow}} Cr^{III} + 3B^{+}$$

Reductant, B[a]	Rate-determining step[b]	References
V^{II}	1	160
V^{III}	1	287
Fe^{II}	1, 2	383, 388, 910
$Fe(phen)_3^{2+}$	1	388
Pu^{IV}	1, 2(?)[d]	805
V^{IV}	2[c]	288, 376, 910
Pu^{III}	1, 2	797
Np^{V}	1, 2	979
$Ce^{III(h)}$	3	1042
Ag^{I}	3	625
Cr^{III}	(e)	29
$Fe(bipy)_3^{2+}$	1	120
$Fe(bipy)_2(CN)_2$	1[f]	120
$Fe(bipy)(CN)_4^{2-}$	1[f]	120
$Fe(CN)_6^{4-}$	1[f]	120
$IrCl_6^{3-}$	2[g]	123
$Mo(CN)_8^{4-}$	1	705
$Ta_6Cl_{12}^{2+}$	1	389
$Ta_6Br_{12}^{2+}$	1[c,i]	389

(a) See *Table 3.1*, footnote (a)
(b) See text, p. 63
(c) Also an analogous term with $(Cr^{VI})^2$ in place of (Cr^{VI})
(d) Pu^{IV} and Pu^{V} dependences apparently not exhaustively studied: hence there may be some Pu^{V} inhibition
(e) Reaction studied is Cr^{III}–Cr^{VI} exchange, Rate = $k[Cr^{VI}]^{2/3}[Cr^{III}]^{4/3}$, implying the rate-determining step $Cr^{V} + Cr^{III} \rightleftarrows 2Cr^{IV}$
(f) Final product is binuclear Cr^{III}–Fe^{III} complex
(g) Mechanism involves more than one form of chromium(V)
(h) Reaction actually studied in reverse
(i) A rate-determining two-electron step is also consistent with the kinetics

summarised in *Table 3.3*. In most cases the evidence is in favour of a stepwise mechanism which may be summarised as

$$Cr^{VI} + B \underset{-1}{\overset{1}{\rightleftharpoons}} Cr^{V} + B^{+} \qquad (3.87)$$

$$Cr^{V} + B \underset{-2}{\overset{2}{\rightleftharpoons}} Cr^{IV} + B^{+} \qquad (3.88)$$

$$Cr^{IV} + B \underset{-3}{\overset{3}{\rightleftharpoons}} Cr^{III} + B^{+} \qquad (3.89)$$

in which either step 1, 2 or 3 may be rate-determining or in certain cases two steps may have comparable rates. In certain systems the observed kinetics are complicated by rapidly established equilibria, especially between the chromium(VI) or chromium(V) and the cation B^+. In most cases the rates are strongly dependent on acidity and from this it has been possible to infer the degree of protonation of the transition state. We shall not discuss these points in detail: a good review is available[382].

The most extensively documented reaction is the reduction by iron(II). Earlier work dating back as far as 1903, has been reviewed by Westheimer[1088] and others[388]. Espenson and King[388] established the rate law

$$\text{Rate} = k[\text{Cr}^{VI}][\text{Fe}^{II}]^2/[\text{Fe}^{III}] \tag{3.90}$$

corresponding to equations (3.87)–(3.89) with step 1 a rapid equilibrium and step 2 rate-determining, so that $k = K_1 k_2$. With lower iron(II) concentrations[383, 910], however, a first-order dependence also becomes apparent, and the full rate law can therefore be written, again in terms of equations (3.87), (3.89), as

$$\text{Rate} = \frac{k_1 k_2 [\text{Cr}^{VI}][\text{Fe}^{II}]^2}{k_2 [\text{Fe}^{II}] + k_{-1}[\text{Fe}^{III}]} \tag{3.91}$$

Participation of lower oxidation states of chromium as intermediates is confirmed by induced oxidation experiments[382, 693, 1088]. When an iodide ion is included in the reaction mixture, under conditions such that the direct reaction of chromium(VI) with the iodide ion does not take place, production of iodine is observed, parallel to the production of iron(III). The ratio of iron(III) to iodine varies according to reagent concentration, but approaches a limiting ratio corresponding to the stoicheiometry.

$$\text{Cr}^{VI} + \text{Fe}^{II} + 2\text{I}^- = \text{Cr}^{III} + \text{Fe}^{III} + \text{I}_2 \tag{3.92}$$

This is consistent with the reaction

$$\text{Cr}^V + 2\text{I}^- \rightarrow \text{Cr}^{III} + \text{I}_2 \tag{3.93}$$

occurring immediately after step 1 of equations (3.87)–(3.89) and so rapidly that the chromium(V) concentration is greatly reduced and steps 2 and 3 cannot effectively compete. The probable mechanism of equation (3.93) is[288]

$$\text{Cr}^V + 2\text{I} \xrightarrow{k_I} \text{Cr}^{III} + \text{I}^+ + \text{I}^- \longrightarrow \text{I}_2 \tag{3.94}$$

where 'I^+' could, for example, be HOI or H_2OI^+. The ratio of iodine to iron(III) product, known as the induction factor[288], F, is given by

$$F = \frac{d[I_2]/dt}{d[\text{Fe}^{III}]/dt} = \frac{k_I[\text{I}^-]}{3k_2[\text{Fe}^{II}] + k_{-1}[\text{Fe}^{III}] + k_I[\text{I}^-]} \tag{3.95}$$

Under conditions where the reverse of step 1 is important, so that in the absence of iodide ion the rate law is equation (3.90), this reduces to

$$F = k_I[I^-]/(k_{-1}[Fe^{III}] + k_I[I^-]) \tag{3.96}$$

which in limits of high $[I^-]$ or low $[Fe^{III}]$ becomes $F = 1$. Various other reagents can undergo induced oxidation, and in all the cases studied, the oxidising intermediate is chromium(V) which reacts by an overall two-electron step. Examples include the following

$$Cr^V + As^{III} \rightarrow Cr^{III} + As^V \tag{3.97}$$

$$Cr^V + S^{IV} \rightarrow Cr^{III} + S^{VI} \tag{3.98}$$

The study of induced oxidation can lead to kinetic information. When the induction factor F varies with reagent concentration, as in equation (3.96), the ratio of rate constants k_{-1}/k_I is obtained. In some instances this ratio has been obtained when both of the overall reactions are too fast to measure. Alternatively, if the overall reaction is slow, but the induced reaction is fast, the latter can be used to obtain k_1, when this cannot be determined directly. For example, vanadium(IV) reduces chromium(VI) only slowly, with the rate law

$$\text{Rate} = \frac{k_1 k_2}{k_{-1}} \frac{[Cr^{VI}][V^{IV}]^2}{[V^V]} \tag{3.99}$$

There is no term first order in vanadium(IV), but the rate constant k_1 has been obtained by performing the reaction $Cr^{VI} + V^{IV}$ in the presence of vanadium(III)[288].

$$Cr^{VI} + V^{IV} \xrightarrow{k_1} Cr^V + V^V \tag{3.100}$$

$$Cr^V + V^{III} \longrightarrow Cr^{III} + V^V \tag{3.101}$$

The induced oxidation, equation (3.101) is fast enough to keep down the chromium(V) concentration so that step -1 is suppressed and step 1 (equation 3.100) becomes rate-determining.

The reaction with silver(I) shown in *Table 3.3*, was actually studied in reverse, as the oxidation of chromium(III) by silver(II), this being generated in a prior reaction of cobalt(III) with silver(I). Thus the overall reaction was the silver(I)-catalysed oxidation of chromium(III) by cobalt(III)

$$Cr^{III} + 3Co^{III} = Cr^{VI} + 3Co^{II} \tag{3.102}$$

with the mechanism

$$Co^{III} + Ag^I \underset{-0}{\overset{0}{\rightleftharpoons}} Co^{II} + Ag^{II} \tag{3.103}$$

$$Cr^{III} + Ag^{II} \underset{3}{\overset{-3}{\rightleftharpoons}} Cr^{IV} + Ag^I \tag{3.104}$$

$$Cr^{IV} + Ag^{II} \xrightleftharpoons[2]{-2} Cr^{V} + Ag^{I} \qquad (3.105)$$

$$Cr^{V} + Ag^{II} \xrightleftharpoons[1]{-1} Cr^{VI} + Ag^{I} \qquad (3.106)$$

(The rate constants here are numbered to conform with equations (3.87)–(3.89). The rate law

$$\text{Rate} = \frac{k_0 k_{-3} [Co^{III}][Ag^{I}][Cr^{III}]}{k_{-0}[Co^{II}] + k_{-3}[Cr^{III}]} \qquad (3.107)$$

indicates that steps (-3) and (-0) are comparable in rate, but among the sequence of steps of oxidation of chromium(III), the first, i.e. step (-3), is rate-determining, with no evidence of any subsequent slow step.

The reaction $Cr^{VI} + Cr^{II}$ is too fast to follow kinetically, but product studies[37] favour the stepwise mechanism.

$$Cr^{VI} + Cr^{II} \rightarrow Cr^{V} + Cr^{III} \qquad (3.108)$$

$$Cr^{V} + Cr^{II} \rightarrow Cr^{IV} + Cr^{III} \qquad (3.109)$$

$$Cr^{IV} + Cr^{II} \rightarrow (Cr^{III})_2 \qquad (3.110)$$

About 50% of chromium(III) appears as the dimer $Cr_2(OH)_2^{4+}$ presumably from the final step of the sequence. This mechanism also requires that the whole of the monomer chromium(III) originates from the chromium(II), and that the dimer should consist of equal parts of chromium from the chromium(VI) and the chromium(III). Actually, tracer studies indicate that only 90% of the monomer originates from chromium(II), and to account for the difference, Hegedus and Haim[530] have suggested an additional pathway with a two-electron step:

$$Cr^{V} + Cr^{II} \rightarrow Cr^{III} + Cr^{IV} \qquad (3.111)$$

Stability of intermediate oxidation states of chromium

Both chromium(IV) and chromium(V) are known in stable compounds but not under the conditions of these experiments. Several authors have attempted to estimate the relevant redox potentials.

For chromium(V), Newton[797] argued as follows. The reaction $Cr^{VI} + Pu^{III}$ induces oxidation of I^-, implying the reactions

$$Cr^{VI} + Pu^{III} \xrightleftharpoons[-1]{1} Cr^{V} + Pu^{IV} \qquad (3.112)$$

$$Cr^{V} + I^- \xrightarrow{k_1} Cr^{III} + I^+ \qquad (3.113)$$

and the induction factor (*see* equation (3.96)) gives $k_{-1}/k_I = 0.24$, in 1.0 M perchloric acid. But k_I cannot exceed the specific rate for a diffusion-controlled encounter (see p. 100). If we assume the reactants to be uncharged H_3CrO_4 and anionic I^-, this gives $k_I \leqslant 1.4 \times 10^{10}\,M^{-1}s^{-1}$; whence $k_{-1} < 0.33 \times 10^{10}\,M^{-1}s^{-1}$, $K_1 \geqslant 3.4 \times 10^{-5}\,M^{-2}$, and from the known potential of the Pu^{IV}/Pu^{III} pair, that of Cr^{VI}/Cr^{V} has a lower limit $E^\ominus > 0.72$ V. An upper limit, $E^\ominus < 1.0$ V is estimated on the argument that if $E^\ominus$ were significantly greater than this, the concentration of chromium(V) would be too great for steady state kinetics to be maintained.

3.3 Reagents of multiple valence

When one or both reagents has more than two readily available oxidation states, we have to consider not only the variety of possible mechanisms, but also the variety of possible stoicheiometries, hence the classification of a given reaction as complementary or non-complementary reactions cannot be applied beforehand, but becomes a matter of experiment.

3.3.1 Reactions of a multi-electron and a one-electron reagent

When one of the reactants has a sequence of well characterised oxidation states, and the other can be securely identified as a one-electron reagent, we expect stepwise mechanisms formally analogous to the non-complementary reactions discussed above. The systems studied are summarised in *Table 3.4*.

Table 3.4 Summary of reactions of multielectron reagents with one-electron reagents

$$A^{n+} + mB = A^{(n-m)+} + mB^{+\,(a)}$$

A^{n+} etc.				B	References
V^{IV}	$\to V^{III\,(d)}$	$\to V^{II}$		Cr^{II}	378, 379
V^{IV}	$\to V^{III}$			Eu^{II}	387
$V^V \to V^{IV\,(c)}$				$Fe(CN)_6^{4-}$	121
$V^V \to V^{IV\,(c)}$				$Fe(bipy)_n(CN)_{6-2n}^{(2n-4)+}$	125
$V^V \to V^{IV}$				$Fe(bipy)_3^{2+}$	125
$V^V \to V^{IV}$	$\to V^{III}$	$\leftarrow V^{II}$		Ti^{III}	368, 124
$V^V \to V^{IV}$	$\leftarrow V^{III}$	$\leftarrow V^{II}$		Fe^{II}	64, 545, 909
↑↑	V^{III}	$\leftarrow V^{II}$		$Ru(NH_3)_6^{2+}$	373
V^{IV}	$\leftarrow V^{III}$	$\leftarrow V^{II}$		$IrCl_6^{3-}$	1033
V^{IV}	$\leftarrow V^{III}$	$\leftarrow V^{II}$		Np^{III}	175
V^{IV}	$\rightleftarrows V^{III}$	$\leftarrow V^{II}$		Cu^{I}	392, 798, 545
$V^V \leftarrow V^{IV}$	$\leftarrow V^{III}$	$\leftarrow V^{II}$		Co^{II}	578, 906

Table 3.4 continued

A^{n+} etc.	B	References
$Ta_6Cl_{12}^{4+} \rightleftarrows Ta_6Cl_{12}^{3+} \rightarrow Ta_6Cl_{12}^{2+}$	Fe^{II}	391
$Mn^{VII} \rightarrow (Mn^{VI}) \twoheadrightarrow Mn^{II}$	$W(CN)_8^{4-}$	541
$Mn^{VII} \rightarrow (Mn^{VI}) \twoheadrightarrow Mn^{II}$	$Mo(CN)_8^{4-}$	1023
$Mn^{VII} \rightarrow (Mn^{VI}) \rightarrow Mn^{II}$	$Fe(CN)_6^{4-}$	875
$Mn^{VII} \rightleftarrows (Mn^{VI}) \rightarrow (Mn^V) \twoheadrightarrow Mn^{II}$	$Fe(phen)_3^{2+}$	542
$Mn^{VII} \rightarrow (Mn^{VI}) \twoheadrightarrow Mn^{V(c)} \rightarrow Mn^{II}$	V^{IV}	65, 762
$Mn^{III} \rightleftarrows Mn^{II}$	Ce^{III}	43
$Mn^{III} \leftarrow Mn^{II}$	Ag^I	560
$U^{VI} \rightarrow U^{V(c)}$	Cr^{II}	798
$U^{VI} \rightarrow U^V$	Eu^{II}	364
$U^{VI} \rightarrow U^V \twoheadrightarrow U^{IV}$	Ti^{III}	689
$\{U^{VI} \twoheadrightarrow (U^V)^{(b)} \twoheadrightarrow U^{IV}$	Fe^{II}	61
$\{U^{VI} \leftarrow (U^V)^{(b)} \leftarrow U^{IV}$	Fe^{II}	118, 418
$U^{VI} \leftarrow (U^V)^{(b)} \leftarrow U^{IV}$	Ce^{III}	66
$Np^{VI} \twoheadrightarrow Np^V \rightleftarrows Np^{IV} \rightarrow Np^{III}$	Fe^{II}	563, 807, 917
$Np^{VII} \rightarrow Np^{VI} \rightarrow Np^V$	V^{IV}	1075, 944
$Np^{VII} \rightarrow Np^{VI} \leftarrow Np^V$	Co^{II}	983, 309
$Np^{VII} \rightarrow Np^{VI}$	Ag^I	309
$Pu^{VI} \rightarrow Pu^V \twoheadrightarrow Pu^{IV} \rightarrow Pu^{III}$	Ti^{III}	869, 870
$Pu^{VI} \rightarrow Pu^V$	Fe^{II}	799

(a) The entries in this first column indicate the principal reaction steps which have been postulated in the references cited. Double headed arrows denote relatively fast steps. For example, the last entry but one implies that the reaction of Pu^{VI} with Ti^{III} leads to Pu^{IV} in a non-complementary reaction, that the step $Pu^{VI} + Ti^{III} \rightarrow Pu^V + Ti^{IV}$ is rate-determining and that the step $Pu^V + Ti^{III} \rightarrow Pu^{IV} + Ti^{IV}$ is fast. The reduction Pu^{IV} to Pu^{III} occurs in a subsequent slower step. Arrows pointing from right to left indicate reactions studied in reverse, e.g. $Ru(NH_3)_6^{3+} + V^{2+} \rightarrow Ru(NH_3)_6^{2+} + V^{3+}$. Intermediates shown in brackets thus (Mn^{VI}) are not directly observed but are deduced from the kinetics unless otherwise stated.
(b) Assumed intermediate
(c) Binuclear complex
(d) Preceded by a binuclear complex, $V^{III} - Cr^{III}$

In the vanadium series, the oxidation potentials are such that a suitably mild oxidant will oxidise vanadium(II) to vanadium(III) but not further; a stronger oxidant will oxidise vanadium(II) to vanadium(III) and on to vanadium(IV), and a still stronger reagent will oxidise any of the lower states to vanadium(V); and conversely with reductants acting initially on vanadium(V). In all cases in which the vanadium changes its oxidation state by more than one unit, the intermediate states have been detected, and usually they are so long-lived that the successive steps have to be studied under different experimental conditions. For example the reaction $VO^{2+} + Cr^{2+}$ proceeds in the stopped-flow time scale, yielding a binuclear intermediate $VOCr^{4+}$, which relatively slowly dissociates to V^{3+} and Cr^{3+}; and the reaction $V^{3+} + Cr^{2+}$ is slower by several orders of magnitude. The reactions $V^V + Ti^{III}$ and $V^{IV} + Ti^{III}$ are also clearly separable, and the reaction $V^{III} + Ti^{III}$ is thermodynamically unfavourable, but have been studied in the reverse direction, $V^{II} + Ti^{IV}$. At the opposite extreme, Co^{3+} oxidises all the lower states of vanadium, V^{2+} on the stopped-flow time scale, the others at 'conventional' rates.

The reaction $Fe^{III} + V^{III}$ illustrates the point that a reaction which appears simple at first sight may have an unexpectedly complicated mechanism. The stoicheiometry and rate law

$$Fe^{III} + V^{III} = Fe^{II} + V^{IV} \tag{3.114}$$

$$\text{Rate} = k_0[Fe^{III}][V^{III}] \tag{3.115}$$

both suggest merely a one-electron process, but when extra concentrations of the products, V^{IV} or Fe^{III}, are added at the start, the rate is respectively increased or decreased; and the full rate law proves to be

$$\text{Rate} = k_1[Fe^{III}][V^{III}] + k_2[Fe^{III}][V^{III}][V^{IV}]/[Fe^{II}] \tag{3.116}$$

When neither V^{IV} nor Fe^{II} is present at the start, they subsequently appear always in equal amounts, and their effects cancel out. The k_1 term is attributed to the complementary one-electron process

$$Fe^{III} + V^{III} \xrightarrow{k_1} Fe^{II} + V^{IV} \tag{3.117}$$

but the second term is attributed to the reactions

$$Fe^{III} + V^{IV} \underset{}{\overset{K}{\rightleftharpoons}} Fe^{II} + V^V \tag{3.118}$$

$$V^{III} + V^V \xrightarrow{k_2'} 2V^{IV} \tag{3.119}$$

where the equilibrium is rapidly attained so that $k_2 = Kk_2'$. The equilibrium constant K was determined in separate experiments[545], and used to calculate the rate constant k_2'. The value so obtained was later satisfactorily confirmed by experiment[278].

The oxidation states of manganese present a more complicated picture. Although the ion MnO_4^- is fairly easily reduced, the immediate product MnO_4^{2-} (or H_2MnO_4) is unstable in acid solutions with respect to disproportionation. Thus most reagents which will reduce manganese(VII), will in practice reduce it directly to manganese(IV) (the sparingly-soluble hydrous oxide) or to manganese(II), depending on the acidity or on the presence of complexing anions. In three cases, (*Table 3.4*), the first step is rate-determining. With $Fe(phen)_3^{3+}$ as reductant, the stoicheiometry is

$$Mn^{VII} + 5Fe^{II} \rightarrow Mn^{II} + 5Fe^{III} \tag{3.120}$$

but the rate equation shows retardation by Fe^{III}, consistent with the mechanism

$$Mn^{VII} + 2Fe^{II} \rightleftarrows Mn^{VI} + Fe^{III} + Fe^{II} \rightarrow Mn^{V} + 2Fe^{III} \tag{3.121}$$

followed by a rapid sequence such as

$$Mn^{V} + 3Fe^{II} \twoheadrightarrow Mn^{IV} + 2Fe^{II} + Fe^{III}$$
$$\twoheadrightarrow Mn^{III} + Fe^{II} + 2Fe^{III} \twoheadrightarrow Mn^{II} + 3Fe^{III} \tag{3.122}$$

The reaction $Mn^{VII} + V^{IV}$ has the overall stoicheiometry

$$Mn^{VII} + 5V^{IV} \rightarrow Mn^{II} + 5V^{V} \tag{3.123}$$

but it occurs in two stages, with an intermediate which can plausibly be identified as manganese(V), possibly stabilised in a binuclear Mn^{V}–V^{V} complex. The rate law is first order in each reagent so the first stage can be written

$$Mn^{VII} + 2V^{IV} \xrightarrow{k_1} Mn^{VI} + V^{V} + V^{IV} \longrightarrow Mn^{V} - V^{V} + V^{V} \tag{3.124}$$

The subsequent reduction of Mn^{V} to Mn^{II} is complicated; among other features, it shows catalysis by the Mn^{II} product[762].

The successive oxidation states of plutonium, from III to VI inclusive, are all well characterised. Of the three one-electron reductions by titanium(III)

$$Pu^{VI} + Ti^{III} \xrightarrow{1} Pu^{V} + Ti^{IV} \tag{3.125}$$

$$Pu^{V} + Ti^{III} \xrightarrow{2} Pu^{IV} + Ti^{IV} \tag{3.126}$$

$$Pu^{IV} + Ti^{III} \xrightarrow{3} Pu^{III} + Ti^{IV} \tag{3.127}$$

Rabideau and Kline found that equation (3.125) proceeded at a measurable rate, but equation (3.126) was rapid. Thus on mixing Pu^{VI} with Ti^{III}, the stoicheiometry was

$$Pu^{VI} + 2Ti^{III} = Pu^{IV} + 2Ti^{IV} \tag{3.128}$$

but the rate law was first order in each reagent[869]. Equation (3.127) was studied separately[870].

Uranium(V) is unstable with respect to the neighbouring oxidation states U^{IV} and U^{VI}, hence reductions of U^{VI} by one-electron reagents tend to be non-complementary. The kinetics of the reaction $U^{VI} + Ti^{III}$ are consistent with the mechanism

$$U^{VI} + Ti^{III} \underset{-1}{\overset{1}{\rightleftarrows}} U^{V} + Ti^{IV} \tag{3.129}$$

$$U^V + Ti^{III} \xrightarrow{k_2} U^{IV} + Ti^{IV} \tag{3.130}$$

with step 1 rate-determining. On adding vanadium(IV), the intermediate uranium(V) is reoxidised

$$V^{IV} + U^V \xrightarrow{k_3} V^{III} + U^{VI} \tag{3.131}$$

thus inhibiting the reaction and leading to the overall rate law

$$\frac{-d[U^{VI}]}{dt} = \frac{k_1 k_2 [Ti^{III}]^2 [U^{VI}]}{k_2 [Ti^{III}] + k_3 [V^{IV}]} \tag{3.132}$$

3.3.2 Reactions of a multi-electron and a two-electron reagent

Reactions of this class are summarised in *Table 3.5*. The main point to be discussed here is the identification of one-electron and two-electron transfer steps.

Table 3.5 Summary of reactions of multi-electron reagents with two-electron reagents

$$2A^{n+} + mB = 2A^{(n-2m)+} + mB^{2+}$$

A^{n+} etc.[a]	B	References
$V^V \longrightarrow V^{III}$	Sn^{II}	344
$V^{IV} \longleftarrow V^{II}$	Tl^I	65
$V^{IV} \longleftarrow V^{III}$	Tl^I	276
$V^{IV} \longleftarrow V^{II}$	Hg^0	483
$V^{III} \longleftarrow V^{II}$	Hg^0	483
$V^V \rightleftarrows V^{IV} \longleftarrow V^{III}$	Hg^0	481
$U^{VI} \longleftarrow U^{IV}$	Tl^I	1077
$U^{VI} \longrightarrow U^{IV}$	Sn^{II}	764
$Pu^{VI} \longrightarrow Pu^{IV}$ $Pu^V \longleftarrow Pu^{IV}$	Sn^{II} Sn^{II}	871
$Np^{VII} \to Np^{VI}$	Tl^I	1032
$Np^{VII} \to Np^{VI}$	Hg^0	1029

[a] For explanation of column 1, see *Table 3.4*

The reaction of vanadium(V) with tin(II) leads to a mixture of vanadium(III) and vanadium(IV) which at first sight might suggest one or more reaction paths with one-electron steps. However, a detailed analysis of the product ratios as functions of reagent concentration, coupled with the failure to detect tin(III) by induced-reduction experiments, render this postulate unnecessary. The stoicheiometry and kinetic data are consistent with the complementary two-electron mechanism

$$V^V + Sn^{II} \rightarrow V^{III} + Sn^{IV} \tag{3.133}$$

with the reaction (3.119) leading to the production of vanadium(IV).

The thallium–vanadium system is more complex. Thallium(III) is capable of oxidising all the lower states of vanadium to vanadium(V), but in practice it does not do so. The reaction $Tl^{III} + V^{III}$ is the non-complementary reaction

$$Tl^{III} + 2V^{III} = Tl^{I} + 2V^{IV} \tag{3.134}$$

and the subsequent oxidation of vanadium(IV) to vanadium(V)—given a sufficient excess of thallium(III)—is slow. The rate law

$$\text{Rate} = k[Tl^{III}][V^{III}] \tag{3.135}$$

is consistent with either of the two mechanisms a or b provided that the first step is rate-determining in either case:

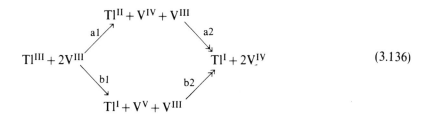

(3.136)

Reaction b2, however, has been measured separately and is too slow to fulfil this condition; hence mechanism a is preferred. Evidence was also obtained of a reactive intermediate, when iron(II) was added to the system it was found that the slow reactions $Tl^{III} + 2Fe^{II}$ and $Tl^{III} + 2V^{IV}$ were enhanced when the reaction (3.134) was in progress, suggesting an increased concentration of Tl^{II} [276].

The reaction $Tl^{III} + V^{II}$ obeys the equation

$$Tl^{III} + V^{II} = Tl^{I} + V^{IV} \tag{3.137}$$

No appreciable amount of vanadium(III) can be detected at the end of the reaction and even when vanadium(II) is initially in excess, the immediate product is the unstable mixture of vanadium(II) and vanadium(IV). The kinetics in the early stages of the reaction obey the rate law

$$\text{Rate} = k[Tl^{III}][V^{II}] \tag{3.138}$$

but the specific rate increases as the reaction proceeds. In spite of this complication, the evidence is generally in favour of the simple two-electron process

$$Tl^{III} + V^{II} \xrightarrow{k} Tl^{I} + V^{IV} \tag{3.139}$$

as the major reaction pathway, since neither thallium(II) nor vanadium(III) can be detected as intermediates. The argument against thallium(II) is that there is no appreciable induction of the reaction $Tl^{III} + V^{IV}$ which is known to involve Tl^{II} as an intermediate (*see* p. 57). The argument against vanadium(III) is more complicated since this species must always be present through the operation of the reaction

$$V^{IV} + V^{II} \rightarrow 2V^{III} \tag{3.119}$$

but its concentration is kept low by the further non-complementary reaction

$$2V^{III} + Tl^{III} = 2V^{IV} + Tl^{I} \tag{3.140}$$

Independent measurements show that reaction (3.119) is sufficiently slow, and reaction (3.140) sufficiently fast, to keep the vanadium(III) concentration below the limits of direct (spectrophotometric) observation. But equations (3.119) and (3.140) together form a parallel pathway with the stoicheiometry given in equation (3.139) and this accounts for the progressive enhancement in rate.

Equations (3.139) and (3.119) together could be described as the non-complementary reaction

$$Tl^{III} + 2V^{II} = Tl^{I} + 2V^{III} \tag{3.141}$$

and the mechanism could be described as mechanism b except for the fact that the two steps are separately observed. It remains to show that there is not also a third reaction path, equation (3.140) with mechanism a, namely

$$Tl^{III} + V^{II} \underset{-a1}{\overset{a1}{\rightleftharpoons}} Tl^{II} + V^{III} \tag{3.142}$$

$$Tl^{II} + V^{II} \xrightarrow{a2} Tl^{I} + V^{III} \tag{3.143}$$

leading to vanadium(III) and thence to vanadium(IV) as before. This pathway, however, must not exceed a certain limiting rate if the vanadium(III) is to be kept below detectable limits, and detailed calculations, using the available known rate constants, have indicated that reaction (3.141) is not effective in comparison with the two-electron reaction, equation (3.137)[65].

Mercury(II)–Vanadium(II). In this system vanadium(II) is thermodynamically capable of acting as either a one-electron reagent

$$2Hg^{II} + 2V^{II} = (Hg^{I})_2 + 2V^{III} \tag{3.144}$$

or a two-electron reagent

$$2Hg^{II} + V^{II} = (Hg^{I})_2 + V^{IV} \tag{3.145}$$

and the evidence is that both pathways occur together. They are distinguishable by their different vanadium products since it is established in separate experiments that further oxidation of vanadium(III) to vanadium(IV) is comparatively slow. The one-electron path may therefore be written as mechanism a (*see* p. 61):

$$Hg^{II} + V^{II} \xrightarrow{a1} Hg^{I} + V^{III} \tag{3.146}$$

$$Hg^{I} + V^{II} \xrightarrow{a2} Hg^{0} + V^{III} \tag{3.147}$$

$$Hg^{II} + Hg^0 \longrightarrow (Hg^I)_2 \tag{3.65}$$

while the two-electron path is

$$Hg^{II} + V^{II} \xrightarrow{b1} Hg^0 + V^{IV} \tag{3.148}$$

$$Hg^{II} + Hg^0 \longrightarrow (Hg^I)_2 \tag{3.65}$$

In both paths the first step must be rate-determining since the rate law is the simple equation

$$\text{Rate} = k[V^{II}][Hg^{II}] \tag{3.149}$$

Mercury(II)–Vanadium(III). This slow reaction has the stoicheiometry

$$2Hg^{II} + 2V^{III} = (Hg^I)_2 + 2V^{IV} \tag{3.150}$$

but the rate law is exceptionally complicated and apparently requires participation of vanadium(V) even though this is not observed as a product.

When the reagents are mixed the kinetics are initially second order, but as the vanadium(IV) product builds up a retardation effect is observed

$$\text{Rate} = A[V^{III}]^2[Hg^{II}]/([V^{III}] + B[V^{IV}]) \tag{3.151}$$

This is consistent with mechanism a'

$$V^{III} + Hg^{II} \underset{-a1}{\overset{a1}{\rightleftarrows}} V^{IV} + Hg^{I} \tag{3.152}$$

$$V^{III} + Hg^{I} \xrightarrow{a2} V^{IV} + Hg^0 \tag{3.153}$$

$$Hg^0 + Hg^{II} \xrightarrow{3} (Hg^I)_2 \tag{3.65}$$

with $A = k_{a1}$, $A/B = K_{a1}k_{a2}$. However, when vanadium(IV) is added at the start of the reaction, another term is found necessary, giving the total rate law as

$$\text{Rate} = \frac{A[V^{III}]^2[Hg^{II}] + C[V^{III}][V^{IV}][Hg^{II}]}{[V^{III}] + B[V^{IV}]} \tag{3.154}$$

An additional parallel path must therefore be included: this can be done in at least two ways consistent with the rate law, namely

$$V^{IV} + Hg^{I} \xrightarrow{5} V^{V} + Hg^0 \tag{3.155}$$

or

$$V^{III} + Hg^{II} \xrightarrow{6} V^{V} + Hg^0 \tag{3.156}$$

and each of these reactions is followed by those of equations (3.119) and (3.65).

$$V^{V} + V^{III} \longrightarrow V^{IV} + V^{IV} \tag{3.119}$$

$$Hg^0 + Hg^{II} \xrightarrow{3} (Hg^I)_2 \tag{3.65}$$

For this purpose step (3.119) must be sufficiently rapid to account for the fact that no vanadium(V) is detected among the reaction products, and this is found to be so. Using equation (3.152) etc., the steady state condition for mercury(I) then gives equation (3.149) with $A = k_{a1}$, $B = (k_{-a1} + k_5)/k_{a2}$, $C = k_1 k_3/k_2$, whence net activation parameters are $k_{a1} = A/2$, $K_{a1}k_{a2} = A^2/(AB-1)$, $K_{a1}k_5 = AC/(AB-1)$. Using the alternative equations (3.155) etc. we have $A = (k_{a1} + k_6)$, $B = k_{a1}/k_{a2}$, $C = k_6 k_{-a1}/k_{a2}$, giving $k_{a1} = (A - C/B)$, $K_{a1}k_2 = (A - C/B)/B$, $K_{a1}^{-1}k_6 = C/B$.

3.3.3 Reactions of a multi-electron and a three-electron reagent

The only reactions in this class involve chromium(VI) as oxidant (*Table 3.6*). In the vanadium and plutonium series, each reductant is oxidised in a series of distinguishable one-electron steps. The individual steps are non-complementary 3:1 reactions, and the kinetics are consistent with stepwise processes involving chromium(IV) and chromium(V) as discussed in Section 3.2.2. The reaction $Cr^{VI} + U^{IV}$ has the 3:2 stoicheiometry expected in view of the instability of uranium(V) and involves two-electron transfers. Induced oxidation experiments indicate chromium(V) as an intermediate, but the lack of inhibition by added iron(III) indicates the absence of uranium(V):

$$Cr^{VI} + U^{IV} \longrightarrow Cr^{IV} + U^{VI} \tag{3.157}$$

$$Cr^{VI} + Cr^{IV} \longrightarrow 2Cr^{V} \tag{3.158}$$

$$Cr^{V} + U^{IV} \longrightarrow Cr^{III} + U^{VI} \tag{3.159}$$

Table 3.6 Reactions of multi-electron reagents with a three-electron reagent

A^{n+} etc.[a]	B	References
$V^V \leftarrow V^{IV} \leftarrow V^{III} \leftarrow V^{II}$	Cr^{III}	160, 288, 376, 910
$Pu^{VI} \leftarrow Pu^{V\,[b]} \leftarrow Pu^{IV} \leftarrow Pu^{III}$	Cr^{III}	805, 797
$U^{VI} \longleftarrow U^{IV}$	Cr^{III}	394
$U^{VI} \leftarrow U^{V}$	Cr^{III}	394

(a) For explanation of column 1, see *Table 4.3*
(b) Binuclear complex

3.3.4 Reactions between two multi-electron reagents

It might be thought that reactions of this class would show every possible variety of non-complementary reaction but in fact most of the possible combinations which have been studied (*Table 3.7*) give complementary one-electron transfer reactions. The main exceptions are some reactions involving the U^{IV}/U^{VI} couple which are non-complementary, the middle valency uranium(V) being relatively unstable. Two-electron transfers might be expected but in fact none have been found. Published redox potential data disclose many instances in which two one-electron transfer reactions could occur spontaneously in such a way as to amount to a net two-electron change, as for example the sequence

$$U^{VI} + V^{II} \rightarrow U^{V} + V^{III} \rightarrow U^{IV} + V^{IV} \tag{3.160}$$

In this and other cases, however, the rates are widely different and the two reactions are easily studied separately. Thermodynamically the most favourable condition for two-electron transfer arises when the intermediate valency is unstable with respect to the initial and final valency, for both reactants. This condition is met in only one case, $Pu^{VI} + U^{IV}$, and in fact the reaction does not occur. When the reagents are mixed in equal amounts, the main reaction is non-complementary:

$$2Pu^{VI} + U^{IV} = 2Pu^{V} + U^{VI} \tag{3.161}$$

A small amount of plutonium(III) is produced, but no detectable plutonium(IV) and when the reagents are mixed in the ratio corresponding to equation (3.161), the stoicheiometry is almost exact. The rate law is consistent with either of the mechanisms a or b discussed above, provided that the first step is rate-determining:

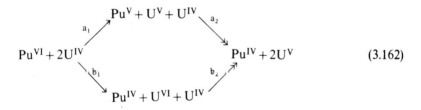

(3.162)

However mechanism b can be ruled out since step b1 leads to stable products and step b2 is thermodynamically unfavourable. Equilibrium constants are $\log K_{a1} \simeq 6.0$ and $\log K_{b1} \simeq 21.0$. It is clear that, in this case at least, two-electron transfer is disfavoured kinetically rather than thermodynamically.

When an element has three consecutive oxidation states differing by one unit, there are several pathways for a net two-electron exchange reaction. Again an actual two-electron process

$$A^{2+} + A = A + A^{2+} \tag{3.163}$$

is unlikely unless the middle valence state is unstable with respect to disproportionation, as with U^{VI}/U^{IV} and Pu^{VI}/Pu^{IV}. There are no data for the plutonium reaction, but the uranium(VI)–uranium(IV) exchange has been thoroughly studied. In hydrochloric acid media, the rate law was found to be[899]

$$\text{Rate} = k[U^{VI}][U^{IV}]^2 \tag{3.164}$$

Table 3.7 Summary of published work on electron transfer reactions between ions of multiple valency. All reactions are 1:1 one-electron transfers, $A^+ + B = A + B^+$, unless otherwise stated

Oxidant (A^+) Reductant (B)	V^{IV}	V^V	Mn^{VI}	Mn^{VII}	U^V	U^{VI}	Np^{IV}	Np^V	Np^{VI}	Pu^{IV}	Pu^V	Pu^{VI}
$Ta_6X_{12}^{2+}$ (c)		381(d)										
$Ta_6X_{12}^{3+}$ (c)		381										
V^{II}	800	390										
V^{III}		278			803	803	175		944			867
V^{IV}				762(d)	803(i)	804	33(b)	33(b)	944(d)			
$Mn^{(VI)}$			992									
U^{III}					1041	1041 (j, e)			984(f)	794(a)		793(a)
U^{IV}					583, 802							
U^V												

Np^{III}		806	
Np^{IV}		806	
Np^{V}	796$^{(h)}$	547, 981	547, 981
Pu^{III}		447	246, 656	868
Pu^{IV}			246	866, 868
Pu^{V}			868	866, 868

(a) Observed stoicheiometry is $2Pu^{VI} + U^{IV} = 2Pu^{V} + U^{VI}$
(b) The reaction $Np^{V} + V^{III} = Np^{IV} + V^{IV}$ has two pathways, $Np^{V} + V^{III} \rightarrow Np^{IV} + V^{IV}$ and $Np^{IV} + V^{III} + Np^{V} \rightleftharpoons Np^{III} + V^{IV} + Np^{V} \rightarrow 2Np^{IV} + V^{IV}$
(c) X is Cl, Br
(d) The reaction $Ta_6X_{12}^{2+} + V^{V} = Ta_6X_{12}^{3+} + V^{IV}$ has two pathways, $Ta_6X_{12}^{2+} + V^{V} \rightarrow Ta_6X_{12}^{3+} + V^{IV}$ and $Ta_6X_{12}^{3+} + V^{V} \rightarrow Ta_6X_{12}^{4+} + Ta_6X_{12}^{2+} + V^{IV}$
$\rightarrow 2Ta_6X_{12}^{3+} + V^{IV}$
(e) $U^{VI} + U^{IV} \rightarrow 2U^{V}$ is preliminary step in the $U^{VI} - U^{IV}$ exchange reaction
(f) Observed stoicheiometry is $2Np^{VI} + U^{IV} = 2Np^{V} + U^{VI}$
(g) Observed stoicheiometry is $2Pu^{IV} + U^{VI} = 2Pu^{III} + U^{VI}$
(h) Observed stoicheiometry is $2Np^{III} + U^{VI} = 2Np^{IV} + U^{IV}$. The reaction paths are (i) $2Np^{III} + U^{VI} \rightarrow Np^{III} + Np^{IV} + U^{V} \rightarrow 2Np^{IV} + U^{IV}$ and
(ii) $2Np^{III} + 2U^{VI} \rightarrow 2Np^{IV} + 2U^{V} \rightarrow 2Np^{IV} + U^{IV} + U^{VI}$
(i) Rapid reaction $V^{IV} + U^{V} = V^{III} + U^{VI}$, mechanism not determined
(j) References 740, 899, 1041

which is interpreted according to the mechanism

$$U^{VI} + U^{IV} \rightleftharpoons U_2^X \tag{3.165}$$

$$U_2^X + U^{IV} \rightarrow 2U^{IV} + U^{VI} \tag{3.166}$$

where U_2^X denotes a binuclear complex, presumably with both uranium atoms in oxidation state V. A similar rate term was found[740] in perchloric acid media, but in addition there is a second-order term

$$\text{Rate} = k[U^{VI}][U^{IV}] \tag{3.167}$$

This, however, is also interpreted in terms of one-electron steps:

$$*U^{VI} + U^{IV} \rightleftharpoons *U^V + U^V \tag{3.168}$$

$$*U^V + U^{VI} \rightleftharpoons *U^{VI} + U^V \tag{3.169}$$

The rates of all the steps in this sequence were known from previous studies, and the rate of exchange thus calculated was in good agreement with the experiments. The importance of uranium(V) as an intermediate is borne out by a recent study of the catalysis of the exchange by the one-electron oxidant, iron(III)[1041].

When the element has four consecutive oxidation states, a curious ambiguity can arise. The reaction $U^{VI} + U^{III}$ for example has the stoicheiometry

$$U^{VI} + U^{III} = U^V + U^{IV} \tag{3.170}$$

which could arise by either two-electron transfer

$$*U^{VI} + U^{III} \rightarrow *U^{IV} + V^V \tag{3.171}$$

or one-electron transfer

$$*U^{VI} + U^{III} \rightarrow *U^V + U^{IV} \tag{3.172}$$

The two possibilities could be distinguished experimentally by isotopic labelling (assuming that there is not a rapid exchange between uranium(V) and uranium(IV)), as shown. No such experiments have been carried out on any of the systems for which this ambiguity arises ($U^{VI} + U^{III}$[806], $Pu^{VI} + Pu^{III}$[868], $Np^{VI} + Np^{III}$[806], $V^V + V^{II}$[390]), but in each case it is assumed that one-electron transfer is the predominant pathway.

3.4 Free energy profiles

Figures 3.1–3.3 show the free energies of transition states, and reaction products, for all the known reactions of thallium(III), tin(II) and mercury(II) which follow mechanism a. For the thallium reactions, the free energies of the intermediates are also shown, calculated from the known redox potentials of the Tl^{2+} ion. For the mercury reactions, also, the stabilities of the intermediate states have been estimated using a semi-theoretical value for the Hg^+/Hg^{2+}

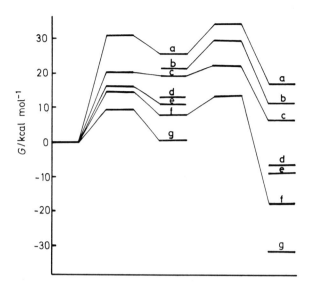

Figure 3.1 Free energy profiles for the reactions

$$Tl^{III} + 2B \rightleftarrows Tl^{II} + B + B^+ \rightleftarrows Tl^I + 2B^+$$

Reductants B are a, Co^{II}; b, Mn^{II}; c, Ce^{III}; d, $Fe(bipy)_3^{2+}$; e, V^{VI}; f, Fe^{II}; g, V^{III}. *(Data from Table 3.1)*

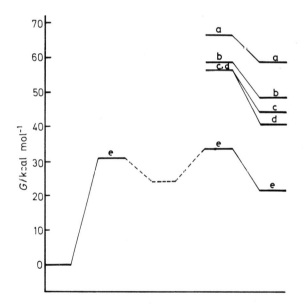

Figure 3.2 Free energy profiles for reactions

$$Sn^{IV} + 2B \rightleftarrows Sn^{III} + B^+ + B \rightleftarrows Sn^{II} + 2B^+$$

Reductants B are a, Co^{II}; b, Mn^{II}; c, Ce^{III}; d, V^{IV}; e, Fe^{II}. *(Data from literature cited in Table 3.1)*

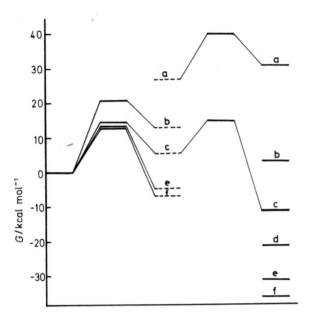

Figure 3.3 Free energy profiles for reactions
$$Hg^{II} + 2B \rightleftarrows Hg^{I} + B + B^+ \rightleftarrows Hg^0 + 2B^+$$
Reductants B are a, Mn^{II}; b, Fe^{II}; c, V^{III}; d, U^V; e, V^{II}; f, Cr^{II}. (Data from literature cited in Table 3.2)

redox potential[1089]. In cases where mass-law retardation effects were observed, it is possible to show complete profiles with both transition states; in the other cases only one transition state is shown, corresponding to the one observed rate-determining step. What emerges clearly is a correlation between activation free energy and standard free energy change, for each of the two reaction steps, and hence a correlation between the form of the rate law and the standard free energy change for the overall, non-complementary process. For the most powerful reductants, step 1 is rate-determining; for the less powerful, step 2; and in between, the two-term rate law is found.

The free energy profiles for reactions of chromium(VI) are shown in *Figure 3.4*. Since neither of the two redox potentials, Cr^{VI}/Cr^V or Cr^V/Cr^{IV} is known, it is not possible to enter the free energy levels for any of the intermediate states with any certainty. Again there is a general correlation between the reaction mechanism and the standard free energy change associated with the overall reaction. In the reaction with the strongest reducing agent, vanadium(III), step 1 is rate-determining; with the milder reductants, iron(II), neptunium(V), vanadium(IV), and plutonium(III), steps 1 and 2 are significant; with cerium(III) only step 2 is significant, and with silver(I) only step 3. When the overall standard free energy change is close to zero, it appears that the rate-determining step is step 2, and it has been argued that this is related to the

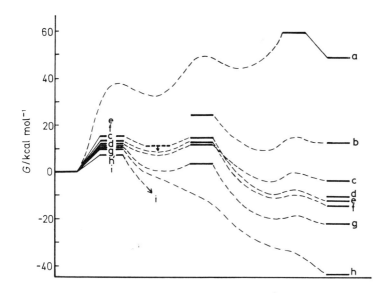

Figure 3.4 Free energy profiles for reactions
$$Cr^{VI} + 3B \rightleftharpoons Cr^{V} + 2B + B^{+} \rightleftharpoons Cr^{IV} + B + 2B^{+} \rightleftharpoons Cr^{III} + 3B^{+}$$

Reductants B are a, Ag^{I}; b, Ce^{III}; c, Np^{V}; d, V^{IV}; e, Pu^{III}; f, Pu^{IV}; g, Fe^{II}; h, V^{III}; i, V^{II}. (Data from literature cited in Table 3.3)

structures of the intermediate oxidation states of chromium. It is generally true that, other things being equal, electron transfer reactions which involve making and breaking of bonds, or changes in the coordination of the reacting centre, are slower than those which do not (see p. 176). It has been suggested therefore that chromium (V) is structurally similar to chromium(VI), while chromium(IV) is structurally similar to chromium(III). Chromium(VI) is well established as the tetrahedral CrO_4^{2-} ion; hence chromium(V) would be written CrO_4^{3-} (and kinetic evidence not discussed here would suggest the protonated form H_3CrO_4, in moderately strongly acidic solution). Chromium(IV), on the other hand, could be written as the octahedral complex $Cr(OH_2)_6^{4+}$ [(c)].

Birk[120] has argued that in the series of reactions $Cr^{VI} + B$, apart from the correlation of rate law with redox potential, there is a tendency for step 1 (equation 3.82) to be rate-determining when B is an outer-sphere reductant, such as $Fe(bipy)_3^{2+}$ or $Fe(phen)_3^{2+}$ (see Chapter 5). It is suggested that inner-sphere processes lead to binuclear intermediates of the type $Cr^{V} \ldots B$ (or $Cr^{III} \ldots B$ at a later stage in the reaction). It is possible that the chromium(V)

(c) The exchange reaction $Cr^{VI} + {}^{*}Cr^{III} = Cr^{III} + {}^{*}Cr^{VI}$ supports this view. The rate law Rate $= k[Cr^{VI}]^{2/3}[Cr^{III}]^{4/3}$ is consistent with the equilibrium $Cr^{VI} + Cr^{III} \rightleftharpoons Cr^{V} + Cr^{IV}$, followed by the rate-determining step $Cr^{V} + Cr^{III} \rightarrow Cr^{IV} + Cr^{IV}$ [29].

3.5 Double electron transfer

3.5.1 General

Several examples of two-electron transfer have been mentioned already, either as elementary steps ($Tl^{III} + Cr^{II}$, $Cr^V + Cr^{II}$, $Hg^{II} + V^{II}$, $Hg^{II} + V^{III}$, $Cr^{VI} + U^{IV}$, $Cr^V + U^{IV}$) or as overall complementary reactions ($V^V + Sn^{II}$, $Tl^{III} + V^{II}$). It is of particular interest, however, to consider reactions between oxidants and reductants each of which is already characterised as a two-electron reagent (see p. 52). All such reactions necessarily conform to the stoicheiometry 3,

$$A^{2+} + B = A + B^{2+} \tag{3.173}$$

and every example so far reported obeys the rate law

$$\text{Rate} = k[A^{2+}][B] \tag{3.174}$$

Equations (3.173) and (3.174) together suggest a two-electron transfer primary step; but they do not constitute proof. They are equally consistent with a number of other mechanisms based on one-electron steps. In some instances these mechanisms have actually been found, in others they have been considered, but rejected for various reasons.

A rational description of possible one-electron mechanisms is based on the reactions

$$A^{2+} + B \underset{-1}{\overset{1}{\rightleftharpoons}} A^+ + B^+ \quad \text{(initiation)} \tag{3.175}$$

$$A^{2+} + B^+ \overset{2}{\rightleftharpoons} A^+ + B^{2+} \tag{3.176}$$

$$A^+ + B \overset{3}{\rightleftharpoons} A + B^+ \tag{3.177}$$

(propagation)

$$A^+ + A^+ \overset{4}{\longrightarrow} A + A^{2+} \tag{3.178}$$

$$B^+ + B^+ \overset{5}{\longrightarrow} B + B^{2+} \tag{3.179}$$

$$A^+ + B^+ \overset{6}{\longrightarrow} A + B^{2+} \tag{3.180}$$

(termination)

Of the two unstable intermediates, B^+ may be further oxidised by A^{2+}, or A^+ may be further reduced by B (steps 2 and 3). Either intermediate may be removed from the system by disproportionation (steps 4 and 5); or by reacting together to give products (step 6) or starting materials (step -1).

If both steps 2 and 3 are kinetically significant, we have a chain mechanism, with step 1 as initiator, steps 2 and 3 as propagators, and steps 4–6 as terminators. This possibility was considered by Harkness and Halpern[517] in a study of the reaction $Tl^{III} + U^{IV} = Tl^{I} + U^{VI}$. In general howeyer, the rate law will not conform to equation (3.174) unless certain relationships between the rate constants are satisfied. These relationships may be worked out in detail by applying the two steady-state conditions $d[A^+]/dt = d[B^+]/dt = 0$. As a particular example it may be shown that if steps 4 and 5 are negligible in both directions, then the reverse steps (-2 or -3) must also be negligible and the overall rate constant (equation 3.174) becomes

$$k = \frac{k_1 k_6}{(k_{-1} + k_6)} \left[1 + \left(\frac{k_2 k_3}{k_1 k_6}\right)^{1/2} \left(1 + \frac{k_{-1}}{k_6}\right)^{1/2} \right] \tag{3.181}$$

This condition for equation (3.174) can be described in two other alternative ways: it could be said that after the two intermediates A^+ and B^+ have been generated in step 1, either (a) B^+ is rapidly oxidised by a second molecule of A^{2+}, while A^+ is rapidly reduced by B in a non-complementary stepwise reaction

$$A^{2+} + B \xrightarrow{1} A^+ + B^+ \tag{3.182}$$

$$A^{2+} + B^+ \xrightarrow{2} A^+ + B^{2+} \tag{3.183}$$

$$A^+ + \tfrac{1}{2}B \xrightarrow{3} \xrightarrow{6} A + \tfrac{1}{2}B^{2+} \tag{3.184}$$

or conversely (b) A^+ is reduced by B and B^+ is oxidised in non-complementary fashion by A^{2+}

$$A^{2+} + B \xrightarrow{1} A^+ + B^+ \tag{3.185}$$

$$B + A \xrightarrow{3} B^+ + A \tag{3.186}$$

$$\tfrac{1}{2}A^{2+} + B \xrightarrow{2} \xrightarrow{6} \tfrac{1}{2}A + B^{2+} \tag{3.187}$$

It follows that if either of the direct reactions $A^+ + \tfrac{1}{2}B$ or $\tfrac{1}{2}A^{2+} + B$ can be measured independently, and turns out to be too slow to generate products at the observed rate, the chain mechanism can be eliminated. Rabideau and Masters[871] applied this argument to the reaction $Pu^{VI} + Sn^{II} = Pu^{IV} + Sn^{IV}$. The reaction $2Pu^{V} + Sn^{II} = 2Pu^{IV} + Sn^{IV}$ could be shown to be too slow to fit into the sequence of equations (3.182)–(3.184).

Inclusion of the backward reactions of steps 1, 2 and 3 makes no difference to these arguments. Generally speaking, since the intermediates A^+ and B^+ are by definition unstable, step -1 will be important; and in addition, either or both of steps -2 and -3 may be important as well, depending on the reactants in question. Clearly, the reverse reactions of steps 4 and 5 do not need to be included, nor the reverse of reaction 6 (unless the overall equilibrium constant K is close to unity, which is not the case for any of the reactions considered). It will be noticed that $k_2 k_3 / k_{-2} k_{-3} = k_1 k_6 / k_{-1} k_{-6} = K$.

Since the steps 4–6 are all bimolecular, between reagents at low concentration, they may be expected to be relatively slow, but they cannot all be neglected without violating the steady-state conditions. If, however, only one of steps 2 or 3 is considered, the chain mechanism no longer applies, and the slowness of steps 4–6 becomes more critical. It is easy to show that if, for example, step 3 is omitted from the scheme, then step 5 at least is required in order to maintain the steady-state condition and the stoicheiometry; but step 5 must be sufficiently rapid also to maintain the rate law. This is generally unlikely, and sometimes can be ruled out by experiment.

If both propagation steps 2 and 3 are negligible, the required stoicheiometry can be maintained only when, by coincidence, $k_4 = k_5$, so that $[A^+] = [B^+]$ throughout the course of the reaction. Then it is easy to show that the rate law is given by equation (3.169) with $k = k_1(k_4 + k_6)/(2k_4 + k_6 + k_{-1})$.

The case usually considered is where $k_4 \cong k_5 \cong 0$, so that the mechanism is reduced to the two steps

$$A^{2+} + B \underset{-1}{\overset{1}{\rightleftharpoons}} A^+ + B^+ \xrightarrow{6} A + B^{2+} \qquad (3.188)$$

For symmetrical exchange reactions $A^{2+} + A$, the above mechanisms all reduce to this one.

The intermediates A^+ or B^+ could, in principle, be detected by appropriate physical methods, but in practice this never seems to have been done. They can, alternatively, be tested for chemically by their reactions with other added reagents. The evidence of such intermediates may be catalysis of some other reaction, the intermediate itself not being used up, or an induced reaction, in which the intermediate is diverted into a different pathway. In the latter case, depending on rate constants and conditions, the original reaction may be entirely suppressed. In either case, an argument for two-electron transfer is based on the *absence* of the effect concerned. The most that can be established is that the concentration of intermediate does not exceed a certain limit, which depends on the (usually unknown) rate constant. It has been pointed out that equation (3.188) is possible even when the concentrations of A^+ and B^+ are too small to be detected by any chemical means. As the reactants A^{2+} and B approach to the distance required for electron transfer, they form a complex enclosed by the solvent molecules:

$$A^{2+} + B \underset{k_{-d1}}{\overset{k_{d1}}{\rightleftharpoons}} (A^{2+} \cdot B) \underset{k'_{-1}}{\overset{k'_1}{\rightleftharpoons}} (A^+ \cdot B^+)$$

$$\underset{k'_{-2}}{\overset{k'_2}{\rightleftarrows}} (\text{A.B}^{2+}) \xrightarrow{k_{d2}} \text{A} + \text{B}^{2+} \tag{3.189}$$

If the electron transfer within the complex is slow compared with the rate of formation and dissociation of the complex, we have in effect a concerted process, and the intermediates A^+ and B^+ do not occur as separate species capable of undergoing independent reactions.

Rich and Taube[888] discussed the problem in terms not of a solvent cage complex, but of the transition state, and following Sykes[997], it may be suggested that this is a third mechanistic possibility: that while still rapid with respect to formation and dissociation of the solvent cage, electron transfer would be slow with respect to the intramolecular motions which define the reaction coordinate in the transition state. This corresponds to the non-adiabatic limit in the gas reactions discussed in chapter 2: it means that the reacting pair may undergo many reaction-like vibrations before electron transfer takes place, and afterwards, it may undergo many more vibrations before the reactants diffuse apart.

The lines of argument which have been used to support the concerted mechanism in particular instances are mentioned in the next section.

3.5.2 Examples

A list of these reactions is given in *Table 3.8*. Some representative examples are discussed here.

Thallium(III)–Thallium(I). The symmetrical exchange process

$$\text{Tl}^{\text{III}} + {}^*\text{Tl}^{\text{I}} = \text{Tl}^{\text{I}} + {}^*\text{Tl}^{\text{III}} \tag{3.190}$$

is now the best understood two-electron transfer reaction. The extensive literature includes kinetic studies in both complexing and non-complexing media, all supporting the second-order rate equation[6, 167, 169, 216, 319, 461, 516, 859, 912, 971, 994, 1096]

$$\text{Rate} = k[\text{Tl}^{\text{III}}][\text{Tl}^{\text{I}}] \tag{3.191}$$

The reaction is catalysed by the hydroxide ion[912], and by other anions,[169, 216, 859] notably halide ions[516] but is retarded by certain organic acids[167]. There is now overwhelming evidence that equation (3.190) represents the mechanism as well as the stoicheiometry of this reaction. The alternative which has been considered is the sequence.

$$\text{Tl}^{\text{III}} + \text{Tl}^{\text{I}} \underset{-1}{\overset{1}{\rightleftarrows}} \text{Tl}^{\text{II}} + \text{Tl}^{\text{II}} \underset{1}{\overset{-1}{\rightleftarrows}} \text{Tl}^{\text{I}} + \text{Tl}^{\text{III}} \tag{3.192}$$

with thallium(II) as intermediate. Tests for thallium(II) in the reacting system have all proved negative; for example, the rate of exchange is not retarded by the addition of reagents such as iron(III)[1069], which rapidly oxidise or reduce

Table 3.8 Summary of complementary two-electron transfer reactions

Reactants	Conditions	Rate law, Notes	References
$T^{III} + Tl^I$			See text, p. 87
$Pt^{IV} + Pt^{II}$			See text, p. 90
$Sb^VCl_6^- + Sb^{III}$	11.7M HCl, 25 °C	$k[Sb^VCl_6^-][Sb^{III}]$	791
$Sn^{IV} + Sn^{II}$	9–11M HCl, 0–25 °C	$k[Sn^{IV}][Sn^{II}]$	166
$Sn^{IV} + Sn^{II}$	10M HCl, 0 °C, UV light	Photochemically induced exchange	260
$Tl^{III} + Sn^{II}$	1.3M HCl, −1.3 °C	No evidence of Sn^{III} intermediate (lack of induced reduction of added cobalt(III) complex)	1089
$Hg^{II} + Sn^{II}$		No evidence of Sn^{III} intermediate (lack of induced reduction of added cobalt(III) complex)	1089
$Tl^{III} + Hg^0$	$HClO_4$, 25 °C	$k[TlOH^{2+}][Hg]$	39, 40
$Tl^{III} + U^{IV}$	75% CH_3OH, 25% H_2O	$k[Tl^{III}][U^{IV}]$	517
$Tl^{III} + U^{IV}$	HCl, aq	$k[Tl^{III}]^{0.67}[U^{IV}]^{0.33}$	599
$U^{VI} + Sn^{II}$	9–12M HCl	$k[U^{VI}][Sn^{II}]$	764
$Mo^{VI} + Sn^{II}$	9–12M HCl	$k[Mo^{VI}][Sn^{II}]$	112
$Mo^V + Sn^{II}$		$k[Mo^V][Sn^{II}]/(1+a[Mo^V]+b[Sn^{II}]+c[Mo^{III}])$	113
$Au^{III}Cl_4^- + Sn^{II}$	0.2M Cl$^-$	No evidence of Au^{II} intermediate	888
$Au^{III}Cl_4^- + Sb^{III}OL^-$	dilute L^{2-}	L^{2-} is tartrate; no evidence of Au^{II} intermediate	888
$Pu^{VI} + Sn^{II}$		Argument against Pu^V intermediate—see text, p. 86	871
$Au^{III}Cl_4^- + PtL_2RCl^{(a)}$	CH_3CN	$k_3[Pt^{II}][Au^{III}][Cl^-] + \dfrac{k_s[Pt^{II}][Au^{III}]}{k+[Au^{III}]}$	837
$Tl^{III} + V^{II}$	1M $HClO_4$	See text, p. 73	65

(a) L is PEt_3, $AsEt_3$; R is various aryl groups

thallium(II). The absence of discrete thallium(II) complexes, at any rate in perchloric acid media, is conclusively proved by the recent measurement of the redox potential and disproportionation rate of the Tl^{2+} ion[320, 398, 930]. From these data, the rate constant k_1 is calculated to be 4.4×10^{-24} $M^{-1} s^{-1}$, whereas the observed rate constant of reaction (3.190) is 7×10^{-5} $M^{-1} s^{-1}$ (at 25°C).

On the other hand, when thallium(II) is introduced into the thallium(III)–thallium(I) system by other means there is a marked catalysis, through the reaction sequence

$$*Tl^I + Tl^{II} \rightleftharpoons *Tl^{II} + Tl^I \quad (3.193)$$

$$Tl^{III} + *Tl^{II} \rightleftharpoons Tl^{II} + *Tl^{III} \quad (3.194)$$

Thus the exchange rate is enhanced by X-radiation[216] through processes such as

$$H_2O \xrightarrow{h\nu} HO + H \quad (3.195)$$

$$Tl^I + OH \rightarrow Tl^{II} + OH^- \quad (3.49)$$

The effect of the radiation is suppressed by Fe^{3+}, Fe^{2+}, Ce^{4+} or Ce^{3+}, all of which can react with both OH and thallium(II). Photochemical studies[971] using UV radiation have been interpreted in terms of the reaction

$$[Tl^{III}OH]^{2+} \xrightarrow{h\nu} Tl^{II} + OH \quad (3.50)$$

In the absence of light, iron(II)[1074] and cerium(IV)[1069] act as catalysts, via the reactions

$$Tl^{III} + Fe^{II} \rightarrow Tl^{II} + Fe^{III} \quad (3.196)$$

$$Tl^I + Ce^{IV} \rightarrow Tl^{II} + Ce^{III} \quad (3.197)$$

The kinetics of these systems have been studied but not yet exhaustively; in particular, the possible inhibitory effects of iron(III) and cerium(III) have apparently not been investigated.

Platinum(IV)–Platinum(II). Exchange reactions between various complexes by the bridging mechanism (see p. 133) appear to be two-electron processes[d]. This conclusion is based on the fact that in certain cases an additional, catalysed pathway can be found, which apparently does involve platinum(III) as an intermediate. For example, the reaction $PtCl_6^{2-} + PtCl_4^{2-}$ is catalysed by light, but this catalysis is inhibited by the oxidant $IrCl_6^{2-}$ [887]. Similar observations apply to the exchange between $PtCl_6^{2-}$ and free Cl^- ion in solution[343]. Platinum(III) has been directly detected in flash photolysis of $PtCl_6^{2-}$ [1110].

---

[d] For a review of complementary and non-complementary reactions of the $Pt^{IV} - Pt^{II}$ system, see reference 836

Other examples. Other complementary two-electron reactions are summarised in *Table 3.8*. The tin(IV)–tin(II) system is of interest since evidence was obtained of a complex between oxidant and reductant species

$$Sn^{IV} + Sn^{II} \rightarrow Sn^{IV}Sn^{II} \tag{3.198}$$

The complex is characterised by an absorption band, probably of the intervalence type and irradiation into this band enhances the rate of exchange. In the notation of Chapter 8 the photo-induced exchange process can be written

$$^{113}Sn^{IV} + Sn^{II} \rightarrow {}^{113}Sn^{IV}Sn^{II} \xrightarrow{h\nu} [^{113}Sn^{III}Sn^{III}]^*$$
$$\rightarrow {}^{113}Sn^{II}Sn^{IV} \rightarrow {}^{113}Sn^{II} + Sn^{IV} \tag{3.199}$$

There is no evidence for or against the participation of isolated tin(III) complexes in the thermal reaction.

The reaction

$$Tl^{III} + Sn^{II} \rightarrow Tl^{I} + Sn^{IV} \tag{3.200}$$

is rapid and complete under the conditions shown in *Table 3.8*, and when $Co(C_2O_4)_3^{3-}$ is included in the mixture, it is reduced to cobalt(II). Evidence that cobalt(III) complexes are good scavengers for tin(III) was obtained from experiments with the $Fe^{III} + Sn^{II}$ system (see p. 59). The reaction.

$$Hg^{II} + Sn^{II} = Hg^{0} + Sn^{IV} \tag{3.201}$$

appears to be a direct process. It is complicated by the appearance of mercury(I) as final product, but the fact that this can be avoided under certain conditions is one of several arguments in favour of two-electron transfer.

In aqueous acidic solution the reaction

$$U^{IV} + Tl^{III} = U^{VI} + Tl^{I} \tag{3.202}$$

obeys the rate law

$$\text{Rate} = k[U^{IV}][Tl^{III}] \tag{3.203}$$

It is strongly catalysed by base, and by sulphate ion, but inhibited by chloride ion, much like the $Tl^{III} - Tl^{I}$ exchange. Other reactions of uranium(IV), with one-electron oxidants and with molecular oxygen, involve uranium(V) as an intermediate. The latter reaction is strongly catalysed by Cu^{2+} and Hg^{2+}, but inhibited by Ag^{+} [510a]. No such effects were observed with reaction (3.197), and it is evident that any uranium(V) which is formed must be short-lived. It seems, however, that the mechanism is sensitive to the conditions, since in a methanol–water mixture, the effects of Cu^{2+}, Hg^{2+} and Ag^{+} reappear[599].

The reactions

$$Au^{III} + Sn^{II} = Au^{I} + Sn^{IV} \tag{3.204}$$

$$Au^{III} + Sb^{III} = Au^{I} + Sb^{V} \tag{3.205}$$

(in chloride media) have not been studied in detail, but Rich and Taube[888] obtained evidence that they proceed by direct two-electron transfer. A kinetic study of the non-complementary reaction

$$AuCl_4^- + 2Fe^{2+} = AuCl_2^- + 2Cl^- + 2Fe^{3+} \qquad (3.206)$$

gave evidence of a gold(II) intermediate, and this in turn was found to catalyse the exchange reaction between $AuCl_4^-$ and free chloride ion. No such catalysis could be detected with reactions (3.204) and (3.205), hence gold(II) is apparently not involved.

3.6 Multiple electron transfer

Simultaneous transfer of more than two electrons between two metal complexes is unknown, but some three-electron oxidations of organic molecules, by chromium(VI), have been characterised[519, 1102, 1103].

3.7 Other reaction pathways

Some examples of electron transfer reactions which do not conform to any of the rate laws already discussed are listed in *Table 3.9*.

3.7.1 Third-order rate laws

A rate law of the form $k[A]^2[B]$ is compatible with several mechanisms of which the two most usually considered can be summarised as

$$A + A + B \rightleftharpoons A_2 + B \rightleftharpoons [AAB]^{\ddagger} \rightarrow \text{products} \qquad (3.207)$$

$$A + A + B \rightleftharpoons A + AB \rightleftharpoons [ABA]^{\ddagger} \rightarrow \text{products} \qquad (3.208)$$

Equation (3.202) is usually postulated in cases where reactant A is known to dimerise under other conditions. Oxidations by chromium(VI) provide several examples of rate laws of the type

$$\text{Rate} = (k[\text{Cr}^{VI}] + k'[\text{Cr}^{VI}]^2)[B] \qquad (3.209)$$

which can be attributed[1088] to parallel pathways involving mononuclear and binuclear chromium(VI) anions, present in equilibrium with each other, for example, the reactions

$$2HCrO_4^- \underset{-1}{\overset{1}{\rightleftharpoons}} Cr_2O_7^{2-} + H_2O \qquad (3.210)$$

$$HCrO_4^- + B \xrightarrow{2} \text{products} \qquad (3.211)$$

$$Cr_2O_7^{2-} + B \xrightarrow{3} \text{products} \qquad (3.212)$$

Table 3.9 Some unusual electron transfer rate laws[a]

Stoicheiometry	Rate expression	References
1. Rate Laws containing squared terms		
$Cr^{VI} + 3Ta_6Br_{12}^{2+} = Cr^{III} + 3Ta_6Br_{12}^{3+}$	$(k[Cr^{VI}] + k'[Cr^{VI}]^2)[Ta_6Br_{12}^{2+}]$	389
$Cr^{VI} + 3V^{IV} = Cr^{III} + 3V^{V}$	$(k[Cr^{VI}] + k'[Cr^{VI}]^2)[V^{IV}]^2[V^{V}]^{-1}$	376
$Cr^{VI} + 3Fe^{II} = Cr^{III} + 3Fe^{III}$	$(k[Cr^{VI}] + k'[Cr^{VI}]^2)[Fe^{II}]^2[Fe^{III}]^{-1}$	388
$Cr^{VI} + 3Fe(phen)_3^{2+} = Cr^{III} + 3Fe(phen)_3^{3+}$	$(k[Cr^{VI}] + k'[Cr^{VI}]^2)[Fe(phen)_3^{2+}]$	388
$U^{VI} + U^{IV} = U^{IV} + U^{VI}$	$k[U^{VI}][U^{IV}]^2$	899
$V^{V} + V^{IV} = V^{IV} + V^{V}$	$k[V^{V}]^2[V^{IV}]$	464
$Ce^{IV} + Ce^{III} = Ce^{III} + Ce^{IV}$	$(k[Ce^{IV}] + k'[Ce^{IV}]^2)[Ce^{III}]$	347
$Mn^{VII} + 4Mn^{II} = 5Mn^{III}$	$k[Mn^{VII}][Mn^{II}]^2$	908
$Co(C_2O_4)_3^{3-} + Fe^{II} = Co^{II} + Fe^{III}$	$(k[Co^{III}] + k'[Co^{III}]^2)[Fe^{II}]$	205
2. Rate Laws with fractional exponents		
$Np^{VI} + Np^{IV} = Np^{IV} + Np^{VI}$ [b]	$k[Np^{VI}][Np^{IV}]^{1/3}$	237
$Np^{V} + Np^{IV} = Np^{IV} + Np^{V}$	$k[Np^{V}]^2 + k'[Np^{V}]^{0.5}[Np^{IV}]^{1.5}$	980
$Cr^{VI} + Cr^{III} = Cr^{III} + Cr^{VI}$	$k[Cr^{VI}]^{2/3}[Cr^{III}]^{4/3}$	29
$Co(edta)^- + \frac{1}{2}Cr_2(OAc)_4 = Co^{II} + Cr^{III}$	$k[Co^{III}][Cr_2(OAc)_4]^{1/2}$	204
3. Non-integral and variable rate laws		
$Sb^{V} + Sb^{III} = Sb^{III} + Sb^{V}$ [c]	$k[Sb^{V}]^{1.1}[Sb^{III}]^{0.6}$	138
$U^{VI} + U^{IV} = U^{IV} + U^{VI}$ [d)(f]	$k[U^{VI}]^{0.2}[U^{IV}]^{0.2}$	742
$Tl^{III} + U^{III} = Tl^{I} + U^{IV}$ [e)(f]	$k[Tl^{III}]^{0.67-}[U^{IV}]^{0.33-1}$	599
$Tl^{III} + U^{IV} = Tl^{I} + U^{V}$ [g]	$k[Tl^{III}]^{0.05}[U^{IV}]^{0.93}$	691

4. *Rate Laws zero-order (or mixed first-order and zero-order) in one reactant*

Reaction	Rate law	Ref.
$Co(C_2O_4)_3^{3-} + \frac{1}{2}Cr_2(OAc)_4 = Co^{II} + Cr^{III}$	$k[Cr_2(OAc)_2]$	204
$Ce^{IV} + Ce^{III'} = Ce^{III} + Ce^{IV}$	$k[Ce^{IV}][Ce^{III}] + k'[Ce^{III}]$	486
$2(Cr^{III}OU^VO)^{4+} + Tl^{III} = 2Cr^{III} + 2U^{VI} + Tl^I$	$k[Cr^{III}OU^VO^{4+}]$	798
$(Cr^{III}OU^VO)^{4+} + V^{IV} = Cr^{III} + U^{VI} + V^V$	$k[Cr^{III}OU^VO^{4+}]$	798
$(Cr^{III}OPu^VO)^{4+} + Np^{VI} = Cr^{III} + Pu^{VI} + Np^V$	$k[Cr^{III}OPu^VO^{4+}]$	805
$Co(NH_3)_4)_2(OH)_2^{4+} + 2Cr^{2+} = 2Co^{II} + 2Cr^{III}$	$(k + k'[Cr^{II}])[(Co^{III})_2]$	549
$Co(NH_3)_4)_2(OH)_2^{4+} + 2V^{2+} = 2Co^{II} + 2V^{III}$	$(k + k'[V^{II}])[(Co^{III})_2]$	549
$(Co(NH_3)_4)_2(NH_2)(OOCH)^{4+} + 2Cr^{2+} = 2Co^{II} + 2Cr^{III}$	$\{k_0[Cr^{II}] + k_1k_2[Cr^{II}]/(k_1 + k_2[Cr^{II}])\}[(Co^{III})_2]$	932

(a) 25 °C, aqueous acidic media (for hydrogen ion dependences see the original references)
(b) 12.1 M ethylene glycol
(c) Concentrated HCl
(d) 0–100% C_2H_5OH
(e) 0–75% CH_3OH
(f) Complicated variation of reaction orders with solvent composition
(g) Catalysed by tartaric acid

with $k = k_2$, $k' = K_1 k_3$. The rate constants k_1, k_{-1} have also been measured[846] and Espenson points out[382, 846] that in principle the mechanism could be checked if a reductant B could be found such that steps 2 and 3 proceeded more rapidly than step -1. In that case it would be possible (using a rapid mixing procedure) to vary the concentrations of $HCrO_4^-$ and $Cr_2O_7^{2-}$ ions independently. So far, no suitable reductant has been found. Most of the other $[A]^2[B]$ type rate laws listed in *Table 3.9* involve highly charged metal ions or oxo-ions, which are prone to hydrolysis and polymerisation[493], and may therefore proceed by mechanism in equation (3.207), but again there is no direct evidence.

The mechanism in equation (3.208) involves as the first step a reaction between different reagents $A + B$; but the transition state has the same composition, A_2B as equation (3.207) and if the equilibrium steps are rapidly established, the two are kinetically indistinguishable. In some cases, however, it seems likely that the two A units in the transition state, while bound to B, are not bound to each other. The reactions of $Co(C_2O_4)_3^{3-}$ with Fe^{2+} (with a net activation process $2Co(C_2O_4)_3^{3-} + Fe^{2+}$) and of MnO_4^- with Mn^{2+} (net activation process $2Mn^{2+} + MnO_4^-$), seem best described in this way.

Some oxidations of halide ions have terms of the type $(A^+)(X^-)^2$, presumably associated with the requirement to form the dimeric X_2 molecule as product. Oxidants A^+ include $VO_2^{+\,214, 936}$, $Fe^{3+\,655}$, $W(CN)_8^{3-\,136}$, and certain binuclear Co^{III} complexes[292, 293] all reacting with an I^- ion; and MnO_4^- with an Br^- ion[657]. Conversely, some reactions in which molecular hydrogen functions as *oxidant*, are second order in reductant. The reaction

$$2Co(CN)_5^{3-} + H_2 = 2Co(CN)_5 H^{3-} \tag{3.213}$$

is of this type[176, 300] and it has been suggested[952] that a binuclear intermediate with Co–H–H–Co bridging is involved.

3.7.2 Fractional rate laws

A rate law in which the concentration dependence is not a whole number may sometimes be adequately represented by an expression with a simple, fractional power. There is an obvious difficulty here: if the dependence found by experiment is $[X]^{0.37 \pm 0.04}$ is this to be interpreted as approximately $[X]^{1/3}$ or approximately $[X]^{2/5}$? The choice becomes convincing only when it can be related to a plausible mechanism usually involving a pre-equilibrium step. The case of the Cr^{VI}–Cr^{III} reaction (p. 87, footnote c) is one example. Another is the series of reactions between the chromous acetate dimer and different oxidants[194, 203, 204]. These evidently involve prior dissociation into a monomeric chromium(II) complex which is the effective reducing agent.

$$Cr_2(OAc)_4 \underset{-1}{\overset{1}{\rightleftharpoons}} 2Cr_2(OAc)_2 \tag{3.214}$$

$$A^+ + Cr(OAc)_2 \overset{2}{\longrightarrow} A + Cr^{III} \tag{3.215}$$

With oxidants $A^+ = Cr(NH_3)_5Cl^{2+}$, $Co(edta)^-$, $Co(NH_3)_5OH_2^{3+}$, the rate law takes the form

$$\text{Rate} = K_1^{1/2}k_2[Cr_2(OAc)_4]^{1/2}[A^+] \quad (3.216)$$

consistent with equations (3.214) and (3.215) provided that step 2 is rate-determining. With more powerful oxidants, $Co(NH_3)_5Cl^{2+}$, $Co(C_2O_4)_3^{3-}$, step 2 is faster, step 1 is rate-determining, and the rate law takes the form (3.212).

$$\text{Rate} = 2k_1[Cr_2(OAc)_4] \quad (3.217)$$

Non-integral rate laws have been reported for a number of reactions in non-aqueous solvents. The list given in *Table 3.9* is not exhaustive. In some cases the reaction order varies drastically with solvent composition, and it is difficult to discuss the reasons without more knowledge of the tendency of the reactants to polymerise in these media.

3.7.3 Zero-order rate laws

Rate laws of zero-order in one component are generally attributable to some reaction which precedes the observed electron transfer, and is rate-determining (*see* equation 3.217). A very curious example is the halide anation of aquoruthenium(III)

$$Ru^{3+} + X^- = RuX^{2+} \quad (3.218)$$

(X = Cl, Br, I), which is catalysed by ruthenium (II) with the rate law

$$\text{Rate} = k[Ru^{2+}][X^-] \quad (3.219)$$

The proposed mechanism involves a rate-determining anation of ruthenium(II) followed by rapid electron transfer[605]

$$Ru^{2+} + X^- \xrightarrow{k} RuX^+ \quad (3.220)$$

$$RuX^+ + Ru^{3+} \longrightarrow RuX^{2+} + Ru^{2+} \quad (3.221)$$

Among other examples, the reaction $(CrOPu^VO)^{4+} + Np^{VI}$ apparently involves prior dissociation of the binuclear complex to yield free PuO_2^+ ion[805]. Reactions of the dimer $[(Co(NH_3)_4)_2(OH)_2]^{4+}$ with Cr^{2+} and V^{2+} proceed partly by a ring-opening step followed by rapid reduction, the first-order rate terms being the same for both reductants[549]. In the reaction of the dimer $[(Co(NH_3)_4)_2(NH_2)(OOCH)]^{4+}$ with Cr^{2+}, Scott and Sykes[932] found the rate law

$$\text{Rate} = k_{Co}[(Co^{III})_2][Cr^{II}] + k_1k_2[(Co^{III})_2][Cr^{II}]/(k_{-1} + k_2[Cr^{III}]) \quad (3.222)$$

the second term of which suggests the mechanism

$$(Co(NH_3)_4)_2(NH_2)(OOCH)^{4+}$$

$$\underset{-1}{\overset{1}{\rightleftharpoons}} H_2OCo(NH_3)_4(NH_2)Co(NH_3)_4OOCH^{4+}$$

$$\xrightarrow{k_2, Cr^{2+}} Co^{III} + Co^{II} + Cr^{III} \xrightarrow{Cr^{2+}} \text{products} \qquad (3.223)$$

in which ring closure (step -1) competes with electron transfer (step 2) (see also reference 1000).

Chapter 4
The reaction path (2): binuclear intermediates

4.1 Introduction

4.1.1 Definitions

The evidence reviewed in the previous chapter makes it clear that most if not all oxidation–reduction reactions between metal ions in solution can be resolved into sequences of bimolecular reactions consisting of single or double electron transfer, and as long as we consider only changes in oxidation states, these are the effective elementary steps. In general, however, the bimolecular process

$$A^+ + B \xrightarrow{k} A + B^+ \tag{4.1}$$

can itself be resolved into a further sequence of steps

$$\underset{i}{A^+ + B} \underset{-1}{\overset{1}{\rightleftarrows}} \underset{p}{A^+ . B} \underset{-2}{\overset{2}{\rightleftarrows}} \underset{s}{A . B^+} \underset{-3}{\overset{3}{\rightleftarrows}} \underset{f}{A + B^+} \tag{4.2}$$

As before, A and B denote reactant ions as complexes with appropriate solvation, and the positive sign denotes the higher of the two valencies. The states i and f are initial and final states, with reactants far apart; states p and s are *precursor* and *successor* states. It is assumed that reactions i ⇌ p and s ⇌ f involve motion of the A and B complexes with little if any transfer of the electron, while the reaction p ⇌ s involves electron transfer with the A–B distance more or less fixed. Depending on the rate constants involved, the states p and s may or may not correspond to well defined chemical species[a].

4.1.2 Historical

The idea of the binuclear intermediate seems to have occurred to several workers at about the same time. In 1947, Whitney and Davidson established

---

(a) The terms *precursor complex* and *successor complex* are due to Sutin[987]. Previously, Hush[570] had used the symbols *p* and *q* to denote '... states immediately preceding and succeeding the electron transfer'

the existence of a coloured $Sb^{III} - Sb^V$ complex in hydrochloric acid[1092] and in 1949 Bonner[138] reported the rate of the $Sb^{III} + Sb^V$ exchange reaction, in which this complex might be expected to participate. From this point of view the result was negative, however: the chloride concentration in which the reaction proceeded most rapidly was not the concentration most favourable to the formation of the complex and Bonner concluded that the known complex was not an intermediate in the reaction. In the following year, Tewes, Ramsey and Garner[1021] measured the kinetics of the $V^{IV} + V^V$ reaction, and undertook a spectrophotometric study 'to check the possibility of formation of some kind of a stable complex between vanadyl and pervanadyl ions, which might account for the apparent rapid exchange'—but again with a negative result.

An intermediate was postulated by Rona[899], in the reaction $U^{IV} + U^{VI}$. Its structure was written $[OU(OH).O.U(OH)_2]^{3+}$ but the valencies were left unspecified. If, as seems likely, it was envisaged as a di-uranium(V) species, then in the more recent terminology it would be called a successor complex.

The role of the intermediate was discussed by several authors at a meeting of the American Chemical Society in 1952[1062]; but in the published transcript it is not always clear which of the various complexes was meant. Thus, in the $Fe^{3+} + Fe^{2+}$ reaction, catalysed by halide ion, Libby[672] postulated a complex $(H_2O)_5Fe^{3+}X^-Fe^{2+}(OH_2)_5$ but described it as a 'transition complex' with a plane of symmetry through the bridging atom.

The definitions now in use were first made explicit in the context of theoretical, not experimental, studies. In papers on the kinetics of electrode processes[569], and later in an analogous treatment of homogeneous reactions[570], Hush used the sequence $i \rightleftharpoons p \rightleftharpoons t \rightleftharpoons s \rightleftharpoons f$ in which t denotes the electron transfer transition state and the other symbols are as above. Previously, Marcus[720] had discussed outer-sphere homogeneous electron transfer in terms of the sequence $i \rightleftharpoons p^\ddagger \rightleftharpoons s^\ddagger \rightarrow f$ in which $p^\ddagger$ and $s^\ddagger$ are resonant forms of the transition state which have identical atomic configurations but different electronic configurations. Although the precursor complex was not considered as a species with a finite lifetime, the work done in forming it from the reactants was included in the calculations.

4.1.3 Outline of mechanisms

Depending on the relative stabilities and rates of formation of the intermediates and products, various mechanisms can be distinguished within this general scheme. A complete formal classification of these mechanisms could be obtained by listing all the permutations of order of magnitude of the six rate constants, but this is unnecessarily elaborate. A broader and more useful classification is as follows. When both precursor and successor states are unstable with respect to the initial and final states, and when both are formed rapidly, the concentrations of precursor and successor remains small throughout the course of the reaction. The reaction is first order with respect to both oxidant and reductant

$$\text{Rate} = k[A^+][B] = k[A^+]_T[B]_T \tag{4.3}$$

where $[A^+]_T$ and $[B]_T$ denote total stoicheiometric concentrations and the steady state conditions give

$$k = k_1 k_2 k_3 / (k_2 k_3 + k_3 k_{-1} + k_{-1} k_{-2}) \tag{4.4}$$

which may be expressed more clearly as

$$k^{-1} = k_1^{-1} + k_{-1} k_1^{-1} k_2^{-1} + k_{-2} k_{-1} k_1^{-1} k_2^{-1} k_3^{-1}$$

$$= k_1^{-1} + (k_2^*)^{-1} + (k_3^*)^{-1} \tag{4.5}$$

where k_2^* and k_3^* are net activation rate constants for the processes i → s and i → f. If any one of the steps 1, 2 or 3 is rate-determining, these equations reduce to $k = k_1$, k_2^* or k_3^* respectively. If the complex $A^+.B$ is of the inner-sphere type (Chapter 5), step 1 can be further subdivided,

$$A^+ + B \underset{}{\overset{0}{\rightleftharpoons}} A^+ \ldots B \underset{}{\overset{1'}{\rightleftharpoons}} A^+.B \tag{4.6}$$

where $A^+ \ldots B$ denotes an outer-sphere complex. In that case, either step 0 or step 1' may be rate-determining, and step 0 in turn may be diffusion-controlled. In the limiting case, where the precursor complex is highly unstable with respect to the reactants, its lifetime may be less than the average time between collisions, and if the same is also true of the successor complex, the entire sequence reduces to a single step i → f, with rate constant k. It is still possible formally to define a precursor and successor state, i.e. configurations of reactants and products poised in position just before and just after the intramolecular electron transfer process, but they can no longer be described as complexes in the chemical sense.

When either the precursor or the successor complex is stable with respect to the reactants there is the possibility of observing it directly, and monitoring its rate of transformation into products. A special case is the symmetrical exchange process ($A \equiv B$), where there is no net reaction, and the precursor and successor complexes are identical, mixed-valence complexes in equilibrium with the reactants. The internal electron transfer could be monitored by resonance methods. A general problem with all labile systems, however, is that although a complex ($A^+ \ldots B$) may be detected and characterised, it cannot in general be proved to be in the direct pathway of the electron transfer reaction. The true precursor or successor may be some other, undetected ($A^+ \ldots B$) species with a structure more favourable to electron transfer (equation 4.89).

When the binuclear intermediate is inert with respect to dissociation or isomerisation this ambiguity is removed. In such a case, the intermediate can be established as a true precursor or successor complex, and its structure to a large extent defines the structure of the electron transfer transition state. Examples of inert binuclear complexes, and other evidence of the structure of the transition state, will be discussed in the next chapter. In this chapter, we review the evidence for the existence of binuclear complexes as kinetically significant intermediates, regardless of structure.

4.2 The diffusion-controlled mechanism

4.2.1 Theory

For this purpose we shall write equation (4.6) in the more explicit form

$$A^{z_1} + B^{z_2} \underset{-0}{\overset{0}{\rightleftarrows}} (A^{z_1})(B^{z_2}) \xrightarrow{1'} P \qquad (4.7)$$

where $(A^{z_1})(B^{z_2})$ denotes the outer-sphere complex or ion-pair; P denotes either the inner-sphere precursor complex, or in the case of an outer-sphere reaction, the final products; and z_1, z_2 are the actual ionic charges of the reactants.

The simplest model of a diffusion-controlled reaction is due to Smoluchowski[959]. The reactants are considered as spheres of equal diameter a, and the solvent as a continuous fluid. For uncharged particles with no interactions, this leads to

$$k_0^0 = 8\pi D_{AB} aL \qquad (4.8)$$

where D_{AB} is the relative diffusion coefficient of the A and B ions. Assuming further that the ions obey Stokes' law this gives

$$k_0^0 = 8RT/3\eta \qquad (4.9)$$

where η is the viscosity of the solvent. For interacting ions, Debye[301] obtained[b]

$$k_0 = k_0^0 \left(a \int_a^\infty \exp[\omega(r)/kT] r^{-2} dr \right)^{-1} \qquad (4.10)$$

where $\omega(r)$ is the mutual energy of the ion pair at distance r. Using the Coulombic potential

$$\omega(r) = (1/4\pi\varepsilon_0) z_1 z_2 e^2/\varepsilon r \qquad (4.11)$$

where ε is the dielectric constant of the medium, this gives the *Debye equation*[301, (b), (c)].

(b) Debye also gives alternative equations for particles of unequal diameter, but the difference is not great, and conventional practice is to use a single parameter, the mean diameter
(c) This and subsequent equations containing the charge-product $z_1 z_2$ are more commonly written using the absolute value $|z_1 z_2|$; but since we are concerned with both like- and unlike-charged pairs, the charge product is given its proper algebraic sign

$$k_0 = k_0^0(z_1z_2s/a)[\exp(z_1z_2s/a) - 1]^{-1} \quad (4.12)$$

where

$$s = (1/4\pi\varepsilon_0)e^2/\varepsilon kT$$

The simplest expression for the stability constant $K_0 = k_0/k_{-0}$ is the Fuoss equation[448(d), (e)]:

$$K_0 = (4/3)\pi a^3 L \exp(-z_1z_2s/a) \quad (4.13)$$

This also is based on the Coulombic potential (equation 4.11) and on the assumption that the majority of outer-sphere complexes consist of pairs of ions in contact at distance a. Previously Bjerrum[128] had considered the thermodynamics of ion pair formation allowing for pairing over a range of distances. Using equation (4.11), but with allowance for the screening effect of the ionic atmosphere as discussed by Debye and Hückel[302, 897], he showed that at equilibrium the radial density of cations around a given anion or vice versa, reached a minimum at a radius $r_{min} = -\tfrac{1}{2}z_1z_2s$ (see footnote(c), p. 100) and he defined an ion-pair as any pair of ions within this distance. Fuoss's treatment, however, makes some allowance for the molecular nature of the solvent. Fuoss pointed out that the range of variation of distance allowed by Bjerrum, from a to r_{min}, is of the order of the diameters of the solvent molecules. Hence, if the solvent in the neighbourhood of each ion is considered to be approximately close-packed, the density of counter-ions at distances intermediate between a and r_{min} will be much less than the density calculated by Bjerrum for a continuous solvent. As an approximation this density may be set equal to zero and all the ion pairs considered as contact pairs.

This argument is intuitively reasonable where oppositely charged ions are concerned, but its application to like-charged ions is not so obvious. However, the form of the result is also reasonable, and for that reason, and for want of anything better, equation (4.13) has been used by several writers to estimate the formation constants of like-charged pairs[(f)].

The above equations refer to ionic species at infinite dilution. To allow for the effects of ionic strength, the Coulombic potential energy is replaced by the

(d) An alternative expression $K_0 = \exp(-z_1z_2s/a)$, due to Denison and Ramsey[308], is dimensionally incorrect, but when K_0 is expressed in the conventional units, the numerical difference between this and equation (4.13) is small; for example, when $a = 7.36$ Å, the pre-exponential factor is $(4/3)\pi a^3 L = 1.0$ dm^3 mol^{-1}

(e) Comparing equation (4.13) with the conventional expression $RT\ln K_0 = \Delta G_0^{\ominus} = \Delta H_0^{\ominus} - T\Delta S_0^{\ominus}$, the pre-exponential factor can be identified as a contribution to the entropy term. It measures the decrease in proper entropy when the set of free-moving ions is transferred from the bulk solvent to a space of volume $(4/3)\pi a^3$. However, it is not in general true to say that the exponent $-z_1z_2s/a$ measures $\Delta H_0^{\ominus}/RT$. It, too, contains an important entropy term, due to the change in ordering of solvent molecules when the charges z_1 and z_2 are combined; expressed in the continuum model by the temperature-dependence of the dielectric constant

(f) Ben-Naim has calculated the free energies of Van der Waals attraction between uncharged molecules in polar solvents[108]

Debye–Hückel energy $\omega(r)$:

$$\omega(r) = \left(\frac{1}{4\pi\varepsilon_0}\right)\left(\frac{z_1 z_2 e^2}{\varepsilon}\right)\left(\frac{\exp(\kappa a)}{1+\kappa a}\right)\left(\frac{\exp(-\kappa r)}{r}\right) \quad (4.14)$$

or in the limit of low ionic strength

$$\omega(r) = \left(\frac{1}{4\pi\varepsilon_0}\right)\left(\frac{z_1 z_2 e^2}{\varepsilon}\right)\left(\frac{\exp(-\kappa r)}{r}\right) \quad (4.15)$$

where $\kappa^2 = 8\pi LsI$, and I is the ionic strength. Hence the Fuoss equation is replaced by

$$K = \tfrac{4}{3}\pi a^3 L \exp[-z_1 z_2 s/a(1+\kappa a)] \quad (4.16)$$

The Coulomb expression (equation 4.11) also breaks down at short interionic distances, since the ions cannot in general be considered as point charges, but the implications of this have not been studied in detail. For spherical ions in close contact, image forces should be considered (see p. 192). Ion–dipole and dipole–dipole forces are also important at short range, as are specific effects such as hydrogen bonding. Some evidence for these effects is considered on p. 127.

The specific rate of dissociation of the ion pair back to reactants may be calculated as $k_{-0} = k_0/K_0$. For the uncharged case this simplifies to $k_{-0} = 6D_{AB}/a^2$, which is also the expression derived by Einstein for the average rate of diffusion of a pair of particles from a separation distance a to a distance $2a$, in a continuous medium[111]. Calculations of this sort are the basis of the *solvent cage* model for reactions in solution. If two non-interacting particles come into contact through random thermal motion, they will remain in contact for an average time $\tau = k_{-0}^{-1}$, during which they may participate in other processes such as repeated collisions and internal rearrangements. The maximum rate for such a process is given in transition state theory by the 'ceiling' unimolecular rate constant $k^{\ddagger} = RT/Lh$. If the lifetime of the precursor complex is long compared with $1/k^{\ddagger}$ it may be considered as a well-defined chemical species.

Some specimen calculations of K_0, k_0 and k_{-0} are shown in *Table 4.1*. The variations in the two rate constants, as a function of the charge-product $z_1 z_2$ are in the expected direction, but what is noteworthy is that for oppositely charged ions k_0 increases only slowly with increased Coulombic attraction, while for like-charged ions k_{-0} increases only slowly with increased repulsion. Consequently not only uncharged reactant pairs, but also some like-charged pairs, can be considered as chemical entities with finite lifetime[534]. Comparing the values of log k_{-0} with the value log $k^{\ddagger} = 12.8$ (at 25°C) it seems that a complex formed from uncharged reactants can survive about 100 collisions before dissociating back to reactants; a complex between two doubly charged cations lasts about one tenth as long, and even a complex of triply charged cations may still be a chemical entity.

Table 4.1 Specific rates of formation and dissociation of ion pairs[a]

$$A^{z_1} + B^{z_2} \underset{0}{\overset{0}{\rightleftharpoons}} (A^{z_1})(B^{z_2})$$

Aqueous solution, 25°C, mean ionic diameter $a = 4.0$ Å.

Charge product z_1z_2	$\log(k_0/M^{-1}s^{-1})$	$\log(k_{-0}/s^{-1})$	$\log(K_0/M^{-1})$
+12	2.1	12.2	−10.0
+9	4.3	12.0	−7.8
+4	7.8	11.7	−3.9
+1	9.5	11.1	−1.6
0	10.0	10.8	−0.8
−1	10.4	10.5	−0.1
−4	10.9	8.6	2.3
−9	11.3	5.1	6.2
−12	11.4	2.8	8.5

(a) Calculated using equations (4.12) and (4.13)

For the general case of an outer-sphere reaction, we may write

$$i \underset{-0}{\overset{0}{\rightleftharpoons}} p \xrightarrow{2} \text{products} \tag{4.17}$$

Applying the steady-state criterion to the precursor complex, this gives

$$\frac{1}{k} = \frac{1}{k_0} + \frac{k_{-0}}{k_0 k_2} \tag{4.18}$$

Using the transition state theory, we have

$$k_2 = k^{\ddagger} K_2^{\ddagger} = k^{\ddagger} \exp(-\Delta G_2^{\ddagger}/RT) \tag{4.19}$$

where $\Delta G_2^{\ddagger}$ is the activation free energy for electron transfer within the precursor complex. But the dissociation rate constant k_{-0} also is of the order of $k^{\ddagger}$ so that

$$k = (k_0 K_2^{\ddagger}/(1 + K_2^{\ddagger})$$
$$\simeq k_0 K_2^{\ddagger} = k_0 \exp(-\Delta G_2^{\ddagger}/RT) \tag{4.20}$$

a relationship first derived by Marcus[720]. When step 2 is sufficiently rapid, $\Delta G_2^{\ddagger} = 0$ and $k = k_0$. An alternative equation sometimes quoted is

$$k = k_0^0 \exp(-[\Delta G_2^{\ddagger} + \omega(a)]/RT) \tag{4.21}$$

which is not strictly consistent with the preceding equations since it lacks the pre-exponential factor $(z_1 z_2 s/a)$.

4.2.2 Examples

The evidence for diffusion controlled mechanisms in general has been reviewed elsewhere[111,182,358]. Some examples of electron transfer reactions which appear to be at or near the diffusion-controlled limit are shown in *Table 4.2*. In some cases the agreement between observed and calculated rate constants is the evidence for the mechanisms. In others there are supporting arguments. The reaction $Co^{3+} + U^{3+}$ is discussed on p. 213. The reaction $Co(NH_3)_5Cl^{2+} + Cr^{2+}$ is one of a large number of Cr^{2+}-reductions which lie in the range $k = 10^5–10^6$ M^{-1} s^{-1} at ionic strength 0.1–1.0 M^{190}. Diffusion control seems likely but other explanations have been considered. Since this is an inner-sphere reaction one possibility[190,209] is that substitution into the Cr^{2+} complex is rate-determining. A number of arguments have been used against this[678,(g)], chiefly the fact that $Cr(H_2O)_6^{2+}$ is exceptionally labile with respect to water exchange[752]. Another possibility[311] is an exceptional contribution to $\Delta G^{\ddagger}$ due to the rearrangement from tetragonal $Cr(H_2O)_6^{2+}$ to octahedral $Cr(H_2O)_6^{3+}$. On the other hand, Fan and Gould[400] have drawn attention to a number of cases in which Eu^{2+} and Cr^{2+}, with the same oxidant, appear to have the same limiting rate, which they attribute to 'joint operation of a diffusion-controlled maximum of $10^{9.8}$ M^{-1} s^{-1} and an activation entropy requirement' (meaning presumably the terms k_0^0 and an entropy contribution to $\Delta G^{\ddagger}$, equation (4.21)[(h)]).

4.3 Rate-determining steps: some indirect arguments

4.3.1 Reaction rates

If the steps of equation (4.2) follow one another rapidly the rate law has the simple form of equation (4.3), and there is no direct evidence for the participation of intermediates. Nevertheless, the rate of reaction itself may be a guide and different mechanisms have been proposed on this basis. The reaction $Co(NH_3)_5Cl^{2+} + Fe^{2+}$ is of the inner-sphere type (see Chapter 5) and step $p \rightarrow s$ may therefore be written

$$(H_3N)_5Co^{III}ClFe^{II}(OH_2)_5^{4+} \rightarrow (H_3N)_5Co^{II}ClFe^{III}(OH_2)_5^{4+} \quad (4.22)$$

From the relevant redox potentials it is clear, however, that this step is highly thermodynamically unfavourable. It is at least possible, therefore, that the rate-determining step is dissociation of the successor complex, presumably at its most labile point, the Co–Cl bond, giving the mechanistic pattern $i \rightleftharpoons s \rightarrow f$. The great majority of the reactions studied are, however, thermodynamically favoured or nearly so, and if the precursor and successor complex are

---

(g) Reference 482, footnote 16
(h) See an argument of Liang and Gould[669] that 'the maximum rate for a diffusion-controlled bimolecular reaction [i.e. $k_0^0 \simeq 7 \times 10^9$ M^{-1} s^{-1} at 25 °C] ... must be corrected for ionic charge and entropy of activation'

Table 4.2 Possible examples of diffusion-controlled electron transfer reactions

$$A^+ + B \rightarrow A + B^+$$

Reactants[a]		Charge product $z_1 z_2$	Ionic strength I/M	Rate constant $k/\text{M}^{-1}\text{s}^{-1}$ [b]		Reference
A^+	B			Observed	Calculated	
$Co(H_2O)_6^{3+}$	U^{3+}	9	0	7.1×10^3	1.7×10^4	362
$Co(NH_3)_5Cl^{2+}$	Cr^{2+}	4	0.1	6×10^5	see text	190
$Fe(CN)_6^{3-}$	H	0	?	4.0×10^9	1.5×10^{9} [c]	787
$FeCl^{2+}$	H	0	var	4.6×10^9	4.8×10^{9} [c]	787
MnO_4^-	Cd^+	−1	?	2×10^{10}		96
$IrCl_6^{2-}$	$Fe(DMP)_3^{2+}$	−4	0.1	1.1×10^{9} [d]		507
$Fe(phen)_3^{3+}$	MnO_4^{2-}	−6	0.45	1.2×10^{11}	7×10^{9} [d] [e]	542
$Mo(cN)_8^{3-}$	$Os(bipy)_3^{2+}$	−6	0	2.0×10^{9} [d] [f]	1×10^{10} [d]	188
$Fe(DMP)_3^{3+}$	$IrCl_6^{3-}$	−9	0.1	1.0×10^{9} [d]	2×10^{10} [d] [e]	507
$Os(bipy)_3^{3+}$	$Mo(CN)_8^{4-}$	−12	0	4.0×10^{9} [d] [f]	2×10^{10} [d]	188

(a) phen is 1:10-phenanthroline, DMP is 4:7-dimethyl-1:10-phenanthroïne; bipy is 2:2′-bipyridyl
(b) Aqueous solutions, 25 °C
(c) Calculated using a different model from that described in the text
(d) 10 °C
(e) $I = 0.05$ M
(f) $I = 0.5$ M

sufficiently labile, as is certainly true for outer-sphere reactions, the mechanism may confidently be assigned as $i \rightleftharpoons p \rightarrow s \rightarrow f$. Examples of substitution-controlled inner-space transfer are also well characterised in certain specific cases, as discussed on p. 147, and these may be written $i \rightarrow p \twoheadrightarrow (s \twoheadrightarrow)f$.

4.3.2 Isotope and non-bridging ligand effects

In the reaction $Co(NH_3)_5Cl^{2+} + Cr^{2+}$, replacement of the normal isotope ^{14}N by ^{15}N makes little difference to the rate[482] and the difference in rate between the reactions $trans$-$Co(en)_2(NH_3)OH^{2+} + Cr^{2+}$ and $trans$-$Co(en)_2(OH_2)OH^{2+} + Cr^{2+}$ is small[200] compared with the difference between $trans$-$Co(en)_2(NH_3)Cl^{2+} + Fe^{2+}$ and $trans$-$Co(en)_2(OH_2)Cl^{2+} + Fe^{2+}$[110]. These and similar observations show that with chromium(II) reductions, there is relatively little Co–N bond stretching, and, therefore, relatively little electron transfer, in the transition state. Hence, it is possible that the rate-limiting step is either diffusion or formation of the precursor complex.

In the reaction $Co(NH_3)_5OH^{2+} + Cr^{2+}$, replacement of ^{16}O by ^{18}O lowers the rate by a significant factor 1.046[313], implying some stretching of the Co–O bond in the transition state. Thus diffusion control can be excluded, and substitution control also seems unlikely. The alternatives are that steps $p \rightarrow s$ or $s \rightarrow f$ are rate-determining

$$Co(NH_3)_5OH_2^{3+} + Cr^{2+} \rightleftharpoons (H_3N)_5Co^{III}OHCr^{II} + H^+$$
$$\rightleftharpoons (H_3N)_5Co^{II}OHCr^{III} + H^+ \rightarrow products \quad (4.23)$$

The reactions $Co(NH_3)_5OH_2^{3+} + M^{II}(edta)^{2-}$ (where M is Eu, Fe, Cr) are unaffected by substitution of ^{18}O. Here either diffusion or substitution control is possible, but the rate constant for the Fe reaction, $k = 1.2 \times 10^6 \, M^{-1} s^{-1}$, argues against diffusion control[313].

4.3.3 Activation enthalpies

In reactions of the type

$$LCo^{III}X + Cr^{2+} \rightarrow L + Co^{2+} + CrX^{2+} \quad (4.24)$$

where X^- is a halide ion and L may be various amine ligands, the enthalpies of activation are frequently low, and in a few cases actually negative; the slow rates observed being due to unfavourable entropy terms. Taube[1010] pointed out that this observation could be rationalised by assuming a pre-equilibrium step with a negative standard enthalpy change, $\Delta H_{ip}^\ominus < 0$;

$$LCo^{III}X + Cr^{2+} \underset{-1}{\overset{1}{\rightleftharpoons}} LCo^{III}XCr^{II\,2+} \overset{2}{\longrightarrow} products \quad (4.25)$$

$$\qquad i \qquad\qquad\qquad\qquad p$$

To fulfil the observed rate law it is necessary to have $k_{ps} \ll k_{ip}$ in spite of the fact that $\Delta H^{\ddagger}_{ps} > \Delta H^{\ddagger}_{pi}$ (it being taken for granted that $\Delta H^{\ddagger}_{ps}$ is positive). This is possible if the activation entropy associated with the electron-transfer step is particularly unfavourable, as is expected if solvent reorganisation is the main factor hindering the reaction. A curious feature is the fact that, whereas activation enthalpies vary over a wide range, activation free energies vary much less, changes in $\Delta H^{\ddagger}$ from one system to another being compensated by opposite changes in $\Delta S^{\ddagger}$. Striking differences occur between systems of apparently similar structure, such as $C_6H_{11}NH_2RCl^{2+} + Cr^{2+}$, with $\Delta H^{\ddagger} \leq -10$ kcal mol^{-1}, and $C_5H_5NRCl^{2+} + Cr^{2+}$ with $\Delta H \geq 1$ kcal mol^{-1} (R is cis-Co^{III}(en)$_2$). Endicott and co-workers[831] have attempted to rationalise these differences in terms of the above mechanism by assuming that the lower activation energies correspond to the case where the electron-transfer step (2) is rate-determining, and the higher ones to the case where the substitution step (1) is rate-determining. The basis of this argument is the assumption that the activation enthalpy of an elementary step must be positive, so that when negative or zero values are observed for the overall activation enthalpy, they must be ascribed to a negative standard enthalpy change ΔH° at some point in the reaction sequence. This point does not seem to have been fully argued, however, and it is by no means intuitively obvious. If an elementary process leads to a concentration of electrostatic charge, then the solvation energy will increase (solvation parameters vary as the square of the total charge) and this change will in general consist of a negative enthalpy change, offset to a greater or lesser extent by a negative entropy change due to the increased ordering of the solvent molecules. As long as the concept of a transition state is retained, as a species in quasi-equilibrium with the reactant particles, the same argument will apply to the transition state as to a precursor complex of the same formula and similar geometry. Low or negative enthalpies of activation may be regarded as a consequence of the concentration of ionic charge in the inner-sphere bridged transition state, as contrasted with the greater charge separation in the outer-sphere transition states of the reactions of $Ru(NH_3)_6^{2+}$. The thermodynamic properties of the transition state, relative to the reactants, are independent of the path by which it is formed. Negative activation enthalpies have since been reported for some outer-sphere reactions, and have been rationalised theoretically (see Chapter 6, p. 216).

Halpern et al.[507] argued similarly for the reaction $Fe(DMP)^{2+} + IrCl_6^{2-}$ (i). The specific rate is close to the limit calculated for diffusion control (Table 4.2) but the activation energy is zero, whereas the activation energy implied by the temperature dependence of the viscosity of water is of the order of 4–5 kcal mol^{-1}. This is taken to imply that $\Delta H_0^{\ddagger}$ (see equation 4.17) is also of the order of 4–5 kcal mol^{-1}. Hence, it is suggested that the rate-determining step is electron transfer, with $\Delta H_0^{\circ} + \Delta H_2^{\ddagger} \simeq 0$.

For the exchange reaction $*Fe^{III}N_3^{2+} + Fe^{2+}$ (where *Fe denotes the isotope ^{52}Fe) Bunn et al.[174] found that the activation enthalpy was strongly

(i) DMP is 4:7-dimethyl-9:10-phenanthroline

temperature-dependent, with $\Delta H^{\ddagger} = 13.3$ kcal mol^{-1} at the lower end of the temperature range (0–13 °C), but $\Delta H^{\ddagger} \leqslant 5$ kcal mol^{-1} at the upper end. This would fit the reaction sequence

$$*FeN_3^{2+} + Fe^{2+} \underset{-1}{\overset{1}{\rightleftharpoons}} *Fe^{III}N_3Fe^{II4+} \underset{-2}{\overset{2}{\rightleftharpoons}} *Fe^{II}N_3Fe^{III4+} \rightleftharpoons *Fe^{2+} + FeN_3^{2+}$$
$$\quad\quad\quad\quad\text{i}\quad\quad\quad\quad\quad\quad\text{p}\quad\quad\quad\quad\quad\quad\text{s}\quad\quad\quad\quad\quad\text{f}$$
(4.26)

with $\Delta H_1^{\ddagger}$ as the higher and $(\Delta H_1^{\ominus} + \Delta H_2^{\ddagger})$ as the lower of the two enthalpies.

4.3.4 Cyclic transition states

There are several instances[495, 574, 960, 1073a] of electron transfer accompanied by simultaneous transfer of two coordinated ligands in a cyclic transition state. These results imply the existence of precursor complexes, if we accept the general principle that ring formation is a stepwise process: prior to ring closure there must be a labile, single-bridged intermediate I

$$A^{+}\!\!\begin{array}{c}X\\ \diagdown\\ Y\end{array} + B \underset{-1}{\overset{1}{\rightleftharpoons}} A^{+}\!\!\begin{array}{c}X-B\\ \diagdown\\ Y\end{array} \underset{-2}{\overset{2}{\rightleftharpoons}} A^{+}\!\!\begin{array}{c}X\\ \diagdown\quad\diagdown\\ Y\end{array}\!B$$

$$\quad\quad\quad\quad\quad\quad\quad\quad\text{I}\Big\downarrow k_S \quad\quad\quad\quad\text{II}\Big\downarrow k_D$$

$$\quad\quad\quad\quad\quad\quad A+Y+X.B^{+}\quad\quad A+XY.B^{+}\quad\quad (4.27)$$

with the metal ions still in the valency states of the reactants. This intermediate must be counted as a chemical entity since its lifetime is long enough to enable it to undergo at least one reaction apart from electron transfer. When there is evidence of parallel pathways, singly and doubly bridged, the intermediate I is the precursor complex of an electron transfer reaction which is actually observed, but when no singly bridged pathway is found, it still means that the lifetime of I with respect to electron transfer is long compared with the lifetime with respect to ring closure, i.e. $k_S < k_2$. Moreover, k_2 can be estimated as being roughly the rate of displacement of solvent molecules from the metal ion B as measured in a solvent exchange reaction and this gives an approximate lower limit for the electron transfer rate.

4.4 Labile intermediates: kinetic evidence from catalysed pathways[j]

4.4.1 General

If neither the precursor nor the successor complex attains measurable concentration during the course of the reaction, the rate law has the form of

---

(j) For a systematic discussion of the rate laws of multiple-path reactions, see Newton and Baker[801]. The authors point out an analogy between rate expressions and expressions for the conductance of electrical circuits

equation (4.3) and the rate constant is given by equation (4.4). This includes terms for the formation and dissociation of both the precursor and the successor complex, but in all the work described in this section it will be sufficient to consider only one of these as kinetically significant. Writing the intermediate as I, the mechanism becomes

$$A^+ + B \underset{-1}{\overset{1}{\rightleftharpoons}} I \overset{2}{\longrightarrow} A + B^+ \tag{4.28}$$

with

$$k = k_1 k_2 / (k_{-1} + k_2) \tag{4.29}$$

which may alternatively be written as

$$\frac{1}{k} = \frac{1}{k_1} + \frac{1}{k_2^*} \tag{4.30}$$

where k_2^* is the net activation rate constant.

The possibility of distinguishing a reaction path kinetically arises when one or more of the steps is subject to catalysis by some other reagent X. Then the terms k_1, k_{-1}, k_2 are dependent on [X] and in favourable cases the form of dependence of k upon [X] proves the existence of the intermediate.

Assuming first that there is just one catalytic pathway for each of the two steps in equation (4.28), we have

$$k_1 = k_1'[X]^p \tag{4.31}$$
$$k_{-1} = k_{-1}'[X]^q \tag{4.32}$$
$$k_2 = k_2'[X]^r \tag{4.33}$$

giving

$$k = \frac{k_1' k_2' [X]^{p+q}}{k_{-1}'[X]^q + k_2'[X]^r} \tag{4.34}$$

or alternatively

$$\frac{1}{k} = \frac{1}{k_1'[X]^p} + \frac{1}{k_2^{*'}[X]^{p+q-r}} \tag{4.35}$$

where $k_2^{*'} = k_1' k_2' / k_{-1}'$. The compositions of the two transition states are $T_1 = A^+ + B + pX$ and $T_2 = A^+ + B + (p+q-r)X$, and that of the intermediate is $I = A^+ + B + (p-q)X$. It is clear from equation (4.35), however, that only the compositions of the transition states, and the rate constants k_1' and $k_2^{*'}$, can be determined by experiment. More generally, each step could consist of parallel paths with different numbers of catalyst molecules. The rate constants k_1, k_{-1}, k_2 would then take the forms

$$\sum_i k_{1i}[X]^{p_i}, \quad \sum_i k_{-1i}[X]^{q_i}, \quad \sum_i k_{-2i}[X]^{r_i}$$

But the law of microscopic reversibility requires that the number of parallel paths forwards and backwards in any step must be the same, hence the parameters must be related by $(p_1 - q_1) = (p_2 - q_2) = \ldots$ and $k'_{11}/k'_{-11} = k'_{12}/k'_{-12} = \ldots$, etc.

The most complicated cases so far met with in practice require only two terms at most in any step. With this restriction, a general mechanism may be written as

$$A^+ + B \begin{Bmatrix} \underset{k_{-1}}{\overset{k_1}{\rightleftarrows}} \\ \underset{k'_{-1}[X]}{\overset{k'_1[X]}{\rightleftarrows}} \end{Bmatrix} I \underset{k'_2[X]}{\overset{k_2}{\longrightarrow}} \text{products} \qquad (4.36)$$

The corresponding rate expression is

$$k = \frac{(k_1 + k'_1[X])(k_2 + k'_2[X])}{(k_{-1} + k_2) + (k'_{-1} + k'_2)[X]} \qquad (4.37)$$

which may alternatively be expressed as

$$\frac{1}{k} = \frac{1}{k_1 + k'_1[X]} + \frac{1}{k_2^* + k_2^{*'}[X]} \qquad (4.38)$$

A set of simpler mechanisms can be derived from equation (4.36) by leaving out one or more steps, and the corresponding rate expressions can be derived from equations (4.37) or (4.38) by setting the corresponding rate constants equal to zero. Before doing this however, we shall introduce a more concise notation, and one further mechanistic complication.

The general mechanism (equation 4.36) will be written $\forall$, where the left-hand diagonal lines denote the reversible steps from reactants to intermediate, and the right-hand lines denote the irreversible steps from intermediate to products. The full lines denote uncatalysed steps, the broken lines denote catalysed steps, first order in [X]. The additional complication is the possibility of a direct reaction path, catalysed and/or uncatalysed, not involving the intermediate, for example, the reaction

$$A^+ + B \underset{k'_0[X]}{\overset{k_0}{\longrightarrow}} \text{products} \qquad (4.39)$$

The mechanism comprising equations (4.36) and (4.39) together may be symbolised $\overline{\forall}$. The rate expression is given by

$$k = \frac{(k_1 + k'_1[X])(k_2 + k'_2[X])}{(k_{-1} + k_2) + (k'_{-1} + k'_2[X])} + k_0 + k'_0[X] \qquad (4.40)$$

Labile intermediates: kinetic evidence from catalysed pathways   111

which is of the second degree in [X] and can be reduced to the four-parameter form

$$k = (a + b[X] + c[X]^2)/(1 + q[X]) \qquad (4.41)$$

The plot of the function $k([X])$ is a hyperbola. By striking out any one of the six rate constants in equation (4.40), six expressions may be obtained, all equally complex but with different parameters. Omitting two rate constants gives 15 expressions, seven of which can be simplified to equation (4.41), two to equation (4.42), two to equation (4.43) and four to equation (4.45).

$$k = (b[X] + c[X]^2)/(1 + q[X]) \qquad (4.42)$$

$$k = (a + b[X])/(1 + q[X]) \qquad (4.43)$$

Continuing in this way, other selections of rate constants give either equations (4.42) or (4.43), the two-parameter hyperbola$^{(k)}$

$$k = b[X]/(1 + Q[X]) \qquad (4.44)$$

or the linear forms

$$k = a + b[X] \qquad (4.45)$$

$$k = b[X] \qquad (4.46)$$

$$k = a \qquad (4.47)$$

The complete picture is summarised in *Table 4.3*, where the mechanisms are classified according to the number of transition states, and the rate expressions according to the number of adjustable parameters. It is the presence of the denominator term $(1 + q[X])$ which diagnoses the intermediate. The uninformative rate laws (equations 4.45–4.47) arise when the pathway through the intermediate consists only of the elements $\vee$ or $\vee$ i.e., a single path from reactants to intermediate and from intermediate to products, of the same order in [X].

We shall not discuss all possible mechanisms and rate laws. The following general points emerge.

(k) Equation (4.44), equivalent to the Michaelis–Menten equation, is usually analysed by plotting k_{obs}^{-1} against $[X]^{-1}$. Alternative forms of plot are k against $(k[X])^{-1}$ and $k^{-1}[X]$ against $[X]$, which also give straight lines. For discussion, see references 316, 504, 551 and others cited therein. The term 'Hofstee plot' has sometimes been used to denote the first of the alternative plots[317]. In the case of hydroxide catalysis, equation (4.44) is more economically written $k = b/(q + [H^+])$ so that the Michaelis–Menten plot is of k^{-1} against $[H^+]$ while the Hofstee plot is k against $k[H^+]$

Table 4.3 Mechanisms and rate expressions for electron transfer reactions

Rate expression	| 1	2	3	4	5	6
$k = a$	—	∨	▽			
$k = b[X]$	----	∨	▽			
$k = a + b[X]$	====		∨ ▽	∨ ▽		
$k = \dfrac{b[X]}{1+q[X]}$		∨∨				
$k = \dfrac{a+b[X]}{1+q[X]}$			∨∨ ▽▽	▽▽		
$k = \dfrac{b[X]+c[X]^2}{1+q[X]}$			∨∨ ▽▽	▽▽		
$k = \dfrac{a+b[X]+c[X]^2}{1+q[X]}$				▽▽ ▽▽ ▽	▽▽ ▽▽	▽ ▽▽ ▽▽

(a) Order of the reaction

All the non-linear expressions are of variable order in [X], but the order varies only between the limits of zero and first order. This can best be expressed by tracing the curves of log k as functions of log [X]. The order with respect to [X] at any given value of [X] is the slope of the log–log curve. The curves are shown in *Figure 4.1*. Each curve has two asymptotes, of slopes 0 or 1 as shown; two of the curves have one point of inflexion and the most complicated plot has two points of inflexion. At all points, the slopes lie in the range 0 to 1. Slopes in this range indicate the participation of transition states containing 0 and 1 molecules of catalytic species. A slope exceeding 1 would necessitate a more complicated mechanism with a higher-order transition state, but the converse is not true; there is an infinity of mechanisms with steps of higher catalytic order, but which still conform to one or other of the equations considered here.

(b) Indeterminate rate laws

The simple second-order rate law (equation (4.3) with equation (4.47)) is consistent with the three mechanisms—, ∨ and ▽ and strictly speaking all of

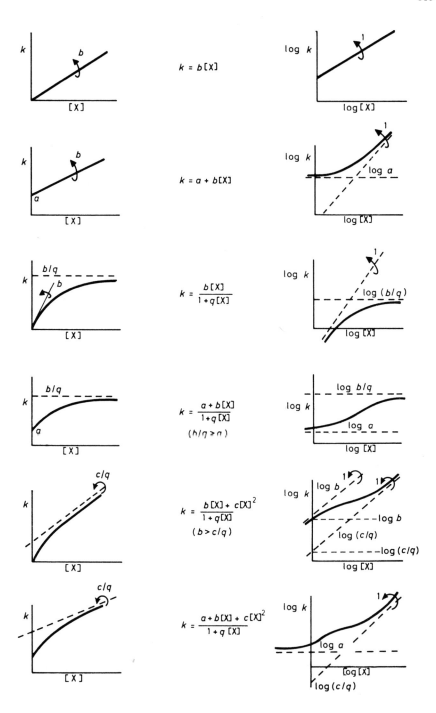

Figure 4.1 Rate expressions derived from equation 4.41. The graphs show k as function of [X] *and* log k *as function of* log [X]. *The curved arrows denote the gradients of the sloping lines.*

these should be considered when discussing any uncatalysed reaction. In practice this is never done, though occasionally there has been some discussion. For example, when it can be shown by other means that parallel outer-sphere and inner-sphere pathways operate, (Chapter 5), these may be assigned the rate constants k_0 and $k_1 k_2/(k_{-1}+k_2)$, corresponding to pathways — and $\vee\!\!/$. Similar remarks apply to the other linear expressions (equations 4.45 and 4.46).

Among the non-linear rate expressions are eleven more cases in which the number of transition states exceeds the number of adjustable parameters. At first glance this suggests that discussions of mechanism will always be impossibly complicated, but this need not be so, since the reasons for indeterminacy are the same as before. In each case there are parallel uncatalysed paths or parallel catalysed paths: the mechanisms contain the elements ∇ or $\vee\!\!/$ or both. In each case there are two choices; either to make some assumption about the values of the rate constants—usually by assigning one of them to zero on mechanistic grounds, or else to extract from the empirical rate law merely those combinations of rate constants which can be determined independently. In fact it is always possible to find combinations of rate constants which have a simple chemical interpretation and are therefore suitable for comparison with other systems. Consider for example the mechanism symbolised by $\nabla\!\!\!/$ The rate expression contains four rate constants, but it reduces to the form

$$k = k_0 + \frac{k_1(k_2 + k'_2[X])}{(k_{-1}+k_2)+k'_2[X]} \qquad (4.48)$$

with only three independent parameters. Two of these may be chosen as the limiting values of k at $[X] = 0$ and $[X] \to \infty$, namely

$$k(0) = a = k_0 + k_1 k_2/(k_{-1}+k_2) \qquad (4.49)$$

and

$$k(\infty) = b/q = k_0 + k_1 \qquad (4.50)$$

In terms of mechanisms, $k(0)$ corresponds to the situation in which the catalysed step is too slow to be effective, i.e. to the mechanism ∇ while $k(\infty)$ corresponds to the situation in which the catalysed step is so fast that, in the indirect pathway, the formation of the intermediate is rate-determining.

(c) Symmetry

Equation (4.37) is symmetrical with respect to interchange of k_1 with $k_2^\ddagger$ and of k'_1 with $k_2'^*$. Thus, for example, the mechanisms symbolised $\vee\!\!/$ and $\backslash\!\!\vee$, which differ only in the order of occurrence of the two transition states

$$A^+ + B \underset{k_{-1}}{\overset{k_1}{\rightleftharpoons}} I \overset{k'_2[X]}{- -\to} A + B^+ \qquad (4.51)$$

$$A^+ + B \underset{k'_{-1}[X]}{\overset{k'_1[X]}{\rightleftharpoons}} I \xrightarrow{k_2} A + B^+ \tag{4.52}$$

have the same rate law, equation (4.44). If we assume mechanism (4.51) operates, the rate constants are obtained from the empirical rate law as $k_1 = b/q$, $k_2^* = k_1 k'_2/k_{-1} = b$; conversely if mechanism (4.52) operates, the rate constants are $k'_1 = b$, $k_2^* = k'_1 k_2/k'_{-1} = b/q$. Thus the same combination of experimental parameters, b and b/q, is required for each mechanism, and they correspond to transition states of the same composition, b to the transition state $[A^+.B.X]$ and b/q to the transition state $[A^+.B]$, regardless of mechanism. Similar remarks apply in other cases. Mechanisms which feature transition states of the same composition and differ only in the order in which they occur are kinetically indistinguishable. The corresponding ambiguity applies also to the general mechanism (equation 4.36). If the rate expression, equation (4.37), were found experimentally it would be possible to calculate activation parameters for four transition states. Then there would be two activation parameters corresponding to the two transition states of composition $[A^+.B]$, (equation 4.36, steps 1 and 2) but it would not be possible to say which parameter corresponded to which transition state; and similarly for the two transition states $(A^+.X.B)$ (steps 1' and 2').

In the face of these ambiguities the generally agreed procedure is to decide, first, what is the simplest empirical rate law which will fit the experimental data, and second, what is the simplest mechanism consistent with the rate law. It is, however, of value to keep in mind the range of possible mechanisms which could be invoked, and some authors cited below have given more or less extensive discussions of this aspect.

4.4.2 Examples

Uranium(VI)–vanadium(III)   The majority of examples studied (*see Table 4.4*) are reactions between metal ions (i.e. aquo or oxo complexes) in aqueous solution, catalysed by the hydroxide ion. In attempting to establish the form of the rate expression, two major ancilliary problems arise: the metal ions may themselves be extensively hydrolysed, and the activity coefficients of the hydrogen ion may vary with concentration. The work of Newton and Baker[804] on the reaction

$$U^{VI} + V^{III} \rightarrow U^V + V^{IV} \tag{4.53}$$

is a good illustration of the difficulties and the way in which they are overcome. This reaction was studied by the device of including iron(III) in the system, to reoxidise uranium(V) to uranium(VI), so that the net reaction amounted to

$$Fe^{III} + V^{III} = Fe^{II} + V^{IV} \tag{4.54}$$

Table 4.4 Kinetic evidence of binuclear intermediate: summary of experimental data according to the rate expression

$$k = (a + b[X] + c[X]^2)/(1 + q[X])$$

Reactants			Ionic strength[a] I/M	Rate parameters[b]				References	Notes
A^+	B	[X]		q	a	b/q	c/q		
PuO_2^{2+}	UOH^{3+}	$[H^+]^{-1}$	2.0	2.5	0	187	0	793	(c)
UO_2^{2+}	VOH^{2+}	$[H^+]^{-1}$	2.0	0.061	0	2.3×10^3	0	804	(d)
V^{3+}	Cr^{2+}	$[H^+]^{-1}$	0.5	0.10	0	5.0	0	15, 378	(e)
$trans$-Co(en)$_2$NH$_3$OH$_2^{3+}$	U^{3+}	$[H^+]^{-1}$	2.0	0.0165	0	248	0	1072	
Co(NH$_3$)$_5$N$_3^{2+}$	Ti^{3+}	$[H^+]^{-1}$	0.5	0.0745	0	29.7	0	122	
Co(NH$_3$)$_5$F^{2+}	Ti^{3+}	$[H^+]^{-1}$	0.5	0.11	0	5.6×10^2	0	1028	
Np^{VII}	Co^{2+}	$[H^+]^{-1}$	1.0	1.83	0	5.4×10^4	0	309	
Np^{VII}	Tl^+	$[H^+]^{-1}$	1.0	0.20	0	26.1	0	1032	(f)
PuO_2^{2+}	Fe^{2+}	$[H^+]^{-1}$	2.0	0.157	1.0×10^3	5.78×10^3	0	799	(g)
(H$_3$N)$_5$CoLH^{2+}	V^{2+}	$[H^+]^{-1}$	1.0	0.035	1.0	7.14	0	813	(h)
Co(NH$_3$)$_5$OH$_2^{3+}$	U^{3+}	$[H^+]^{-1}$	0.2	0.015	23.8	54.7	0	1072	
cis-Co(NH$_3$)$_4$(H$_2$O)$_2^{3+}$	U^{3+}	$[H^+]^{-1}$	0.2	0.041	202	34.1	0	1072	
$trans$-CoL(H$_2$O)$_2^{3+}$	Cr^{2+}	$[H^+]^{-1}$	1.0	0.033	0	9×10^3	1.2×10^2	683, 684	(i)
$trans$-CoL(H$_2$O)OH^{2+}	Cr^{2+}	$[H^+]^{-1}$	1.0	0.033	31×10^6	1.3×10^6	0	683, 684	(j)

(a) In all cases the medium is acidic aqueous solution, with ClO_4^- as the principal anion. Temperature 25 °C. For data at other temperatures and ionic strengths, see the original references
(b) Units as implied by the rate law, e.g. a in units of M^{-1}s^{-1}, q in units of M when $[X] = [H^+]^{-1}$ or in units of M^{-1} when $[X] = [H^+]$
(c) The reaction is the first step of a two-step non-complementary reaction $2Pu^{VI} + U^{III} \rightarrow 2Pu^V + U^{VI}$
(d) The reaction in is $U^{VI} + V^{III} \rightarrow U^V + V^{IV}$. The U^V product is reoxidised in a further reaction $U^V + Fe^{III} \rightarrow U^{VII} + Fe^{II}$
(e) For discussion, see also references 995; 496 and text
(f) Data of reference 1032 recalculated in reference 309
(g) The reaction is the first step of a two-step non-complementary reaction $Pu^{VI} + 2Fe^{II} = Pu + 2Fe^{III}$
(h) H_2L = nicotinic acid
(i) L $\equiv$ 5, 7, 7, 12, 14, 14-hexamethyl-1, 4, 8, 11-tetraazacyclotetradeca-4, 11-diene
(j) Calculated from data in the line above

with uranium (VI) catalysis, and allowance had also to be made for the uncatalysed and vanadium(IV)-catalysed reactions. Since both UO_2^{2+} and H^+ concentrations were to be varied over a wide range the double medium technique was employed: variations in $[H^+]$ were counterbalanced by adding Li^+ ion, and variations in $[UO_2^{2+}]$ were counterbalanced with Mg^{2+} [1]. The hydrogen ion activity coefficient was still not assumed to remain constant over the range of concentration used (0.04–1.67 M), but it was assumed that any change would be consistent with Harned's rule[897]. The UO_2^{2+} ion is not appreciably hydrolysed under the conditions of the experiments but the V^{3+} ion is hydrolysed according to the reaction

$$V^{3+} \xrightleftharpoons{K_a} VOH^{2+} + H^+ \qquad (4.55)$$

Allowing for this the rate law could be expressed as

$$k_{obs} = \text{Rate}/[UO_2^{2+}][V^{3+}] = (1 + K_a[H^+]^{-1})^{-1} \exp(\alpha[H^+])k'([H^+]) \qquad (4.56)$$

where $k([H^+])$ is the function of $[H^+]$ remaining to be determined. The value of K_a was taken from other experiments, but the Harned term α was regarded as one of the adjustable parameters. Two alternative functions were considered for $k'([H^+])$:

$$k' = (k_I[H^+]^{-1} + k_{II}[H^+]^{-2}) \qquad (4.57)$$

$$k' = \frac{k'_I k'_{II}[H^+]^{-2}}{k'_I + k'_{II}[H^+]^{-1}} \qquad (4.58)$$

Equation (4.57), equivalent to equation (4.45) with $k = k'[H^+]$, $k_I = a$, $k_{II} = b$, requires parallel steps with transition states UVO_3H^{4+} and UVO_3^{3+}; equation (4.58), equivalent to equation (4.44) with $k = k'[H^+]$, $k_{II} = b$, $k_{II}/k_I = q$, requires consecutive steps with the same two transition states in either order. A careful statistical analysis favoured equation (4.58).

From this example it is clear that the data required to establish such complex rate laws must be of good quality and must cover a wide range of hydrogen ion concentrations. Care and discretion need to be exercised also: a slight deviation from a simple first or second power dependence on $[H^+]$ may indicate not an additional rate term, but merely a small systematic error or a medium effect.

---

(1) In earlier work it was common practice to counterbalance the H^+ ion with Na^+, but several studies have shown that this can lead to artificial terms in the rate law, which disappear when Li^+ is used. Examples include the reactions $Co(NH_3)_5H_2O^{3+} + Cr^{2+}$ (Chapter 5, footnote (e)), $Cr(NH_3)_5Cl^{2+} + Cr^{2+}$ [203], and $(Co(NH_3)_5)_2NH_2^{6+} + V^{2+}$ [304a]. The problems associated with medium effects in these systems are generally similar to those encountered in studies of hydrolysis and polymerisation of metal ions in aqueous solutions[949].

Vanadium(III)–chromium(II) Both the kinetics and the mechanism of this reaction have been the subject of some controversy, and the discussion illustrates usefully the distinction between these two aspects of the subject. The reaction

$$V^{3+} + Cr^{2+} = V^{2+} + Cr^{3+} \tag{4.59}$$

proceeds to completion, with the rate law

$$\text{Rate} = k[V^{3+}][Cr^{2+}] \tag{4.60}$$

and with strong base catalysis. Comparing two possible hydrogen ion dependences

$$k = a + b/[H^+] \tag{4.61}$$

$$k = b/(q + [H^+]) \tag{4.62}$$

equivalent to equations (4.45) and (4.44) respectively, Espenson[378] showed that a significantly better fit was obtained using equation (4.62). The decision was made on the basis of a least squares fitting, and illustrated by plotting a graph of $\log k$ against $\log [H^+]$. Sykes[995] pointed out that a still better fit, as judged by least squares, could be obtained by using the equation

$$k_{obs} = (b + c[H^+]^{-1})/(q + [H^+]) \tag{4.63}$$

(which is equivalent to equation (4.43)) and illustrated the point by plotting graphs of $k_{obs}(q + [H^+])$ against $[H^+]^{-1}$, for different values of q. Espenson replied[386] that although the equation of Sykes did represent the data better than equation (4.62), it did not follow that equation (4.63) was to be preferred. The mean deviation of observed rate constants from those calculated by equation (4.62) was 3.7%, while the mean deviation from equation (4.63) was 1.4%, but the mean deviation of data measured at the same H^+ concentration was 3.0%, and it could be argued that it was unjustified to seek any closer agreement than that. A further study by Adin and Sykes[15] confirmed the validity of equation (4.62).

As to the mechanism, Espenson[378] originally proposed the mechanism \/ , but Sykes[995], and Adin and Sykes[15] argued that a more plausible mechanism would assign a direct role to the process $VOH^{2+} + Cr^{2+}$, by analogy with the chromium(II) reductions of other aquoions. This question is independent of the question of the rate law, and accepting the rate law (equation 4.62) it is possible to invoke the desired process by reversing the order of the transition states. The mechanism then becomes \/ or, specifically

$$V^{3+} \underset{}{\overset{K_a}{\rightleftharpoons}} VOH^{2+} + H^+ \tag{4.64}$$

$$VOH^{2+} + Cr^{2+} \rightleftharpoons VOHCr^{4+} \tag{4.65}$$

$$VOHCr^{4+} + H^+ \rightarrow V^{2+} + Cr^{3+} \tag{4.66}$$

Strong base catalysis suggests an inner-sphere mechanism with the hydroxide ion as bridging group (see p. 149), hence, the intermediate is formulated as a μ-hydroxy binuclear complex, analogous to other well-characterised species such as $Cr^{III}OHCr^{III\,5+}$. The valencies are assigned as V^{II}–Cr^{III}, being more stable than V^{III}–Cr^{II}. This gives both metal ions the configuration $3d^3$, consistent with the slow rate of hydrolysis to products. The first-order acid dependence in the hydrolysis step (equation 4.66) is characteristic of such reactions.

Plutonium(VI)–iron(II) The redox potentials are strongly favourable for the process $Pu^{VI} + Fe^{II} \rightarrow Pu^V + Fe^{III}$, and the binuclear intermediate can plausibly be formulated as the oxygen-bridged species $OPu^VOFe^{III}(OH_2)_5^{4+}$. Similar intermediates have been detected by direct means in other reactions[780, 805, 977, 978] and it is well established that the pentavalent actinide ions MO_2^+ have a greater affinity for other cations than the corresponding hexavalent ions MO_2^{2+}.

With this intermediate the mechanism becomes

$$PuO_2^{2+} + Fe^{2+} \xrightarrow{k_0} PuO_2^+ + Fe^{3+}$$

$$\Bigg\downarrow k_1 \quad \Bigg\uparrow k_{-1} \quad \nearrow \begin{array}{c} k_2 \\ k_2'[H^+]^{-1} \end{array} \qquad (4.67)$$

$$OPuOFe^{4+}$$

The independently measurable constants are (see equations (4.48)–(4.50)) $(k_0 + k_1) = b/q$, $k_0 + k_1k_2/(k_{-1} + k_2) = a$, and $K_1 k_2'/(k_1 + K_1 k_2) = q$. Newton and Baker calculated activation parameters for the appropriate transition states with the aid of two alternative assumptions—that $k_0 = 0$ or that $k_2 = 0$.

The base-catalysed pathway k_2' is analogous to the base-catalysed substitution reactions of other iron(III) complexes. If $k_0 = 0$, then

$$k_2'/k_2 = q/a = 1.6 \times 10^{-4} \, M^{-1}$$

but this seems unreasonably small. (The corresponding ratios for the complexes $FeFH^{3+}$, $FeCl^{2+}$, $FeBr^{2+}$, $FeNCS^{2+}$, FeN_3H^{3+} lie in the range $10^{\pm 1.5} \, M^{-1}$ (m).) The assumption $k_2 \simeq 0$ is therefore preferred.

Cobalt(III)–chromium(II) Liteplo and Endicott[683, 684] reported the kinetics of reactions of the form

$$CoL^{3+} + Cr^{2+} = CoL^{2+} + Cr^{2+} \qquad (4.68)$$

(m) Calculated from rate data tabulated in reference 359; using $K(Fe^{3+} \rightleftharpoons FeOH^{2+} + H^+) = 10^{3.1}$ (reference 950).

where L denotes either of the macrocyclic tetraammine ligands I, II. In contrast to most cobaltammine–chromium(II) reactions, the cobalt(II) product is low-spin and retains the chelate ligand. The strong base catalysis is evidence of a bridged mechanism; this has yet to be confirmed by oxygen tracer experiments. The rate law is of the form of equation (4.42), consistent with any of six mechanisms involving only steps of first order in the OH ion (Table 4.3), but if we exclude the possibility of a 'direct' k'_0 path, the number is reduced to two: in our notation $\sqrt{}$ and $\sqrt{}$, that is equations of the form,

$$CoL^{3+} \underset{}{\overset{K_a}{\rightleftharpoons}} CoLOH^{2+} + H^+ \tag{4.69}$$

$$CoLOH^{2+} + Cr^{2+} \underset{-1}{\overset{1''}{\rightleftharpoons}} CoLOHCr^{4+} \tag{4.70}$$

$$CoLOHCr^{4+} + H^+ \xrightarrow{2} products \tag{4.71}$$

$$CoLOHCr^{4+} \xrightarrow{2'} products \tag{4.72}$$

and

$$CoL^{3+} \underset{}{\overset{K_a}{\rightleftharpoons}} CoLOH^{2+} + H^+ \tag{4.73}$$

$$CoL^{3+} + Cr^{2+} \underset{-1}{\overset{1}{\rightleftharpoons}} CoLOHCr^{4+} + H^+ \tag{4.74}$$

$$CoLOH^{2+} + Cr^2 \underset{-1}{\overset{1''}{\rightleftharpoons}} CoLOHCr^{4+} \tag{4.75}$$

$$CoLOHCr^{4+} \xrightarrow{2} products \tag{4.76}$$

with $k''_1 = k'_1 K_a$.

The intermediate has been written here as a hydroxy-bridged species, but Liteplo and Endicott[684], in their second paper on these reactions, preferred to postulate an oxo-bridged complex, (corresponding in our notation to the uptake of two catalyst species; i.e. $(A^+.B.X_2)$). This is plausible by analogy with other hydrolysed dimers of these macrocyclic complexes, e.g. $CoLOLCr^{4+}$.

The intermediate could be written with valencies Co^{III}–Cr^{II} or Co^{II}–Cr^{III}. To identify it as the precursor complex we note the very large driving force associated with the overall reaction. The equilibrium constant is approximately 10^{17}. If, therefore, the intermediate were of the form Co^{II}–Cr^{III}, the backward steps of equations (4.70) and (4.74) would be too slow to be kinetically significant.

Cobalt(III)–vanadium(II) Norris and Nordmeyer[313] reported the kinetics of the reaction:

$$Co(NH_3)_5 N\langle\rangle\text{–COOH} + V^{2+} \to Co^{II} + V^{III} \quad (4.78)$$

The rate expression is analogous to that of the $PuO_2^{2+} + Fe^{2+}$ reaction and leads to the same ambiguity: there are two symmetrically related mechanisms each with four transition states, and two pairs of mechanisms each with three transition states. The authors considered the pair $\bigtriangledown$ and $\bigtriangledown$, with equations of the form

(i) $CoLH^{3+} + V^{2+} \xrightarrow{k_0}$ products (4.79)

$CoLH^{3+} \underset{}{\overset{K_a}{\rightleftharpoons}} CoL^{2+} + H^+$ (4.80)

$CoL^{2+} + V^{2+} \underset{-1'}{\overset{1''}{\rightleftharpoons}} CoLV^{4+}$ (4.81)

$CoLV^{4+} + H^+ \xrightarrow{2}$ products (4.82)

and

(ii) $CoLH^{3+} + V^{2+} \xrightarrow{k_0}$ products (4.83)

$CoLH^{3+} + V^{2+} \underset{-1}{\overset{1}{\rightleftharpoons}} CoLV^{4+} + H^+$ (4.84)

$CoLV^{4+} \xrightarrow{2'}$ products (4.85)

with $L^- = C_6H_4O_2N^-$; $k_1'' = k_1'/K_a$. In this case it is possible to decide between the two alternatives. In mechanism (i), $k_1'' = 200\ M^{-1}s^{-1}$ and in mechanism (ii), $k_1 = 5.3\ M^{-1}s^{-1}$. Since these are substitution reactions in the $V(H_2O)_6^{2+}$ complex, the rate should not exceed $\simeq 30\ M^{-1}s^{-1}$ (see p. 147) hence mechanism (ii) is preferred.

The intermediate is assumed to be the precursor complex and this is reasonable in view of the overall driving force.

4.4.3 Free-energy profiles

Newton and Baker[801] have reviewed a series of reactions of the type

$$MO_2^{2+} + M'^{4+} + 2H_2O = MO_2^+ + M'O_2^+ + 4H^+ \quad (4.86)$$

122 The reaction path (2): binuclear intermediates

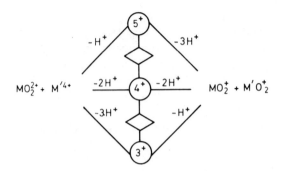

Figure 4.2 (above) Summary of reaction mechanisms for reactions of the type

$$MO_2^{2+} + M'^{4+} = MO_2^+ + M'O_2^+ + 4H^+$$

(where M, M' are actinide metals). The circles denote transition states with overall ionic charges as shown. The diamonds denote intermediates

Figure 4.3 (right) Reaction profiles for reactions of the type

$$MO_2^{2+} + M'^{4+} = MO_2^+ + M'O_2^+ + 4H^+$$

The reaction coordinate x is drawn on an arbitrary scale. The transition states at $x = 1, 2, 3$ have formulae $[MM'O_2H_{-y}]^{(6-y)+}$, which may alternatively be expressed $[MM'O_2(OH)_y]^{(6-y)+}$. The metals M, M' are as indicated

Table 4.5 Summary of mechanisms of reactions

$$MO_2^{2+} + M'^{4+} + 2H_2O = MO_2^+ + M'O_2^+ + 4H^+ \text{ (a)}$$

Reactants	$\Delta G°/$ kcal mol^{-1}	Mechanism	$\Delta G^{\ddagger}$/kcal mol^{-1}		
			(5+)	(4+)	(3+)
$NpO_2^{2+} + U^{4+}$	−12.9	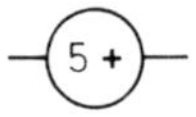 —(5+)—	16.0		
$AmO_2^{2+} + Am^{4+}$	−12.9	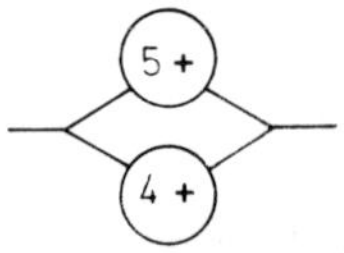 (5+)/(4+)	11.6		
$NpO_2^{2+} + Np^{4+}$	−9.2	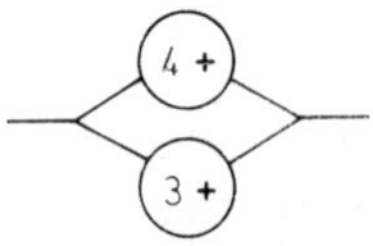 (4+)/(3+)		19.2	19.2
$PuO_2^{2+} + U^{4+}$	−7.8	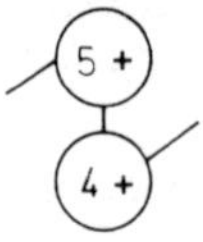 (5+)/(4+)	16.6	16.0	
$PuO_2^{2+} + Pu^{4+}$	+5.8	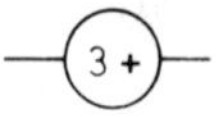 —(3+)—			26.6
$UO_2^{2+} + U^{4+}$	+11.9	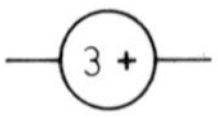 —(3+)—			26.8

(a) Reference 801

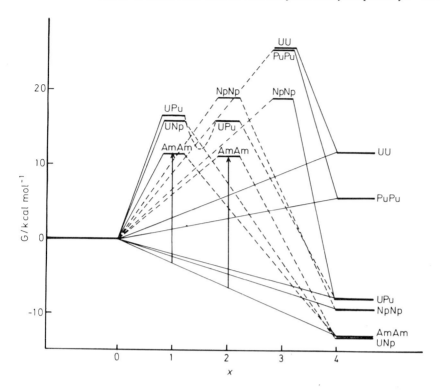

involving the actinide metals U, Np, Pu, Am. Of the 16 possible reactions, six have been studied kinetically. The mechanisms include apparent single-step processes, parallel paths with two transition states and, in one case, consecutive transition states. The rate terms are of the form $[MO_2^{2+}][M'^{4+}][H^+]^{-n}$ where n is 1, 2 or 3. A comprehensive mechanistic scheme is summarised in *Figure 4.2*, in which the circles denote transition states, numbered according to the overall charges, and the diamonds denote presumed intermediates. Possible reaction paths are represented by the various routes which can be traced from left to right of the diagram.

The actual mechanisms deduced from the rate expressions are shown, with the same formalism, in *Table 4.5*. The reactions are arranged in descending order of the thermodynamic 'driving force', and it can be seen that a pattern has emerged. The net reaction involves the release of four protons, and in general some of these are lost before the transition state(s), and some after. In the reaction $NpO_2^{2+} + U^{4+}$ which is strongly favoured thermodynamically, one proton is lost before the transition state and three afterwards; in the reaction $UO_2^{2+} + U^{4+}$ which is almost equally strongly disfavoured, the reverse is true. Reactions with intermediate driving force show intermediate laws, three of them having two transition states. These are instances of the 'principle of similitude'. The transition state resembles reactants or products, according to whether the free energy change is negative or positive, but here the principle applies not merely to the detailed structure of the transition state, but to the composition.

Reaction profiles calculated from the data of *Table 4.5* are shown in *Figure 4.3*. Intermediates are detected only when the two transition states appear at nearly the same free energy level. Analogous intermediates may occur in other reactions, but the reaction profile is so steep that the retarding back-reaction is insignificant.

The free energies of the non-observed transition states can be guessed from the diagrams in *Figure 4.3*. Taking the horizontal coordinate x as a reaction coordinate which increases directly with successive removals of H^+, an 'intrinsic' free energy barrier can be estimated for each transition state by subtracting from the observed free energy the appropriate fraction of $\Delta G^\ominus$. For example, the two intrinsic barriers for the $AmO_2^{2+} + Am^{4+}$ reaction are marked with vertical arrows. When this is done, each of the three transition states, defined by charges $5+$, $4+$ and $3+$, shows a consistent trend of intrinsic barriers, descending in the order $Np > Pu > U > Am$. On this basis it is possible to interpolate other transition states—$[Np_2O_2(OH)^{5+}]^{\ddagger}$ at $\Delta G^{\ddagger} \geqslant 13.6 \text{ kcal mol}^{-1}$ and $[Am_2O_2(OH)^{5+}]^{\ddagger}$ at $\Delta G^{\ddagger} < 5.0 \text{ kcal mol}^{-1}$. The former might possibly be detected in experiments at sufficiently high acidity, but the latter is well below the limits of observation.

4.5 Labile intermediates: direct observation

If the binuclear intermediate is both labile and fairly stable with respect to reactants, and if the final decomposition to products is sufficiently slow, the mechanism may be written

$$A^+ + B \underset{}{\overset{K}{\rightleftharpoons}} I \overset{k}{\longrightarrow} A + B^+ \tag{4.87}$$

where I is either the precursor complex ($K = K_{ip}$, $k = k_{ps}$) or the successor complex ($K = K_{ip}k_{ps}$, $k = k_{sf}$). With one reagent (say B) in large excess, the specific rate of reaction is then given by the Michaelis–Menten equation[753]:

$$k_{obs} = -\left(\frac{1}{[A^+]_T}\right)\left(\frac{d[A^+]_T}{dt}\right) = \frac{kK[B]}{1+K[B]} \tag{4.88}$$

where $[A^+]_T$ denotes the total concentration of $[A^+]+[I]$. Thus, a plot of k_{obs} against $[B]$ is curved: at the limit of low concentrations of $[B]$ it approaches the pseudo-first-order rate law $k_{obs} = k_{ps}K_{ip}[B]$; at the limit of high concentrations it approaches pseudo-zero-order behaviour, $k_{obs} = k_{ps}$. In effect, the method is an analytical technique for simultaneously measuring the concentration of the complex and the rate of the reaction. As already mentioned, a general limitation of the method is that although the complex can be identified the kinetics do not prove it to be an intermediate. It could alternatively lie 'to one side' of the reaction path as in the reaction

$$I \underset{K^{-1}}{\rightleftharpoons} A^+ + B \xrightarrow{k_2} A + B^+ \qquad (4.89)$$

in which case the experimental constant K has the same significance, but the quantity kK denotes k_2, the second-order rate constant of the direct reaction. This point recurs several times in the discussion of individual systems.

4.5.1 Precursor complexes *(Table 4.6)*

These studies are of particular interest in that they give the rate constant for the intramolecular electron transfer process, p → s. Most of the work reported

Table 4.6 Electron transfer reactions involving labile precursor complexes, and some related systems[a]

Reactions	K/M^{-1}	k/s^{-1}	References
$Co(NH_3)_5OOCCH_2N(CH_2COO)_2^-$ + Fe^{2+}	1.15×10^6	0.115	201
$Co(NH_3)_5OH_2^{3+} + Fe(CN)_6^{4-}$	1.5×10^3	1.9×10^{-1} [b]	452
$Co(NH_3)_5OOCCH:CH_2^{2+} + Cu^+$	1.4×10^3	$< 10^{-9}$ [c]	566
$Co(NH_3)_5NH_2CH_2CH=CH_2^{2+} + Cu^+$	—	—	402
$Co(en)_3^{3+} + Fe(CN)_6^{4-}$	95 [d]	—	452, 652, 608
$Co(phen)_3^{3+} + Fe(CN)_6^{4-}$	—	$> 3 \times 10^3$	452
$Cr(C_2O_4)_2^- + Cr^{2+}$	15 [e]	$< 3 \times 10^{-5}$ [e]	963
$HCr(C_2O_4)_2 + Cr^{2+}$	—	$\geq 3.6 \times 10^{-2}$ [f]	963
$HCrO_4^- + IrCl_6^{3-}$ [g]	1.2×10^2	—	123
$Co(C_2O_4)_3^{3-} + Fe^{2+}$	≤ 10	≥ 2	205
$Co(NH_3)_5F^{2+} + Fe^{II}$	10.6 [h]	3.2 [h]	743
$Co(NH_3)_5Cl^{2+} + Fe^{II}$	19.1 [h]	1.6×10^{-3} [h]	743
$Co(NH_3)_5Br^{2+} + Fe^{II}$	12.4 [h]	3.4×10^{-4} [h]	743
$V^{3+} + Fe(CN)_6^{4-}$	(i)_	—	121
$NpO_2^{2+} + NpO_2^+$	0.7 [k]	1.6×10^2 [j]	239, 982
$VO_2^{2+} + UO_2^+$	16 [m]	3.3 [l]	361, 472, 802

(a) At 25°C, ionic strength 1.0 M, except as otherwise stated
(b) $I = 0.1$ M
(c) Taking $kK < 138$ M^{-1} s^{-1} from a single experiment in which no electron transfer reaction was detected
(d) $I = 0.2$ M
(e) $I = 2.0$ M, $T = 20°C$; $kK < 0.5 \times 10^{-3}$ M^{-1} s^{-1}
(f) Calculated from the specific rate term $k[H^+ + Cr(C_2O_4)_2^- + Cr^{2+}] = 2.7 \times 10^{-2}$ M^{-2} s^{-1}, using $K_1[Cr(C_2O_4)_2^- + Cr^{2+}] = 15$ M^{-1} and an assumed value $K_2[Cr_2(C_2O_4)_2^+ + H^+] \leq 5 \times 10^{-2}$ M^{-1}. Authors obtained $k_{ps} \simeq 1$ s^{-1} at 25°C, using $k = 4 \times 10^{-2}$ M^{-1} s^{-1} and $K_2 \simeq 10^{-3}$ M^{-1}
(g) Non-complementary reaction
(h) 25.8°C, solvent dimethylformamide, $I = 0.34$ M
(i) Purple complex not yet characterised
(j) Using $kK = 1.1 \times 10^2$ M^{-1} s^{-1}, extrapolated from data at 0°C and 10°C
(k) $I = 3.0$ M. K_{1p} assumed same as for $UO_2^{2+} + NpO_2^{4+}$
(l) Using $kK \geq 52$ M^{-1} s^{-1}, from experiments at $H^+ = 1.0$ M, assuming zero-order $[H^+]$-dependence, as found for the reaction $NpO_2^{2+} + NpO_2^+$
(m) $I = 2.1$ M, $[H^+] = 0.08$ M

(a) Ion-pair complexes

Outer-sphere complex formation between oppositely charged ions is well established, and when oxidant and reductant are oppositely charged, the ion-pair is a possible precursor complex. Gaswick and Haim[452] have reported kinetics of the reaction $Co(NH_3)_5OH_2^{3+} + Fe(CN)_6^{4-}$ consistent with the mechanism

$$Co(NH_3)_5OH_2^{3+} + Fe(CN)_6^{4-} \underset{}{\overset{K}{\rightleftharpoons}} [Co(NH_3)_5OH_2^{3+}.Fe(CN)_6^{4-}]$$

$$\overset{k}{\longrightarrow} Co^{II} + Fe(CN)_6^{3-} \qquad (4.90)$$

(To avoid precipitation of cobaltous ferricyanide, the chelating agent edta was included in the reaction mixture.)

The reaction $Co(CN)_5SCN^{3-} + V^{2+}$ involves ion pairing followed by a substitution process. The reaction obeys the Michaelis–Menten rate law (V^{2+} ion in excess), and the product $VNCS^{2+}$ can be detected by observing the rate of subsequent dissociation to V^{3+} and NCS^- ions. Thus the overall sequence can be written

$$Co(CN)_5SCN^{3-} + V^{2+}\cdot aq \overset{1}{\rightleftharpoons} [Co(CN)_5SCN^{3-}.V(H_2O)_6^{2+}]$$

$$\overset{1'}{\longrightarrow} [Co(CN)_5SCNV(H_2O)_5]^- + H_2O$$

$$\longrightarrow Co(CN)_5^{3-} + VNCS^{2+}$$

$$\longrightarrow Co(CN)_5^{3-} + V^{2+} + NCS^- \qquad (4.91)$$

As with many other electron transfer reactions involving the V^{2+} ion, the rate of step $1'$ is characteristic of substitution into the inner coordination sphere of $V(H_2O)_6^{2+}$. Most probably, therefore, the immediate substitution product has the valency states $Co^{III}\ldots V^{II}$ and then rearranges to $Co^{II}\ldots V^{III}$. On this basis, the substitution product, not the ion pair, is the true precursor complex.

Some negative results are also noteworthy. In a number of cases, authors have been aware of the possibility of formation of precursor complexes, and have remarked on their inability to detect them. One reason must be that the stability of an opposite-charged ion pair falls off with increasing ionic strength; hence the attempt to detect the complex by increasing the reagent concentrations can be self-defeating. This may account for Thorneley and Syke's[1033] failure to detect a complex between $IrCl_6^{2-}$ and V^{3+} ions, and it almost certainly accounts for the failure of Cannon and Stillman[205] to detect a complex between $Co(ox)_3^{3-}$ and Fe^{2+}. Clearly the ionic charges and radii are by no means the only factors affecting complex stability: others such as the surface charge distribution and the local interaction with water molecules, are relevant as well. For example, the ion pair $Co(NH_3)_5OH_2^{3+}\cdot Fe(CN)_6^{4-}$ has a formation constant $K_{ip} = 1500$ M^{-1} at 25 °C, ionic strength 0.1 M; but the pair $Co(phen)_3^{3+}\cdot Fe(CN)_6^{4-}$ is undetectable under the same conditions; $K_{ip} \leqslant 100$ M^{-1}. The interionic distance is smaller in the former complex than in the latter, but not smaller by an amount sufficient to account for this difference. Formation constants predicted (for zero ionic strength) using equation (4.13) are 440 and 114 M^{-1} respectively. It has been suggested that the extra stability of $Co(NH_3)_5OH_2^{3+}\cdot Fe(CN)_6^{4-}$ is due to hydrogen bonding between the nitrogen atoms of the ammine and cyano groups.

A group of reactions worth studying further is exemplified by the system $Fe(DMP)_3^{3+} + IrCl_6^{3-}$ (see footnote (i), p. 107). The charges on the reactants are equal and opposite both before and after the reaction, and the redox potentials are so finely balanced that both reactants and products can coexist at appreciable concentration. It should, therefore, be possible, by excluding all other anions and cations, to prepare solutions containing both the precursor and the successor complexes at appreciable concentrations.

(b) Chelate complexes

Cannon and Gardiner[201] studied the reaction $Co(NH_3)_5LH_2^{2+} + Fe^{2+}$, where L^{3-} is the anion $N(CH_2COO)_3^{3-}$, initially linked to cobalt through one carboxyl group, with a tridentate side chain for attachment to the reducing ion. The rate law supported the mechanism

$$Co(NH_3)_5LH^{2+} + Fe^{2+} \underset{}{\overset{K}{\rightleftharpoons}} Co(NH_3)_5LFe^{2+} + 2H^+ \quad (4.92)$$

$$Co(NH_3)_5LFe^{2+} \xrightarrow{k} products \quad (4.93)$$

Both the numerical value and the hydrogen ion dependence of the equilibrium constant K confirm the structure of the intermediate as complex III, where Ro is $Co(NH_3)_5$.

RoOOCCH$_2$N——Fe(OH$_2$)$_3$

III

with the coordination of the Fe^{II} analogous to that of the known mononuclear complex $CH_3N(CH_2COO)_2Fe$.

The formation of a multi-chelate complex such as complex III is generally accepted to be a stepwise process, displacing one water molecule from the Fe^{II} centre at a time. If the driving force for the redox reaction were very strong, it might be expected that electron transfer would occur at an earlier stage, for example, when only one ring had formed in complex IV, but in that case the apparent formation constant K would be similar to that of the iron(II)–glycine complex; i.e. at least two orders of magnitude less than is observed.

Evidence of this kind establishes the nature of the predominant Co^{III} – Fe^{II} complex in the solution, but it does not prove that the predominant complex is also the precursor complex. That is, it does not prove that the electron transfer step is merely a redistribution of charges in complex III with no additional making or breaking of bonds. The possibility that one of the iron–ligand bonds has to be broken to reform the mono-chelate complex IV can presumably be discounted, but the alternative, that an additional iron–ligand bond must be formed to give complex V, cannot be ruled out.

IV V

(c) Other systems

Also shown in *Table 4.7* are some complexes which contain potentially oxidisable and reducible metal centres, but for which no electron transfer has been detected. Presumably the step p → s is highly unfavourable. In most cases, however, alternative redox reaction paths exist.

The ion pair $Co(en)_3^{3+}.Fe(CN)_6^{4-}$ is well characterised and its association constant has been determined[104]. From redox potential data[950], the reaction to form $Co(en)_3^{2+}.Fe(CN)_6^{3-}$ is opposed by an equilibrium constant of $\simeq 10^{-12}$, and, in fact, it is kinetically undetectable. Irradiation in the visible leads to a cherry-red binuclear complex also containing Co^{III} and Fe^{II} [652]. Larsson suggested that this arose by way of a photo-induced electron transfer, followed by substitution of $Fe(CN)_6^{4-}$ in the labile cobalt(II) centre, and reverse thermal electron–electron transfer to give Co^{III} – Fe^{II}. Kane-Maguire and Langford[608], however, have shown that this is preceded by a ligand field transition in the $Co(en)_3^{3+}$ and they suggest that the excited state has a lifetime long enough to permit a further, photochemical electron transfer from $Fe(CN)_6^{4-}$:

$$Fe(CN)_6^{4-} + Co(en)_3^{3+}[^1A_{1g}] \xrightarrow{h\nu} Fe(CN)_6^{4-} + Co(en)_3^{3+}[^1T_{1g}]$$

$$\xrightarrow{h\nu} Fe(CN)_6^{3-} + Co(en)_3^{2+} \longrightarrow (NC)_5Fe^{II}CNCo^{III}(en)_x^- \quad (4.94)$$

An analogous photochemical reaction of $Co(phen)_3^{3+}$ and $C_2O_4^{2-}$ leads to net reduction to cobalt(II)[651].

Farr et al.[402] have synthesised cobalt ammine complexes with olefinic side chains and have found stable binuclear complexes with copper(I), presumably attached to the double bond

$$(NH_3)_5CoOOC\underset{CH_2}{\overset{CH}{\diagup\!\!\!\diagdown}}Cu^+$$

There is no doubt that the overall reactions of these complexes, in acid solution, to form Co^{2+} and Cu^{2+}, are thermodynamically favourable; but it is most unlikely that electron transfer will be the initial step in such a process, since this would entail an organometallic π-complex of copper(II) for which there is no known precedent, as well as the reduction of cobalt(III) in an environment of nitrogen ligands. Direct photochemical electron transfer is well established, however, as discussed in Chapter 8.

The complex trans-$Cr(C_2O_4)_2^-$ forms an ion pair or complex with Cr^{2+}, and also undergoes a Cr^{2+}-catalysed aquation. The latter, however, is also acid-catalysed, so the measured parameters (K and $KK_a^{-1}k'$) are insufficient to yield either of the two possible intramolecular electron transfer rates:

$$H^+ + Cr(C_2O_4)_2^- + Cr^{2+}$$
$$\Updownarrow K$$
$$H^+ + (C_2O_4)Cr(C_2O_4)Cr^+ \xrightarrow{k} H^+ + CrC_2O_4 + CrC_2O_4^+$$
$$\Updownarrow K_a^{-1}$$
$$(HC_2O_4)CrC_2O_4Cr^{2+} \xrightarrow{k'} HC_2O_4 + Cr^2 + CrC_2O_4^+ \quad (4.95)$$

From the kinetic data, however, one may estimate a lower limit for the non-acid catalysed pathway Kk, and by analogy with other complexes, the authors have proposed upper limits for K_a^{-1}. These estimates give the limiting values of k and k' shown in Table 4.6.

4.5.2 Successor complexes (Table 4.7)

The discovery of labile successor complexes preceded that of precursor complexes by a considerable time, beginning with the system

Table 4.7 Electron transfer reactions involving labile successor complexes[a]

Reaction[b]	K/M^{-1}	k/s^{-1}	Notes	References
Fe(CN)$_6^{3-}$ + Co(edta)$^{2-}$	6.7 × 10²	6.2 × 10⁻³	(c)	5
	1.5 × 10³		(d) (e)	561
	7.4 × 10²	6.0 × 10⁻³	(f)	901
	8.3 × 10²	5.4 × 10⁻³	(e)	558
Fe(CN)$_6^{3-}$ + Co(dtpa)$^{3-}$	1.1 × 10²	1.35 × 10⁻³	(g) (h)	559
Fe(CN)$_6^{3-}$ + Co(cydta)$^{2-}$	4.4 × 10¹	2.2 × 10⁻²	(e)	558
	3.3 × 10¹	2.7 × 10⁻²	(f)	818
Fe(CN)$_6^{3-}$ + Co(pdta)$^{2-}$	2.4 × 10³	3.0 × 10⁻³	(f)	818
Fe(CN)$_6^{3-}$ + Co(trdta)$^{2-}$	6.2 × 10²	2.0 × 10⁻²	(f) (i)	818
Cr^{VI} + Ti^{III}	—	—	—	451
Ce^{IV} + IrCl$_6^{3-}$	2.2 × 10²	62	(j)	1116
cis-Ru(NH$_3$)$_4$Cl$_2^+$ + Cr^{2+}	4.6 × 10²	1.5 × 10²	(k)	770, 771
Ru(NH$_3$)$_5$Cl^{2+} + Cr^{2+}	7.0 × 10¹	4.6 × 10²	(k)	771
cis-Ru(NH$_3$)$_4$(H$_2$O)Cl^{2+} + Cr^{2+}	3.6 × 10²	1.2 × 10²	(k)	771
RuCl^{2+} + Cr^{2+}	7.7 × 10²	1.2	—	937

(a) 25 °C, $I = 1.0$ M
(b) edta^{4-} is ethylenediaminetetraacetate, dtpa^{5-} is diethylenetriaminpentacetate, cydta^{4-} is trans-1,2-cyclohexanediaminetetraacetate, pdta$^-$ is 1,2-diaminopropanetetraacetate, prdta^{2-} is 1,3-diaminopropanetetraacetate
(c) At low ionic strength
(d) $I = 0.6$ M
(e) The rate constant for the steps i $\rightleftarrows$ s were also determined
(f) $I = 0.66$ M
(g) The Co(dtpa)$^{3-}$ shown here is the hexa-coordinate form, present in equilibrium with penta-coordinate Co(dtpaH)(OH$_2$)$^{2-}$
(h) 30 °C
(i) 5 °C
(j) 0.1 M H$_2$SO$_4$
(k) 0.1 p-toluenesulphonic acid

Fe(CN)$_6^{3-}$ + Co(edta)$^{2-}$ described by Adamson and Gonick in 1963[5], though not fully described until 1963. Both the absorption spectrum and the observed diamagnetism confirmed the assignment of valencies as Fe^{II} – Co^{III}. Later, Huchital and Wilkins[561] obtained the separate rate constants k_1 and k_{-1} (equation 4.96) by temperature-jump measurements. The complex decomposes to mononuclear products at measurable rate, and it was originally assumed that this rate denoted the rates of dissociation, k_2 (equation 4.96). The cobalt (III) complex immediately produced by such dissociation would be the pentadentate form, but under the conditions of the experiments it was known that this would yield the hexadentate form relatively rapidly (step 4). Hain and coworkers[901] have since presented strong evidence against this, arguing that the dissociation involves the reverse of step 1, followed by the direct, outer-sphere reaction, step 3.

$$\begin{array}{c}
\text{Fe(CN)}_6^{3-} + \text{Co(edta)}^{2-} \underset{-1}{\overset{1}{\rightleftharpoons}} (\text{NC})_5\text{Fe}^{II}\text{CNCo}^{III}(\text{edta})^{5-} \\
{\scriptstyle -3}\updownarrow{\scriptstyle 3} \qquad\qquad\qquad {\scriptstyle 5}\qquad\qquad\qquad {\scriptstyle -2}\updownarrow{\scriptstyle 2} \\
\text{Fe(CN)}_6^{4-} + \text{Co(edta)}^{-} \underset{-4}{\overset{4}{\rightleftharpoons}} \text{Fe(CN)}_6^{4-} + \text{Co(edta)(OH}_2)^{-}
\end{array} \qquad (4.96)$$

The most direct proof is the fact that under alkaline conditions, where step 4 is known to be relatively slow, the hexadentate product is still formed. Secondly, the specific rate k' of the reverse reaction $\text{Co(edta)}^- + \text{Fe(CN)}_6^{4-}$ has been measured, using the hexadentate form of the cobalt complex. If the parameter Kk (equation 4.88) of the forward reaction is interpreted as the factor K_1k_2 of equation (4.96), then k' must be interpreted either as $K_4^{-1}k_{-2}$ with step -2 rate-determining, or as k_5, the rate of a direct substitution process. The former interpretation is ruled out since k_{-4} is known and is too small, while the latter is considered unlikely since substitution reactions of Co(edta)^- in general proceed by prior ring-opening. The rates of known substitution reactions are all much less than the observed rate constant k'. Finally, the value k' is in good agreement with the value of k_{-3} calculated theoretically using the Marcus cross-relation (see p. 205).

These ambiguities do not apply to the various $\text{Ru}^{III} + \text{Cr}^{II}$ reactions listed in *Table 4.3*, since the transfer of the chloride ion proves that the overall reaction is by way of the inner-sphere pathway.

The reaction $\text{RuCl}^{2+} + \text{Cr}^{2+}$ differs from the others in that the equilibrium is proceeds to completion on mixing, but when V^{2+} is added, the equilibrium is driven backwards by the operation of the reaction

$$\text{RuCl}^{2+} + \text{Cr}^{2+} \rightleftharpoons (\text{Ru}^{II}\text{ClCr}^{III})^{4+} \xrightarrow{\text{slow}} \text{Ru}^{2+} + \text{CrCl}^{2+} \qquad (4.97)$$

$$\text{RuCl}^{2+} + \text{V}^{2+} \rightarrow \text{RuCl}^{+} + \text{V}^{2+} \qquad (4.98)$$

Equations (4.97) and (4.98) together constitute two alternative modes of decomposition of the binuclear complex, i.e. spontaneous dissociation and vanadium(II)-catalysed dissociation with and without transfer of chloride ion.

Chapter 5
The mechanism of electron transfer

5.1 Introduction

The information reviewed in the previous two chapters consisted mainly of kinetic studies, supplemented where possible by the direct detection of reaction intermediates; when successfully applied it led to the elucidation of the reaction path, as a sequence or network of elementary reactions. Concerning any individual rate-determining step, the most that can be deduced from kinetics is the specific rate and the composition of the transition state. The next stage in defining the mechanism of electron transfer is to determine as far as possible the structure of the transition state, and for this purpose a wide variety of arguments have been deployed. In this chapter we classify and review both the mechanisms and the lines of argument which have been used.

5.1.1 Historical

The idea of direct transfer of an electron from one atom to another in a chemical reaction seems to have first arisen in the context of isotopic exchange studies, when it was realised that some of the elementary steps involve no net chemical change other than the change of valency of the central atom. For example, the equation

$$Mn^{3+} + {}^*Mn^{2+} \rightarrow Mn^{2+} + {}^*Mn^{3+} \tag{5.1}$$

was postulated in 1936 [853] as a step in the exchange between MnO_4^- and Mn^{2+}. Although there was little discussion in the literature, the earliest workers seem to have assumed that such reactions were indeed 'pure' electron transfer, the arrangement of atoms around the central atom being the same not only before and after, but during, the redox process. The first definite evidence for such a mechanism was provided in 1950 by Hornig, Zimmerman and Libby[557], who established that the reaction

$$^*MnO_4^- + MnO_4^{2-} = {}^*MnO_4^{2-} + MnO_4^- \tag{5.2}$$

is more rapid, by many orders of magnitude, than the exchange of oxygen between MnO_4^- and solvent water. Hence the electron transfer mechanism was favoured, 'the argument being that exchange of the manganese atom could hardly occur without the surrounding oxygen atoms exchanging with water in the process'[557]. From the choice of words (expressed parenthetically at the end of a short communication), it is not clear precisely what alternative mechanisms the authors had in view; but, if it is accepted that the MnO_4^- ion is a rigid structure unable to gain or lose oxygen atoms within the time of the experiment, that the same is true of the MnO_4^{2-} ion, and that these properties are not affected by the transfer of the electron, then Libby's mechanism is established.

The alternative mechanism, with an atom or group directly linking the oxidising and reducing centres, was recognised as a possibility by several groups of workers in 1951–52 and first definitely established in 1953. Several symmetrical exchange processes ($Eu^{3+} + Eu^{2+}$, $Fe^{3+} + Fe^{2+}$, $Ce^{4+} + Ce^{3+}$) were shown to be catalysed by halide ions. Meier and Garner[748] conjectured 'that an electron and a chloride ion are interchanged between reactants in the activated complex, either simultaneously or in two steps', and Libby[672] suggested as an example of the structure of a transition state the bridged formulation $((H_2O)_5FeXFe(OH_2)_5)^{4+}$. The first direct proof of such a mechanism was obtained by Taube, Myers and Rich[1017], using the reaction

$$Co(NH_3)_5Cl^{2+} + Cr^{2+} + 5H^+ = Co^{2+} + 5NH_4^+ + CrCl^{2+} \quad (5.3)$$

The method depends on the fact that ligand replacement reactions of cobalt(III) and chromium(III) are characteristically slow while those of cobalt(II) and chromium(II) are much more rapid. In a review published the previous year Taube[1006] had surveyed what little was then known of the rates of substitution reactions, and had pointed out the distinction between complexes which reacted at measurable rates, and those which reacted within the time of mixing of two solutions. His terms 'inert' and 'labile' for the two categories have come into general use. In equation (5.3), the cobalt complex changes from inert to labile, while the chromium changes from labile to inert. The fact that the chloride ion is transferred from one to the other proves that at no point in the reaction event did it become free in solution; hence, the structure of the transition state can only be the bridged binuclear species $[(H_3N)_5CoClCr(H_2O)_5]^{4+}$. In a further experiment, radioactive chloride ion was included in the solution and after separation of the products it was shown that none of the radioactivity was taken up into the $CrCl^{2+}$ complex.

This observation initiated a period of rapid advances, and the distinction between *outer-sphere* and *inner-sphere*, or *non-bridged* and *bridged* electron transfer mechanisms came to be accepted as fundamental. The subject has been reviewed many times [a], and a comprehensive survey would exceed the limits imposed on this book. It has been shown that many other ligands besides the chloride ion can function as bridges, and that a number of other oxidants can take the place of cobalt(III). Attempts to test other reductants besides

(a) References 506, 679, 985, 986, 987, 989, 996, 1007, 1008, 1009, 1011, 1012, 1013, 1015.

chromium(II) were frustrated at first by the lability of the corresponding oxidised product, but in more recent years measurements by fast reaction methods have made it possible to work with other metals. Equally important, however, has been the recognition of other mechanisms, or of subdivisions within the two main categories already recognised. The varieties of inner–sphere mechanism which can be distinguished are summarised in section 5.1.2, and the experimental evidence is reviewed in sections 5.2 and 5.3. Outer-sphere mechanisms are treated in more detail in section 5.4.

5.1.2 Summary of mechanisms (inner-sphere)

The varieties of inner-sphere mechanism which have been clearly established are set out in *Table 5.1*, together with a classification of the lines of argument on which they are based. The evidence varies from direct proof to more or less strong inference by analogy with other systems.

What we have called 'direct' electron transfer, has been very little studied, and rarely discussed. It seems that early workers assumed that reactions of the type of equation (5.1) were of this kind, analogous to electron transfer between atoms in the gas phase, but the subsequent recognition that all metal ions in solution are coordinated with water molecules in definite orientations led to the abandonment of this idea. It seems possible that the rapid exchange [1104] between Hg^{2+} and Hg^0 in solution involves the Hg_2^{2+} ion as intermediate, and this reaction is accordingly entered in *Table 5.1*. Other binuclear systems are known which can be regarded formally as adducts of two different oxidation states, but they are not necessarily electron transfer intermediates. For example, the species H_2^+.aq has been postulated [890, 255], though the evidence is not conclusive [440]. It might be expected to be an intermediate in the reaction

$$H^+.aq + H.aq = H.aq + H^+.aq \tag{5.4}$$

but the measured rate is consistent with an outer-spere mechanism[638, 788].

Inner-sphere mechanisms can be subdivided according to the distance between the reacting centres. The only strictly monatomic bridges known for metal–metal electron transfer reactions are the halide ions, and possibly the oxide ion in some reactions of MnO_4^- and CrO_4^{2-} [473]. Monatomic bridges with side groups attached include OH^-, $SC(NH_2)_2$ [863] and $H_2NC_2H_4S^-$ [1086].

Electron transfer through extended bridging groups is amply documented, the distance between centres ranging from $\simeq 4$ Å, with the CN^- ion as bridge, to over 11 Å, with 4-4'-bipyridyl. Most of the experiments are analogues of reaction (5.1), with the bridging group initially attached to the inert oxidant. Thus, when the bridging group offers more than one site for the attachment of the reductant, the distinction between *adjacent attack* and *remote attack* becomes possible. In some cases, such as the reaction $Co(NH_3)_5N_3^{2+} + Cr^{2+}$, the preferred point of attack has not been ascertained (compare p. 157); in other cases, such as $Co(NH_3)_5SCN^{2+} + Cr^{2+}$ [943], adjacent and remote attack proceed at comparable rates. Some of the early reports of remote attack

Table 5.1 Classification of electron transfer mechanisms, and of experimental evidence of mechanism[a]

Mechanism		Evidence					
		Long-lived precursor or successor complex	Transfer or non-transfer of bridging group	Electron transfer rate compared with substitution rate	Catalysis		Assignment based on LFER or other rate comparisons
					Kinetically significant intermediate	Structure of transition state guessed from rate law	
Direct	$A^+ . B$	Hg_2^{2+} $Cr^{2+} + H$	[c]		[c]		
Inner-sphere:							
monatomic bridge	$A^+ . X . B$	$Cr^{III}ClIr^{III}$	$Cr^{III}(OH)Cr^{II}$	$Co^{III}FV^{II}$	$V^{II}(OH)Cr^{III}$	$Fe^{III}(OH)Cr^{II}$	$Co^{III}FTi^{III}$
remote attack (small molecule)	$A^+ . XY . B$	$Co^{III}NCFe^{II}$	$Co^{III}SCNCr^{II}$	$Co^{III}SCNV^{II}$			
remote attack (large molecule)	$A^+ . X\text{---}Y . B$	$Co^{III}(ina)Ru^{II}$	$Co^{III}(ina)Cr^{II}$	$Co^{III}(ina)V^{II}$	$Co^{III}(ina)V^{II}$		$Co^{III}(ina)Eu^{II}$
double bridge	$A^+ \genfrac{}{}{0pt}{}{X}{Y} B$		$Cr^{III}(N_3)_2Cr^{II}$			$Cr^{III}\genfrac{}{}{0pt}{}{(OH)}{N(CS)}Cr^{II}$	
Linked	$A^+ . \widehat{XY} . B$	$Co^{III}(na)Ru^{II}$				$\begin{array}{c}CH_2\text{---}CO\\ \vert \quad\quad \vert \\ Co^{III}NH_2 \quad OCr^{II}\end{array}$	
Outer-sphere[b]	(A⁺) (B)	[c]	$Co^{III}(iz) + Cr^{2+}$	$Co^{3+} + V^{2+}$		$Co^{3+} + Cr^{2+}$	$Co^{III}Cl + V^{2+}$

(a) Non-bridging ligands not shown. *ina* is the *iso*-nicotinate ion $(OOCC_5H_4N\text{-}p)^-$; *na* is the nicotinate ion $(OOCC_5H_4N\text{-}m)^-$; *iz* = imidazole
(b) For further subdivisions see *Table 5.7*
(c) Not applicable

involving conjugated organic bridging groups were later found to be incorrect, but the confusion has since been dispelled[b].
involving conjugated organic bridging groups were later found to be incorrect, but the confusion has since been dispelled[b].

The term *linked electron transfer* is used here to denote a remote attack in which the oxidant and reductant centres are joined by a molecular chain, but the electron transfer is a 'through space' interaction, i.e. an outer-sphere reaction in which the linking group merely serves to hold the ions at the requisite distance. Strictly speaking, this definition presupposes an understanding of how the bridging group serves to carry the electron between the reacting centres, a question which is more conveniently discussed later on (chapter 7). For the present it is sufficient to say that if the reacting centres have inner coordination spheres similar to those of other systems which readily undergo outer-sphere transfer, and if the connecting chain is sufficiently flexible to allow them to approach closely, the linked structure is sufficient to explain the observed reaction.

Experimental tests for inner- and outer-sphere reactions, are reviewed in section 5.2, and the examples of remote attack are discussed in section 5.3. The wealth of data should not be allowed to obscure the fact that many fundamental questions remain unanswered. There are still comparatively few examples in which the specific rate of the elementary electron transfer reaction, and the structure of the transition state, are both known; certainly not enough for systematic comparison between different metals, and none at all with simple monatomic bridges. It has not yet been possible to vary systematically the distance between oxidant and reductant centres without substantially altering the chemical nature of the bridge. We have no information on the importance or otherwise of the relative orientations of the coordination polyhedra surrounding the reactive centres.

(b) It was initially[434] reported that chromium(II) reduction of the half-ester complex $[Co(NH_3)_5OOC\sim COOCH_3]^{2+}$ was accompanied by ester hydrolysis, giving the product $[CrOOC\sim COOH]^{2+}$. This was taken as evidence of the formation of the intermediates, followed by ester hydrolysis (by analogy with the catalysis of ester hydrolysis by metal ions), and hence as evidence of remote attack. This and analogous[432, 433, 435, 436] claims have not, however, been verified, and later, more extensive work has shown[567, 821] that in fact ester hydrolysis is negligible. The claim that not only chromium(II) but also vanadium(II)[431, 435-437] and europium(II)[437] led to ester hydrolysis introduced a new factor, for it implied that hydrolysis occurred not after but during the electron transfer process, i.e. that 'activation for hydrolysis taken place in an intermediate which contains the Co centre as well as the reducing cations associates with the ligand'[437]. A similar explanation seemed called for by reports that reductions of the maleate complex by both chromium(II) and vanadium(II) led to *cis–trans* isomerisation of the ligand[436, 438]. Both claims have now been disproved[821], but it is worth noticing that even at the time the latter two results were recognised as anomalous 'it is difficult to see why electron conduction to the ligand *requires* such a dramatic redistribution of electrons in the alkoxy bond as to produce loss of the alkyl groups'[438]

5.2 Inner- and outer-sphere mechanisms: experimental evidence

5.2.1 Isolation of binuclear complex[c]

The most direct evidence of the inner-sphere mechanism is the isolation of the precursor or successor complex, $[A.X.B.]^+$, provided that the reacting centres are inert to substitution. This condition is necessary in the case of the precursor complex in order to be sure that the preferred reaction path does not involve dissociation followed by an outer-sphere reaction, $A^+.X.B \rightarrow A^+.X + B \rightarrow A + X + B^+$, and conversely in the case of the successor complex, to eliminate pathways such as $A^+.X + B \rightarrow A.X + B^+ \rightarrow A.X.B^+$.

(a) Precursor complexes

Complexes in which two 'inert' metal cations (cobalt(III), low-spin iron(III) or iron(II), ruthenium(III), etc.) are linked by a bridging group have been known since 1961, but the problem of synthesising such a complex with two ions metastable with respect to electron transfer has been solved only recently. Isied and Taube[590] have described a series of complexes containing cobalt(III) and ruthenium(II), linked by organic groups of varying rigidity. They could not be made by condensing units containing cobalt(III) and ruthenium(II), since the outer-sphere electron transfer was too rapid, and the alternative synthetic route, of condensing cobalt(III) and ruthenium(III) complexes, followed by reduction to ruthenium(II), also failed in these cases. The following ingenious synthesis relies on the properties of ruthenium(II) complexes with the sulphite ion as ligand; they are exceptionally stable, so that sulphitopentammine ruthenium(II) does not readily reduce cobalt(III); and the ligand *trans* to the sulphur atom is labilised by a strong kinetic *trans* effect.

$$(H_3N)_5Co^{III}OOC\text{-}\langle\_\rangle\text{-}NH^{3+} + trans\text{-}ClRu^{II}(NH_3)_4SO_3^+$$
$$\downarrow$$
$$(H_3N)_5Co^{III}OOC\text{-}\langle\_\rangle\text{-}N-Ru^{II}(NH_3)_4SO_3^{3+}$$
$$\downarrow H_2O_2$$
$$(H_3N)_5Co^{III}OOC\text{-}\langle\_\rangle\text{-}N-Ru^{III}(NH_3)_4SO_4^{3+} \quad (5.5)$$

The $Co^{III}-Ru^{III}$ complexes could be isolated and purified; then on treatment with a strong outer-sphere reducing agent such as Eu^{2+} or $Ru(NH_3)_6^{2+}$, the rapid reduction to $Co^{III}-Ru^{II}$ ensued, followed by the desired electron transfer reaction:

---

[c] For discussion of the important class of symmetric complexes, $A^+.X.A$, see p. 287

$$(H_3N)_5Co^{III}OOC\text{-}\langle\rangle\text{-}NRu^{II}(NH_3)_4SO_4^{2+}$$
$$\xrightarrow{k_{ps}} (H_3N)_5Co^{II}OOC\text{-}\langle\rangle\text{-}NRu^{III}(NH_3)_4SO_4^{2+}$$
$$\rightarrow \text{decomposition products} \qquad (5.6)$$

The rate constants k_{ps} are listed in *Table 5.2*.
Related to these systems is the reaction

$$(NH_3)_5Co^{III}N\text{-}\langle\rangle\text{-}\langle\rangle\text{-}N^{3+} + (H_2O)Fe^{II}(CN)_5^{3-} \underset{k_{pi}}{\overset{k_{ip}}{\rightleftharpoons}}$$

$$(NH_3)_5Co^{III}N\text{-}\langle\rangle\text{-}\langle\rangle\text{-}N\text{-}Fe^{II}(CN)_5 \xrightarrow{k_{ps}}$$

Table 5.2 Substitutionally inert electron transfer precursor complexes

Complex[a] $A^{III}.X.B^{II}$	k_{ps}/s^{-1} [b]	Reference
$(NH_3)_5Co^{III}OOC\text{-}\langle O\rangle\text{-}N\text{-}Ru^{II}(NH_3)_4SO_4$	$\simeq 1 \times 10^2$ [b]	590
$(NH_3)_5Co^{III}OOC\text{-}CH_2\text{-}\langle O\rangle\text{-}N\text{-}Ru^{II}(NH_3)_4H_2O$	1.3×10^{-2}	590
$(NH_3)_5Co^{III}OOC\text{-}\langle O\rangle\text{-}N\text{-}Ru^{II}(NH_3)_4H_2O$	1.7×10^{-3}	590
$(NH_3)_5Co^{III}OOCCH_2\text{-}\langle O\rangle\text{-}N\text{-}Ru^{II}(NH_3)_4H_2O$	0.86×10^{-3}	590
$(NH_3)_5Co^{III}N\text{-}\langle\rangle\text{-}\langle\rangle\text{-}NFe^{II}(CN)_5$	2.7×10^{-3} [c]	453
$(NH_3)_5Co^{III}OOC\text{-}\langle\rangle\text{-}NFe^{II}(CN)_5$	1.75×10^{-4}	604

(a) Overall ionic charges not shown
(b) 25 °C, $I = 1.0$ M
(c) $I = 0.1$ M

$$(NH_3)_5Co^{II}N\underset{}{\diagdown}\underset{}{\diagup}N\text{-}Fe^{III}(CN)_5 \longrightarrow$$

$$Co^{2+} + N\underset{}{\diagdown}\underset{}{\diagup}N\text{-}Fe^{III}(CN)_5^{2-} \quad (5.7)$$

studied by Gaswick and Haim[453]. Here the complex was not isolated, but it was detected as an intermediate, and by suitable choice of initial conditions it was possible to obtain all three rate constants. The rate constant $k_{ip} = 5.5 \times 10^3 \text{ M}^{-1}\text{s}^{-1}$ is characteristic of an inner-sphere substitution reaction at the Fe^{II} centre; hence the structure of the intermediate is unambiguously determined. Jwo and Haim[604] have since reported reactions of $Fe^{II}(CN)_5H_2O^{3-}$ with the isomeric oxygen-bonded pyridinecarboxylato-pentammine cobalt(III) complexes. With the 2-isomer, there is no reaction, presumably for steric reasons; with the 3-isomer a binuclear complex is produced which shows no tendency to react further ($k_{ps} \leq 3 \times 10^{-5}\text{s}^{-1}$), but with the 4-isomer, the reaction sequence is

$$Co(NH_3)_5L^{2+} + Fe(CN)_5H_2O^{3-} \underset{k_{pi}}{\overset{k_{ip}}{\rightleftharpoons}}$$

$$Co^{III}(NH_3)_5OOC\underset{}{\diagdown}\underset{}{\diagup}N\text{-}Fe^{II}(CN)_5^{-} \overset{k_{ps}}{\longrightarrow} Co^{II} + Fe^{III}L \quad (5.8)$$

with all steps proceeding at measurable rates ($k_{ip} = 1.5 \times 10^3 \text{ M}^{-1}\text{s}^{-1}$; $k_{pi} = 2.5 \times 10^{-3}\text{s}^{-1}$; $k_{ps} = 1.75 \times 10^{-4}\text{s}^{-1}$).

(b) Successor complexes

When the reaction sequence $i \rightleftharpoons p \to s$ is fast and thermodynamically favourable, but the step $s \to f$ is slow, the successor complex may be isolated in high yield. This has formed the basis of some interesting syntheses, a representative selection of which is shown in *Table 5.3*. The binuclear product of reaction 13 in *Table 5.3* was isolated as the barium salt. It dissociates to $Fe(CN)_6^{4-}$ and $Co(CN)_5OH_2^{2-}$ only on prolonged heating in aqueous solution, and undergoes reversible redox reactions characteristic of the $Fe^{II}(CN)_6^{4-}$ group. The oxidised form of the complex can react with another molecule of the cobalt(II) complex to give a trinuclear species.

$$(NC)_5Co^{III}NCFe^{III}(CN)_5^{5-} + Co(CN)_5^{3-} \to$$
$$(NC)_5Co^{III}NC(Fe^{II}(CN)_4)CNCo^{III}(CN)_5^{8-} \quad (5.9)$$

The product of reaction 1, *Table 5.3* was not isolated, but it was identified on the basis of its spectrum and reactions, characteristic of uranium(V).

In other cases the binuclear product has not been isolated but has been observed directly. Either the electron transfer reaction is rapid and the reaction studied kinetically is the subsequent decomposition of the intermediate, or else both reactions are of comparable rate, and the rise and fall of concentration of

Table 5.3 Electron transfer reactions leading to substitutionally inert successor complexes

No.	Reaction	References
1	$UO_2^{2+} + Cr^{2+} \to OU^VOCr^{III}(OH_2)_5^{4+} \to UO_2^+ + Cr^{3+}$ (f)	363, 471, 798
2	$V^V + Fe(CN)_6^{4-} \to V^{IV}-Fe^{III} \to V^{IV} + Fe^{III}$	121
3	$V^V + Fe(bipy)(CN)_4^{2-} \to V^{IV}-Fe^{III} \to VO^{2+} + Fe(bipy)(CN)_4^-$	125
4	$V^V + Fe(bipy)_2(CN)_2 \to V^{IV}-Fe^{III} \to VO^{2+} + Fe(bipy)_2(CN)_2^+$	125
5	$VO^{2+} + V^{2+} \to V^{III}(OH)_2V^{III4+}$	800
6	$Cr^{VI} + Pu^{IV} \to \to Cr^{III}OPu^VO^{4+}$ (g)	805
7	$Cr^{VI} + Mo(CN)_8^{4-} \to \to Cr^{III}-Mo^V$ (not fully characterised) (g)	705
8	$Cr^{VI} + Fe^{II} \to (Cr^V-Fe^{III}$ or $Cr^{III}-Fe^{III}) \to$ products	383
9	$Cr^{VI} + Fe(CN)_6^{4-} \to \to Cr^{III}-Fe^{III}$ (g)	120
10	$Cr^{VI} + Fe(bipy)(CN)_4^{2-} \to \to Cr^{III}-Fe^{III}$ (g)	120
11	$Cr^{VI} + Fe(bipy)_2(CN)_2 \to \to Cr^{III}-Fe^{III}$ (g)	120
12	$Cr^{VI} + Co(edta)^{2-} \to Cr-Co$	105
13	$Fe^{III}(CN)_6^{3-} + Co^{II}(CN)_5^{3-} \to (NC)_5Fe^{II}CNCo^{III}(CN)_5^{6-}$	503
14	$Fe(CN)_6^{3-} + Cr^{2+} \to (NC)_5Fe^{II}CNCr^{III}(OH_2)_5^-$	1035
15	$RuCl^{2+} + Cr^{2+} \rightleftharpoons (H_2O)_5Ru^{II}ClCr^{III}(OH_2)_5^{4+}$ $\to Ru(OH_2)_6^{2+} + CrCl(OH_2)_5^{2+}$ (d)	937
16	$Ru^{III}LCl + Cr^{2+} \rightleftharpoons Ru^{II} LClCr^{III} \to Ru^{II}L + CrCl^{2+}$ (e)	771
17	$Ru(NH_3)_5OOCR^2 + Cr^{2+}$ $\to (H_3N)_5Ru^{II}O(CR)OCr^{III}(OH_2)_5^{4+}$ (c)	974
18	$Ru(NH_3)_5OH_2^{3+} + Cr^{2+} \to (H_3N)_5Ru^{II}OHCr^{III}(OH_2)_5^{5+}$	974
19	$Ru^{III}(NH_3)_5L + Cr^{2+} \to (H_3N)_5Ru^{II}LCr^{III}$ (a)	454
20	$Ru(NH_3)_5NC_5H_4CSNH_2 + Cr^{2+}$ $\to Ru^{II}(NH_3)_5NC_5H_4CS(NH_2)Cr^{III}$	259
21	$Ru^{III}(NH_3)_4L_2 + Cr^{2+} \to L[Ru^{II}(NH_3)_4]LCr^{III}$ (b)	383
22	$Ru^{III}(bipy)_2Cl_2^+ + Cr^{2+} \to cis\text{-}Cl[Ru^{II}(bipy)_2]ClCr^{III}$ $\to cis\text{-}Ru^{II}(bipy)_2Cl(H_2O)^+ + CrCl^{2+}$	771
23	$IrCl_6^{2-} + Cr^{2+} \to Cl_5Ir^{III}ClCr^{III}(OH_2)_5$ $\to IrCl_5OH_2^{2-} + CrCl(OH_2)_5^{2+}$	999, 1034
24	$IrCl_6^{2-} + Co(CN)_5^{3-} \to Cl_5Ir^{III}ClCo^{III}(CN)_5^-$ $\to IrCl_6^{3-} + Co(CN)_5(H_2O)^{2-}$	484

(a) L is nicotinamide, *iso*-nicotinamide or methyl-*iso*-nicotinate
(b) L is *iso*-nicotinamide or methyl-*iso*-nicotinate
(c) R is H, CH_3 or CF_3
(d) The hydrolysis step also has a base-catalysed path, in which the Cr–Cl and Ru–Cl bonds break at comparable rates
(e) $Ru^{III}LCl$ is cis-$Ru(NH_3)_4Cl_2^+$, cis-$Ru(NH_3)_4ClH_2O^{2+}$ or $Ru(NH_3)_5Cl^{2+}$
(f) Oxygen is transferred from U to Cr
(g) In these expressions $\to \to$ denotes a multistep reaction

the intermediate can both be monitored. When the decomposition rate is sufficiently slow it is possible to distinguish inner- and outer-sphere pathways. This may be done on the basis of identifying the different inert products of the two paths, or on the basis of the kinetics, when the outer-sphere reaction leads to the same products as arise from the break-up of the binuclear intermediate, but more rapidly. In the reaction $IrCl_6^{2-} + Cr^{2+}$ (reaction 23, *Table 5.3*) the outer-sphere path gives the inert complexes $IrCl_6^{3-}$ and $Cr(H_2O)_6^{3+}$, but the inner-sphere path leads to the chloride-bridged intermediate and then to $IrCl_5(OH_2)^{2-}$ and $Cr(H_2O)_5Cl^{2+}$. The two paths operate simultaneously. In the reaction $VO^{2+} + V^{2+}$ (reaction 5, *Table 5.3*) the inner-sphere path leads to a dark brown intermediate VOV^{4+} (or more probably, $V_2(OH)_2^{4+}$) which decomposes at a measurable rate to give the blue-black V^{3+} ion. The kinetics of the reaction, however, show that never more than a small fraction of vanadium(III) is present as the binuclear complex. Most of the V^{3+} product must arise by a direct outer-sphere process.

Reactions of the type

$$M^{II} + M^{IV} \rightarrow M_2^{III}(OH)_2^{4+} \qquad (5.10)$$

(where M is Cr [37, 102, 530], or Fe [247]) which lead to binuclear products, have been cited as evidence of the unstable M^{IV} intermediate in non-complementary reactions (*see* chapter 3).

5.2.2 Transfer of bridging group (inert reactants)

The argument of Taube, Myers and Rich has been successfully applied to many other reactions. Only a representative selection can be discussed here. With chromium(III) as oxidant and chromium(II) as reductant, the net reaction is a chromium(II)-catalysed substitution, such as aquation[817]:

$$Cr(NH_3)_5Cl^{2+} + Cr^{2+} + 5H^+ = Cr^{2+} + 5NH_4^+ + CrCl^{2+} \qquad (5.11)$$

or isotopic exchange

$$CrCl^{2+} + {}^*Cr^{2+} = Cr^{2+} + {}^*Cr^{III}Cl \qquad (5.12)$$

Some oxidants retain their non-bridging ligands after reduction by reason of either stability or inertness, but the bridging group is sufficiently labile in the reduced form to be effectively transferred[372]

$$Ru(NH_3)_5Cl^{2+} + Cr^{2+} = Ru(NH_3)_5OH_2^{2+} + CrCl^{2+} \qquad (5.13)$$

Complexes of iron(III) are generally too labile to be studied in this way. An exception is FeLCl (where L is tetraphenylporphinate) which in benzene solution is relatively inert[241]:

$$FeLCl + Cr(acac)_2 \rightarrow FeL + Cr(acac)_2Cl \qquad (5.14)$$

The chromium(III) product has not been identified fully, but transfer of the Cl^- ion was demonstrated by radioactive tracer experiments.

Other reductants used include pentaammine ruthenium(II)[374, (d)] and low-spin cobalt(II) in the dimethylglyoximato[13] and pentacyano[1] complexes

$$Co(NH_3)_5I^{2+} + Ru(NH_3)_5OH_2^{2+} + 5H^+ \\ = Co^{2+} + 5NH_4^+ + Ru(NH_3)_5I^{2+} \quad (5.15)$$

$$Co(NH_3)_5Cl^{2+} + Co(dmg)_2 + 5H^+ \\ = Co^{2+} + 5NH_4 + Co(dmg)_2(H_2O)Cl \quad (5.16)$$

$$Co(NH_3)_5Br^{2+} + Co(CN)_5^{3-} = Co^{2+} + 5NH_3 + Co(CN)_5Br^{3-} \quad (5.17)$$

The discovery of reaction (5.17) stemmed from the remarkable observation[876, 877] that acidopentaammine cobalt(III) complexes, when dissolved in aqueous cyanide solution, apparently suffered rapid replacement of the ammonia but not of the halide ion. The effect was explained as catalysis by cobalt(II), present in trace quantities in the cobalt(III) samples used. Curiously, high-spin cobalt(II) as a reductant has only rarely been studied[704, 812, 1093]:

$$Co^{III}(py)_4Cl_2^+ + Co^{II}(edta)^{2-} \rightarrow \\ Co^{2+} + 4py + Cl^- + Co^{III}(edta)Cl^{2-} \quad (5.18)$$

$$trans\text{-}Co^{III}(en)_2Cl_2^+ + Co^{II}(en)(NO_2)_3^- \rightarrow \\ Co^{2+} + 2en + Cl^- + Co^{III}(en)(NO_2)_3Cl^- \quad (5.19)$$

$$Co(acac)_3 + {}^*Co(acac)_2 \rightarrow Co(acac)_2 + {}^*Co(acac)_3 \quad (5.20)$$

Transfer of oxygen, generally as hydroxide, can usually only be detected by isotopic labelling experiments. Classic examples are cobalt(III)–chromium(II) reactions such as

$$cis\text{-}Co(en)_2(^*OH_2)_2^{3+} + Cr^{2+} + 4H^+ = \\ Co^{2+} + 2enH_2^{2+} + Cr(^*OH_2)_6^{3+} \quad (5.21)$$

and equation (5.38) below[634]

A rather limited amount of work has been done on oxyanions of metals. Examples include the reactions

$$4H^+ + 3U^{4+} + 2CrO_4^{2-} = 3UO_2^{2+} + 2Cr^{3+} + 2H_2O \quad (5.22)$$

$$2H_2O + 5U^{4+} + 2MnO_4^- = 5UO_2^{2+} + 2Mn^{2+} + 4H^+ \quad (5.23)$$

which proceed with extensive transfer of oxygen to uranium[473].

Other non-complementary reactions with bridging mechanisms are[102, 1016]

(d) Ford[420] has pointed out that the identification of the product $Ru(NH_3)_5I^{2+}$ in this case is not conclusive proof of the inner-sphere mechanism. The same product could arise from the anation reaction $I^- + Ru(NH_3)_5OH_2^{3+}$, catalysed by ruthenium(II). Endicott and Taube[374] had considered this possibility, and they calculated that the yield of $Ru(NH_3)_5I^{2+}$ observed was too high to be accounted for in this way. The argument depends, however, on the magnitude of the equilibrium constant for the aquation reaction, which according to Ford is subject to some uncertainty

$$AuCl_4^- + 2Cr^{2+} = AuCl_2^- + 2CrCl^{2+} \quad (5.24)$$

$$Pt(NH_3)_5Cl^{3+} + 2Cr^{2+} + H^+ = Pt(NH_3)_4^{2+} + NH_4^+ + Cr^{3+} + CrCl^{2+} \quad (5.25)$$

A number of two-electron transfer processes have been shown to be bridged, the first to be recognised were platinum(IV) and platinum(II) exchange reactions[734, 836]. A typical example is[80, 735, 914] the reaction

$$Pt(NH_3)_5Cl^{3+} + Pt(NH_3)_4^{2+} + Cl^- + H^+ \quad (5.26)$$
$$= NH_4^+ + Pt(NH_3)_4^{2+} + \textit{trans-}Pt(NH_3)_4Cl_2^{2+} \quad (5.27)$$

The net chemical change is a ligand-substitution reaction

$$Pt(NH_3)_5Cl^{3+} + Cl^- = \textit{trans-}Pt(NH_3)_4Cl_2^{2+} \quad (5.28)$$

The product is specifically the *trans*-isomer and for many years reactions of this type were discussed as examples of substitution, controlled by a *trans*-directing effect of the chloride ion attached to the platinum(IV)[83]. The electron transfer mechanism was postulated by Basolo, Pearson et al.[84, 735], and is confirmed by a number of kinetic studies showing first-order dependence on both platinum(IV) and platinum(II). In the typical case of equation (5.28) the rate law is[735]

$$\text{Rate} = k[Pt(NH_3)_5Cl^{3+}][Pt(NH_3)_4^{2+}][Cl^-] \quad (5.29)$$

When the initial platinum(IV) complex is a dichloro complex, the reaction is a symmetrical exchange process[84],

$$\textit{trans-}Pt(en)_2Cl_2^+ + Pt(en)_2^{2+} + {}^*Cl^-$$
$$= Cl^- + Pt(en)_2^{2+} + \textit{trans-}Pt(en)_2Cl^*Cl^{2+} \quad (5.30)$$

By using isotopically labelled platinum[258], and also using ligands labelled either isotopically[80] or as optical isomers[84], it was shown that the rate of exchange of chloride ions is the same as the rate of exchange of valency states.

In all these reactions, the rate law indicates the presence of an additional, non-bridging ion (usually Cl^-) which moves into the vacant space on the platinum(II) to complete the octahedral geometry. The importance of this term reflects the strength of the Pt–Cl bond in the platinum(IV) product. The converse effect has also been observed: removal of the axial non-bridging halide ion from the oxidant, by the Hg^{2+} ion

$$Hg^{2+} + Pt(en)(NO_2)_2Br_2 + Pt(en)(NO_2)_2 + H_2O$$
$$= HgBr^+ + Pt(en)(NO_2)_2 + Pt(en)(NO_2)_2(H_2O)Br^+ \quad (5.31)$$

The net reaction is dissociation of the bromide ions from the platinum(IV) complex[312], catalysed by the Hg^{2+} ion

$$\text{Rate} = k[Hg^{2+}][Pt^{IV}][Pt^{II}] \quad (5.32)$$

The oxidation of platinum(II) by gold(III) is believed, though not actually proved, to involve chloride ion transfer[837]

$$AuCl_4^- + trans\text{-}PtL_2Cl_2 + Cl^- = Cl^- + AuCl_2^- + trans\text{-}PtL_2Cl_4 \quad (5.33)$$

where L denotes various tertiary phosphine or arsine ligands. One term in the rate law involves a catalytic chloride ion:

$$\text{Rate} = k[AuCl_4^-][trans\text{-}PtL_2Cl_2][Cl^-] \quad (5.34)$$

Since the chloride ion moves in the opposite direction to the two electrons, these reactions can formally be described as transfer of the 'chloronium' ion Cl^{+} [79]. They can equally well be written in the manner of the non-metal chemist, with curved arrows to denote a mesomeric shift (for clarity, the equatorial NH_3 molecules are omitted):

$$H_3N^+\!\!\frown\!\!Pt^{IV}\!\!\frown\!\!Cl \ldots Pt^{II} \ldots Cl^- \rightarrow H_3N + Pt^{II} + Cl\!-\!Pt^{IV}\!-\!Cl \quad (5.35)$$

Analogous mechanisms have been established involving non-metals[215], as typified by the reactions[181,337]

$$\begin{aligned} trans\text{-}PtL_2Cl_2^{2+} + SCN^- &\rightarrow [Cl(PtL_2)Cl \ldots SCN]^+ \\ &\rightarrow Cl^- + PtL_2^+ + ClSCN \end{aligned} \quad (5.36)$$

(where L is 1:2-bis (dimethylarsino) benzene) and

$$ROO^*H + NO_2^- \rightarrow ROH + O^*NO_2^- \quad (5.37)$$

(where O* is isotopically labelled oxygen).

Equally important are demonstrations that a bridging mechanism does not occur, when at first sight it might have been expected. For example, it is known from x-ray crystallography that the water molecule can bridge two metal ions, as in $LiClO_4 \cdot 3H_2O$[1084a], and electron transfer bridging has therefore also been considered. In all cases where a direct test has been made, however, the result is negative. Thus the reaction

$$Co(NH_3)_5OH_2^{3+} + Cr^{2+} + 5H^+ = Co^{2+} + 5NH_4^+ + Cr^{3+} + H_2O \quad (5.38)$$

proceeds with the transfer of oxygen from cobalt to chromium, but the rate law is wholly of the base-catalysed type[734,836,(e)], implying that hydroxide ion, not water, is the bridging group

$$\text{Rate} = k[Co(NH_3)_5OH_2^{3+}][Cr^{2+}][OH^-] \quad (5.39)$$

(e) Earlier work [779,1128] suggested a detectable, non-base-catalysed path, but this is now thought to be a spurious effect arising from the use of $NaClO_4$ to maintain ionic strength over a wide range of (H^+). The later result is based on the use of $LiClO_4$

and similar comments apply to the exchange reaction $Cr(H_2O)_6^{3-} + Cr^{2+}$ [31, 310]. The chromium(II) reductions of the complexes I[475] and II[932] are outer-sphere, although attack at the hetero-nitrogen and carboxy-oxygen atoms might have been predicted.

$$Co(NH_3)_5-N\underset{}{\overset{}{\diagup}}NH \qquad (H_3N)_4Co\underset{CH_3}{\overset{NH_2}{\diagup\!\!\diagdown}}Co(NH_3)_4 \qquad (5.40)$$

$$\text{I} \qquad\qquad\qquad \text{II}$$

In many instances, an outer-sphere mechanism can confidently be assigned, simply because, from the structure of the reactants, there is apparently no way of forming a bridge. Bridging by the NH_3 molecule is generally ruled out on this basis, though in one case

$$Co(NH_3)_6^{3+} + Cr^{2+} + 6H^+ = Co^{2+} + Cr^{3+} + 6NH_4^+ \qquad (5.41)$$

it has also been demonstrated by experiment that no ammonia is transferred[318a].

5.2.3 Transfer of bridging group (labile reactants)

The advent of rapid reaction techniques made it possible to identify reaction products hitherto classed as 'labile'. Sutin and co-workers[248, 500] proved the existence of the bridge mechanism in the reactions

$$CoCl^{2+} + Fe^{2+} \rightarrow Co^{2+} + FeCl^{2+} \qquad (5.42)$$

$$Co(C_2O_4)_3^{3-} + Fe^{2+} \rightarrow Co^{2+} + 2C_2O_4^{2-} + FeC_2O_4^+ \qquad (5.43)$$

observing both formation and dissociation of the primary iron(III) product. Analogous experiments have demonstrated transfer from chromium(III) to vanadium(II)[1031], and from cobalt(III) to vanadium(II) in several instances[286, 380, 543, 860]. In the case of the reaction

$$Co(NH_3)_5OOCCOOH^{2+} + V^{2+} \xrightarrow{H^+} Co^{2+} + 5NH_4^+ + VC_2O_4^+ \qquad (5.44)$$

the product $VC_2O_4^+$ is thermodynamically stable, hence an alternative possible mechanism would be the two-stage process

$$Co(NH_3)_5OOCCOOH^{2+} + V^{2+} \xrightarrow{H^+}$$
$$Co^{2+} + 5NH_4^+ + V^{3+} + HC_2O_4^- \qquad (5.45)$$
$$V^{3+} + HC_2O_4^- \rightarrow VC_2O_4^+ + H^+ \qquad (5.46)$$

but from the known rate of reaction 5.46 it could be shown that the kinetics of formation of $VC_2O_4^+$ agree with the mechanism of equation (5.44)[860].

When the reaction products are inert but the reactants are labile, difficulties arise in establishing bridging or non-bridging mechanisms, but these too have been overcome in some cases with the aid of fast reaction techniques.

Taube and Myers[1016] had observed early on that the reaction between chromium(II) and iron(III) in solutions containing chloride ions produced a high yield of $Cr(H_2O)_5Cl^{2+}$ and it was tempting to suppose that this arose by transfer of the chloride ion from the chloroiron(III) complex which is known to be present

$$FeCl^{2+} + Cr^{2+} \rightarrow Fe^{2+} + CrCl^{2+} \tag{5.47}$$

At this stage, however, it was not possible to exclude the alternative, that the chloride ion might be a non-bridging ligand, bound initially to chromium(II)

$$Cr^{2+} + Cl^- \rightleftharpoons CrCl^+ \tag{5.48}$$

$$Fe^{3+} + CrCl^+ \rightarrow Fe^{2+} + CrCl^{2+} \tag{5.49}$$

The methods used to distinguish these mechanisms rely on the slowness of the reaction

$$Fe^{3+} + Cl \rightleftharpoons FeCl^{2+} \tag{5.50}$$

compared with the reaction (5.47). Ardon, Levitan and Taube[36] carried out a sequential mixing experiment at $-50°C$ (the solvent was a water–$HClO_4$ mixture). Solutions containing Fe^{3+} and Cl^- ion were pumped from separate vessels, through a mixing junction, into a third solution of Cr^{2+} ions. The temperature and pumping rate were such that no appreciable amount of $FeCl^{2+}$ was produced, yet the chromium(III) was mainly $CrCl^{2+}$ and the reaction was noticeably faster than the direct reaction

$$Fe^{3+} + Cr^{2+} = Fe^{2+} + Cr^{3+} \tag{5.51}$$

carried out in the absence of chloride ions. Hence, the reaction under these conditions was equation (5.49).

This does not exclude the possibility that the inner-sphere reaction (equation 5.47) might provide a still more favourable reaction pathway if the $FeCl^{2+}$ had time to form, and this was verified by Dulz and Sutin[348], by rapid mixing experiments in aqueous solution at room temperature. In one set of measurements Cr^{2+} and Cl^- were mixed, then combined with a solution containing Fe^{3+} and no chloride ion. The rate of reduction of iron(III) conformed to the equation

$$\text{Rate} = (k_1 + k_2 K_h/[H^+])[Fe^{3+}][Cr^{2+}] + k_5[Fe^{3+}][Cl^-][Cr^{2+}] + k_3[Fe^{3+}][Cl^-] \tag{5.52}$$

where $(k_1 + k_2 K_h/[H^+])$ agreed with the known rate constant for equation (5.51), and k_3 with the known rate constant for equation (5.50), (corresponding to a pathway $Fe^{3+} + Cl^- + Cr^{2+} \rightarrow FeCl^{2+} + Cr^{2+} \rightarrow Fe^{2+} + CrCl^{2+}$).

From the yield of chromium(II) product, the k_5 term corresponded to production of $CrCl^{2+}$, not Cr^{3+}. This term could thus only be attributed to the pathway with non-bridging chloride attached to chromium(II). In a second set of measurements, a solution containing Fe^{3+} and Cl^- at equilibrium was mixed with a solution containing Cr^{2+} and no Cl^- ion. The rate of reduction of iron(III) was such that no displacement of the equilibrium (equation 5.50) could occur during reaction, and the rate of disappearance of $FeCl^{2+}$ conformed to

$$\frac{-d[FeCl^{2+}]}{dt} = k_4[FeCl^{2+}][Cr^{2+}] \tag{5.53}$$

The chromium(III) product was entirely $CrCl^{2+}$ and thus the bridging mechanism (equation 5.47) was established.

5.2.4 Isomeric reaction products

When the reaction product can exist in more than one isomeric form, the structure actually produced may give information about the mechanism. In general, metal thiocyanato complexes have two possible structures, MNCS and MSCN, but most metal ions show a strong preference for either one or the other. The labile, blood-red iron(III) complex has the nitrogen-bonded structure (confirmed by comparison of the electronic spectrum with that of the purple species $FeSCNH^{3+}$ formed in strongly acid solution[852]). When the equilibrium mixture of iron(III) thiocyanate complexes reacts with chromium(II) the chromium(III) product is a green complex, which subsequently rearranges to the purple nitrogen-bonded species $CrNCS^{2+}$ [498]. Thus the green complex is identified as the sulphur-bonded form and the primary process is most probably

$$FeNCS^{2+} + Cr^{2+} \rightarrow Fe^{2+} + NCSCr^{2+} \tag{5.54}$$

A similar argument has been applied to reactions of $Co(edta)^{2-}$ with oxidising aquo ions. The reductant is labile and could exist in a five-coordinate or a six-coordinate form. It has been suggested that outer-sphere oxidation favours production of the six-coordinate $Co(edta)^-$ complex, while inner-sphere reductants lead to the five-coordinate $Co(edtaH)OH_2$ [352, 1098].

5.2.5 Substitution-controlled electron transfer

With certain reductants, notably V^{2+}, it has been found that the rates and activation parameters of electron transfer are similar to those of ligand replacement. The point is brought out best when the oxidant is varied in a systematic way. Thus Grossman and Haim[485] compared the rates of reduction of series of oxidants with vanadium(II) and with iron(II), and Chen and Gould[221] similarly compared vanadium(II) with chromium(II). It is concluded that the electron transfer reaction is preceded by a substitution process of the form

$$V(H_2O)_6^{2+} + NCSCo(NH_3)_5^{2+} \rightarrow [(H_2O)_5 VNCS Co(NH_3)_5^{4+}] + H_2O$$
$$\rightarrow (H_2O)_5 VNCS^{2+} + Co^{2+} + 5NH_3 \quad (5.55)$$

In some cases, this has also been confirmed, by identification of product.

A general review of oxidation reactions involving the $CoOH^{2+}$ ion has led to a similar conclusion[283]. For a series of reductants with a wide range of reducing power (Cl^-, HNO_2, H_2O_2, Br^-, SCN^-, $H_2C_2O_4$, HN_3, $N_2H_5^+$, NH_3OH^+), the rate and activation parameters are similar to each other and to the rate of water exchange[249], hence the common mechanism

$$Co(H_2O)_5 OH^{2+} + X \rightarrow Co(H_2O)_4(OH)X + H_2O$$
$$\rightarrow Co^{2+} + \text{oxidation products} \quad (5.56)$$

is proposed[283-285]. It is noted further[284] that where significant rate differences do occur, as in the comparison of reactions where X is HN_3, Br^-, NCS^-, analogous differences occur in rates of substitution of the same ligands on $FeOH^{2+}$: presumably, therefore, the same differences will occur in the substitution step of equation (5.56), consistent with the proposed mechanisms.

The possibility that $MnOH^{2+}$ reacts in the same way has been discussed, and has been used in turn to obtain estimates of the rate of water exchange in $Mn(H_2O)_5 OH^{2+}$ [281].

Similarly, the oxidation of Ru^{2+} by ClO_4^-

$$2Ru^{2+} + ClO_4^- + 2H^+ = 2Ru^{3+} + ClO_3^- + H_2O \quad (5.57)$$
$$\text{Rate} = k[Ru^{2+}][ClO_4^-] \quad (5.58)$$

is believed to be substitution-controlled, the rate constant k and activation parameters being similar to those of halide/anion reactions[606].

The converse of this argument is widely used to establish outer-sphere mechanisms. The example $MnO_4^- + MnO_4^{2-}$ has already been mentioned. Other reactions between complexes inert to substitution in both their oxidised and reduced forms, include $Fe(CN)_6^{4-} + Fe(CN)_6^{3-}$ [1030]; $Mo(CN)_8^{3-} + Mo(CN)_8^{4-}$ [188]; $W(CN)_8^{3-} + W(CN)_8^{4-}$ [56]; L-Os(bipy)$_3^{3+}$ + D-OS(bipy)$_3^{3+}$ [351]; and metal cluster complexes as in the reaction[250]

$$Ta_6 Cl_{12}^{4+} + Ta_6 Cl_{12}^{2+} \rightleftharpoons 2Ta_6 Cl_{12}^{3+} \quad (5.58)$$

The extension to less inert metal ions has been made possible by increased knowledge of the rates of substitution. Thus, in the reaction $Ru(NH_3)_5 Cl^{2+} + V^{2+}$, the chloride ion might be expected to furnish an effecting bridging group, but evidently it does not, since the rate is faster than expected for substitution at $V(H_2O)_6^{2+}$ [974].

5.2.6 Catalysis

As mentioned above (p.98) catalysis by halide ions was one of the first criteria of mechanism to be suggested, and since then various strong and specific catalytic effects which have been observed in electron transfer reactions have been taken as evidence for inner- or outer-sphere mechanisms. This argument is never conclusive, since catalysis in itself merely defines the composition of the transition state and not its structure.

The most firmly established case is catalysis by hydroxide ion[282, 987, 1073]. If a reaction between aquo metal ions, such as equation (5.1) which involves no net gain or loss of protons, is strongly base-catalysed, this seems to imply an inner-sphere reaction with hydroxide ion as bridge, and conversely, an uncatalysed rate law implies an outer-sphere mechanism. This criterion can sometimes be checked independently, and it holds good in every case. Thus, the reaction $Co(NH_3)_5OH_2^{3+} + Cr^{2+}$ is known to be inner-sphere with transfer of oxygen from cobalt to chromium[779], and is base-catalysed[1043] (see footnote (e), p. 144); while the reaction $Co(NH_3)_5OH_2^{3+} + Cr(bipy)_3^{2+}$ which is necessarily outer-sphere, is predominantly uncatalysed[1129]. The reaction $Co(NH_3)_5OH_2^{3+} + V^{2+}$, which might be expected to have base-catalysed pathways, is in fact entirely uncatalysed[80], and from kinetic isotope measurements seems to be outer-sphere[1130]. The reaction $Co^{3+} + Cr^{2+}$ has a specific rate close to the diffusion-controlled limit, which suggests an outer-sphere reaction, and the rate law contains an uncatalysed as well as a base-catalysed term[83]. When the form of the catalytic term implies a kinetically significant intermediate (see chapter 4), this strongly suggests an inner-sphere mechanism, since the formation and dissociation of an outer-sphere intermediate would certainly be diffusion-controlled.

Halpern and co-workers found an interesting example where catalysis implies the outer-spere mechanism[192]. Several reactions of the type $Co^{III}X + Co^{II}(CN)_5^{3-}$ (see equation 5.17) proceed with transfer of the bridging group and with the expected second-order rate law; others such as $Co(NH_3)_6^{3+} + Co^{II}(CN)_5^{3-}$ in which no bridging group is available are catalysed by cyanide ion and yield the product $Co(CN)_6^{3-}$. Evidently the complex $Co(CN)_6^{4-}$ is a more powerful outer-sphere reductant than $Co(CN)_5^{3-}$.

5.2.7 Linear free energy relationships (LFER)

It is often found that in a series of reactions the rates are a smooth function of one or more parameters related to the structure or stability of the reactants. Most commonly there is a linear relation between the activation free energy $\Delta G^{\ddagger}$ and the overall free energy change $\Delta G^{\ominus}$, and, for electron transfer reactions in particular, a linear relation between the values of $\Delta G^{\ddagger}$ or $\log k$ for the reactions of a series of oxidants with a given pair of reductants, or vice versa. The significance of these relationships has been explored in depth for both organic and inorganic substitution reactions[(f)]. For electron transfer reactions,

(f) References 106, 217, 356, 660, 1085

a basis is provided by the theories reviewed in chapter 6. Regardless of theory, however, empirical correlations can be used to draw conclusions about mechanisms. Studies of this kind are summarised in *Table 5.4*. Guenther and Linck[488] found a linear correlation between the (log) rates of reduction of various chloropentammine cobalt(III) complexes by the three reductants $Ru(NH_3)_6^{2+}$, V^{2+} and Fe^{2+}. The first of these three must react by outer-sphere mechanism in all the reactions considered. It does not follow that the other two reductants also react by outer-sphere mechanisms. What can be inferred is that, since all of the $Ru(NH_3)_6^{2+}$ reductions have the same mechanism, then to the extent that the LFER is valid, all the V^{2+} reactions likewise have a common mechanism, and similarly all of the Fe^{2+} have a common mechanism, be it outer-sphere or inner-sphere. In fact, several of the V^{2+} reactions are known, from their rapid rates, to be outer-sphere, and one of the iron(II) reactions is known, from detection of the product $FeCl^{2+}$, to be inner-sphere. Comparisons of the reactions of V^{2+} and Cr^{2+} with cobalt(III) oxidants have established that the ratio of rate constants $k_V/k_{Cr} \simeq 0.02$ is characteristic of outer-sphere mechanisms[579, 931, 1043]. This has been used as a criterion for the structural identification of cobalt(III) complexes[933].

Table 5.4 Summary of linear free energy relationships and similar rate comparisons

Reagents compared[a]	Factors discussed[b]	References
V^{2+}, Fe^{2+}, $Ru(NH_3)_6^{2+}$	i.s. and o.s.; effect of ΔG°	488
V^{2+}, $Cr(bipy)_3^{2+}$, $Ru(NH_3)_6^{2+}$	i.s. and o.s.	579
Cr^{2+}, V^{2+}, $Ru(NH_3)_6^{2+}$	critical assessment of LFER	891, 893
Cr^{2+}, Eu^{2+}, $Ru(NH_3)_6^{2+}$; V^{2+}	o.s. rates of unknown reactions	401
Cr^{2+}, V^{2+}	rate-ratio for o.s. mechanism	931, 932
		1043, 1090
Cr^{2+}, V^{2+}	i.s. (V^{2+}); steric factors	221
Cr^{2+}, V^{2+}	i.s. (Cr^{2+} + cobaloximes)	861
Cr^{2+}, V^{2+}	non-linear relationship	862
Eu^{2+}, V^{2+}		577
Cr^{2+}, Eu^{2+}	i.s. (Eu^{2+}); steric factors	400
Fe^{2+}, VO^{2+}	exhaustive review	904
Cr^{2+}, Cu^+	i.s. (Cu^+)	317
Fe^{3+}, $(Co(NH_3)_5)_2O_2^{5+}$ (oxidants)	mechanism of Fe^{2+}/Fe^{3+} exchange	511
Co^{3+}, Ag^{2+} (oxidants)		560
Ce^{IV}, Cr^{VI} (oxidants)	i.s. and o.s.	120
Ce^{IV}, V^V, Cr^{VI} (oxidants)	i.s. and o.s.	125
Cr^{III}, Co^{III} (oxidants)	Yb^{2+} reductant i.s. and o.s.	227

(a) The reagents listed in this column are the reductants, studied with a common sequence of oxidants, unless otherwise indicated
(b) i.s., inner-sphere mechanism; o.s., outer-sphere mechanism

If the series of oxidants is sufficiently long and varied, then the mere existence of a correlation does become, if not conclusive, at least a persuasive argument for a common mechanism. Thus, Gould et al.[221] have found correlations between the rates of reaction of V^{2+} and Cr^{2+} with various cobalt(III) complexes. The Cr^{2+} reactions are known to be inner-sphere, and it is concluded that the same is true of the V^{2+} reactions. Similar evidence points to inner-sphere mechanisms for Eu^{2+} reduction[400]. By extrapolating such correlations it is possible to estimate rates for hypothetical unobserved reaction paths, i.e. outer-sphere rates for systems which mainly react by the inner-sphere path or vice versa. In this way Fan and Gould[401] estimated the ratios of rates of outer-sphere to inner-sphere paths for Eu^{2+} and Cr^{2+} reductions of carboxylato cobalt(III) complexes; some of these ratios being as low as $10^{-6} M^{-1} s^{-1}$.

The two reductants need not necessarily be closely comparable in structure. In an exhaustive review of the reactions of Fe^{2+} and VO^{2+}, with a wide diversity of oxidants, Rosseinsky found good correlations of the rates for sequences of common mechanisms, and was able to draw a 'mechanistic boundary' on the plot of $\log k_V$ against $\log k_{Fe}$. Points lying above a certain line of unit slope corresponded to inner-sphere reactions, points below to outer-sphere reactions[904].

Possibly, however, the most reliable, and certainly the most interesting, applications of LFER arise when certain points deviate markedly from the expected sloping line. (Deviations due to substitution control in reactions of V^{2+} have already been mentioned.) In the reactions of various cobalt oxidants with $Ru(NH_3)_6^{2+}$ and V^{2+}, a good correlation is found when the vanadium(II) reactions are known to be outer-sphere, but several vanadium reactions known to be inner-sphere deviate markedly; others which deviate are, therefore, held to be inner-sphere as well. In this way, it is concluded that the reactions $Co(NH_3)_5 X^{2+} + V^{2+}$ are outer-sphere when X is Cl, Br, I, but inner-sphere when X is F [579].

A more sophisticated, multiple application of LFER has been offered by Hand, Hyde and Sykes[511]. Reaction rates are available for the two oxidants Fe^{3+} aq and $[Co(NH_3)_5]_2 O_2^{5+}$ with a series of five reductants $Fe^{2+}, Cr^{2+}, Eu^{2+}, V^{2+}$ and $Ru(NH_3)_6^{2+}$. Let us call the rate constants for the two oxidants k_{Fe} and k_S respectively. The plot of $\log k_S$ against $\log k_{Fe}$ is not linear, but in view of the small number of points it would be difficult to decide whether the deviation from linearity is due to some systematic effect leading to curvature, or to scatter. It is argued, however, that the appropriate plot is a straight line of slope 1.0 passing through or near to the points from Cr^{2+} to $Ru(NH_3)_6^{2+}$, and that the point for Fe^{2+} is aberrant. It is shown, first, that the rates of reaction of V^{2+} and Eu^{2+} with a common series of oxidants, and likewise of V^{2+} and $Ru(NH_3)_6^{2+}$, and V^{2+} and Cr^{2+}, all correlate on straight lines with slopes close to 1.0, and in each case some or all of the reactions are outer-sphere. Hence, in a plot of $\log k_S$ against $\log k_{Fe}$ the reductants $Cr^{2+}, Eu^{2+}, V^{2+}, Ru(NH_3)_6^{2+}$ should also correlate on a straight line with a slope of

$1.0^{(g)}$. The eight reactions so far mentioned are taken to be outer-sphere either from direct evidence or from the correlations. Presumably, therefore, the deviation of the Fe^{2+} point implies that one of the two reactions, $(S + Fe^{2+})$ or $(Fe^{3+} + Fe^{2+})$ has some other mechanism. The deviation is such that the reaction $Fe^{3+} + Fe^{2+}$ is more rapid than expected, and it is suggested that this reaction may proceed by a more favourable inner-sphere pathway.

Finally, it should be noted that if the two series of reactions cover a very wide range of ΔG° values, systematic curvature is expected even when the mechanism is the same throughout both series. This point is discussed more fully in the next chapter (p. 213).

5.2.8 Other rate comparisons: bridging groups

The equation

$$A^+.X + B \rightarrow A + X.B^+ \tag{5.60}$$

denotes an inner-sphere reaction with transfer of the bridging group from oxidant to reductant; and the equation

$$A^+.X + B \rightarrow A.X + B^+ \tag{5.61}$$

denotes the alternative outer-sphere process involving the same reactants. As noted above, other things being equal, the rates of reaction tend to correlate with the thermodynamic driving force. Thus, if the group X forms a particularly stable complex with B^+, in preference to A, a bridged mechanism will be favoured; conversely, if, in a series of suitably comparable systems, rapid rates are observed whenever the bridging group specifically favours B^+ rather than A, this is evidence of a bridging mechanism.

(a) Azide and thiocyanate ions

$$CrNNN^{2+} + Cr^{2+} \rightarrow Cr^{2+} + CrNNN^{2+} \tag{5.62}$$

$$CrNCS^{2+} + Cr^{2+} \rightarrow Cr^{2+} + CrSCN^{2+} \tag{5.63}$$

$$CrNCS^{2+} + Cr^{2+} \rightarrow Cr^{2+} + CrNCS^{2+} \tag{5.64}$$

Equation (5.62) is a symmetrical exchange with $\Delta G^\circ = 0$, but equation (5.63) which assumes remote attack, leads to the unstable S-bonded complex. Alternatively, equation (5.64), with adjacent attack, is symmetrical, but presumably slower than equation (5.62), since the metal atoms must be

---

(g) If the correlations for pairs of reductants V^{2+}, Eu^{2+}, V^{2+}, Cr^{2+}, etc. were *exact*, the correlations for pairs of oxidants would be exact also. Consider for example six reactions involving the oxidants A_1^+, A_2^+ and reductants B_1, B_2, B_3. It is easily shown that if the pairs of reactions $(A_1^+ + B_1)$, $(A_2^+ + B_2)$ correlate with the pairs $(A_2^+ + B_1)$ and $(A_2^+ + B_2)$ and also with the pairs $(A_1^+ + B_3)$, $(A_2^+ + B_3)$ then the series $(A_1^+ + B_1)$, $(A_1^+ + B_2)$, $(A_1^+ + B_3)$ correlates with the series $(A_2^+ + B_1)$, $(A_2^+ + B_2)$, $(A_2^+ + B_3)$.

brought closer together. Thus, both equations (5.63) and (5.64) are expected to be slower than equation (5.62), and this is observed[69]. On the other hand, the reactions $Co(NH_3)_5N_3^{2+} + Cr(bipy)_3^{2+}$ and $Co(NH_3)_5NCS^{2+} + Cr(bipy)_3^{2+}$, which are necessarily outer-sphere, are comparable in rate. It has, therefore, been argued that the ratio of rates, $r = k_{NNN}/k_{NCS}$ is a criterion for the mechanism[377, 987]. Provided that the reductant is of the type which, when oxidised, bonds nitrogen in preference to sulphur, a value $r \simeq 1$ implies outer-sphere mechanisms for both systems, and a value $r \gg 1$ implies the contrary. The latter may mean that both azide and thiocyanate systems react by the inner-sphere path, or that azide is inner-sphere and thiocyanate outer-sphere[543].

A similar argument has been founded on the differences in the rates of reduction of N-bonded and S-bonded thiocyanato cobalt(III) complexes[9], but here the difference in thermodynamic stability of the two oxidants is a complicating factor. The rate order observed with reductants Fe^{2+}, Cr^{2+} and $Co(CN)_5^{3-}$ is $Co(NH_3)_5SCN^{2+} > Co(NH_3)_5NCS^{2+}$, but the same order might be expected with outer-sphere reductants.

(b) Halide ions

It is well known that most metal ions exhibit a marked trend in affinity for the halide ions[21], and the terms 'hard' and 'soft' have been used[834] to denote cations which bond preferentially in the order $F > Cl > Br > I$ and vice versa. From equation (5.61), it is clear that for an outer-sphere mechanism, if the oxidant A^+ is 'harder' than the corresponding reduced form A, then rates with a common reductant B will fall in the sequence $I > Br > Cl > F$. Most metal ions are indeed 'harder' in the higher valencies than in the lower[21], and the sequence just mentioned is the one commonly observed[497]. Since the same sequence is observed in a number of systems which are unquestionably bridged (e.g. $CrX^{2+} + Cr^{2+}$, $Co(NH_3)_5X^{2+} + Cr^{2+}$) nothing so far can be deduced about the mechanism. When the reverse sequence, $I < Br < Cl < F$, is found, however, this can reasonably be attributed to a bridging mechanism. The preference for F^- as bridge reflects the fact that the oxidised ion B^+ is a harder acid than B. The reversed sequence has in fact been found in a number of systems, which are discussed more fully on p. 257.

(c) Chelating ligands

There have been many studies of systems involving multidentate organic groups as bridging ligands[180, 733, 1022]. With chromium(II) as reductant, there is abundant evidence, from structure analysis of the product, of inner-sphere reactions. With more labile reductants, evidence comes from the comparison of rates between systems with and without the chelating function, as, for example, the lactato and propionato complexes, $Co(NH_3)_5OOC(CHOH)CH_3^{2+}$, $Co(NH_3)_5OOCC_2H_5^{2+}$. The ratio of rates of reduction $k(\text{lactate})/k(\text{propionate})$ increases[733] according to the reducing metal ion, $Cu^+ \ll Cr^{2+} < Eu^{2+} \ll TiOH^{2+}$ and this is the expected sequence based on the increasing stability of chelated transition state or product.

5.2.9 Other criteria

(a) Volume of activation

Candlin and Halpern[191] suggested that there might be a systematic difference in $\Delta V^\ddagger$ between inner- and outer-sphere reactions. An inner-sphere mechanism $A^+ \cdot X + B(H_2O)_n$ involves the release of a water molecule and, since the molecular volume of free water is greater than that of the coordinated value, $\Delta V^\ddagger$ should be more positive for inner-sphere than for outer-sphere reactions, other things being equal. Stranks[970a] has put forward quantitative predictions for a number of reactions, and experimental work is now in progress. Measurements of a number of known or probable inner-sphere reactions, such as $CrOH^{2+} + Cr^{2+}$ [970a] or $Co(NH_3)_5Cl^{2+} + Cr^{2+}$ [191], have yielded positive values for $\Delta V^\ddagger$, whereas the outer-sphere reaction $Co(en)_3^{3+} + Co(en)_3^{2+}$ gives a negative value[970a].

(b) Solvent isotope effect

With the V^{2+}.aq ion as reductant, comparisons of rates in H_2O and D_2O media have been used to differentiate mechanisms. Rate ratios $k(H_2O)/k(D_2O)$ appear to be higher for outer-sphere[1130] than for inner-sphere reactions[860]. This does not apply to chromium(II) reductions, however[1130].

(c) Bridge isotope effect

Taube and co-workers[313, 779] compared the rates of reduction of $Co(NH_3)_5OH_2^{3+}$ (and/or its deprotonated form $Co(NH_3)_5OH^{2+}$) containing ^{16}O and ^{18}O isotopes, with various reductants. The inner-sphere reaction of $Co(NH_3)_5OH^{2+}$ with Cr^{2+} was characterised by a higher fractionation factor (fractionation factor $f = d \ln[^{16}O]/d \ln[^{18}O]$) than the outer-sphere reaction with $Ru(NH_3)_6^{2+}$, and this is consistent with a specific stretching of the Co–O bond in the bridged activated complex. Comparisons of fractionation factors were used to assign outer-sphere mechanisms to the reactions of $Co(NH_3)_5OH_2^{3+}$ with V^{2+} and Eu^{2+}. Reactions with $Cu(NH_3)_n^+$, $Eu(edta)^{2-}$, $Fe(edta)^{2-}$ and $Cr(edta)^{2-}$ showed no appreciable isotope fractionation. For the copper complex, the rate is high enough to suggest diffusion control, but with the other reductants some other mechanism must be invoked, such as control by substitution at the reductant – again, a bridged mechanism[313].

(d) Activation entropy

Since the difference of mechanism is a difference of geometry and of the number of solvent molecules in the transition state, it might be expected that inner- and outer-sphere mechanisms should show characteristically different entropies of activation. The factors affecting entropies of activation have been discussed by several authors[505, 506, 545, 808] but it does not appear that there is any simple correlation with mechanism. Rather, $\Delta S^{\ddagger}$ is determined primarily by the total charge on the activated complex, whether bridged or unbridged

5.3 'Remote attack': experimental evidence

Continuing the plan of this chapter, the experimental evidence for remote attack will be reviewed according the nature of the arguments involved. There is not space to discuss all the well-authenticated examples, but they are listed in the tables. More details of the mechanisms, and especially the role of the bridging groups, will be discussed in chapter 7.

5.3.1 Inert binuclear complexes

The most direct proof of remote attack is the isolation of a binuclear precursor or successor complex. This approach has already been discussed and the known examples are listed in *Tables 5.2 and 5.3*. All that need be added here is that for remote attack the determination of the structure of products is not always straightforward. In some cases the structure may reasonably be assumed, simply because it is difficult to envisage any alternative, as in $(NC)_5FeCNCo(CN)_5^{6-}$. In other cases, e.g. $(NH_3)_5RuOC(OCH_3)Cr(OH_2)_5^{3+}$, there are good arguments but not direct proof.

5.3.2 Transfer of bridging group

(a) Inorganic systems

Examples of electron transfer through small polyatomic bridges are summarised in *Table 5.5*. Bridging through the cyano group is proved by the detection of an unstable intermediate which undergoes intramolecular rearrangement (in the following equations, the atom underlined is the one bonded to the metal).

$$Co(NH_3)_5\underline{C}N^{2+} + Co(CN)_5^{3-} \to \underline{C}NCo(CN)_5^{3-} \to N\underline{C}Co(CN)_5^{3-} \quad (5.65)$$

$$Co(NH_3)_5\underline{C}N^{2+} + Cr^{2+} \to Cr\underline{N}C \to Cr\underline{C}N^{2+} \quad (5.66)$$

The rearrangement step of equation (5.65) has both a spontaneous and a chromium(II)-catalysed pathway, the latter attributable to a second electron transfer step with remote attack:

$$Cr\underline{N}C^{2+} + Cr^{2+} \to Cr^{2+} + N\underline{C}Cr^{2+} \quad (5.67)$$

The N-bonded intermediates are identified partly by their UV spectra, but again mainly on the argument that it is difficult to envisage any other reasonable structure.

Table 5.5 Examples of remote attack (small inorganic ligands)

System	Remarks	References
$(NH_3)_5Co-N{\overset{O}{\underset{O}{\diagup}}}Co(CN)_5^{2-}$	Primary product detected (rearranges to form $Co(CN)_5NO_2^{3-}$)	508
$(NH_3)_5Co-C{\equiv}N\ Co(CN)_5^{2-}$	Primary product detected (rearranges to form $Co(CN)_6^{3-}$)	508
$(NH_3)_5Co-O{\underset{O}{\overset{\diagdown}{\diagup}}}N\ Co(CN)_5^{3-}$	Primary product identified as $Co(CN)_5NO_2^{3-}$	508
$(NH_3)_5CoNCS\ Co(CN)_5^{3-}$	Primary product identified as $Co(CN)_5SCN^{3-}$	508
$Co(NH_3)_5CN\ Cr^{2+}$	Primary product $CrNC^{2+}$ detected (rearranges to form $CrCN^{2+}$)	384
$FeNCS^{2+} + Cr^{2+}$	Primary product $CrSCN^{2+}$ detected (rearranges to form $CrNCS^{2+}$)	498
$CrSCN^{2+}\ Cr^{2+}$	Evidence of simultaneous remote attack (leading to $CrNCS^{2+}$) and adjacent attack (leading to Cr exchange)	498, 499
$Co(NH_3)_5SCN^{2+} + Cr^{2+}$		498, 499
$Co(NH_3)_5SSO_3^+ + Cr^{2+}$	No evidence of intermediate sulphur-bonded chromium(III) product	882
$Co(NH_3)_5O(CO)CH_3^{2+} + Cr^{2+}$	Indirect evidence. Arguments summarised in the references cited. See also *Table 5.3*, entry 17.	932

The proposed structure of the transition state is supported by the discovery of the analogous dicobalt(III) complex $(H_3N)_5CoNCCo(CN)_5$[1071] and the linkage isomer $(H_3N)_5CoCNCo(CN)_5$[297, 443]. Several other CN-bridged complexes have been characterised[211, 603].

The same arguments apply to the reaction of the nitro complex

$$Co(NH_3)_5\underline{N}O_2^{2+} + Co(CN)_5^{3-} \rightarrow (NC)_5Co\underline{O}NO^{3-} \rightarrow (NC)_5Co\underline{N}O_2^{2-} \quad (5.68)$$

and in addition there is the contrasting behaviour of the nitrite isomer

$$Co(NH_3)_5\underline{O}NO^{2+} + Co(CN)_5^{3-} \rightarrow (NC)_5Co\underline{N}O_2^{3-} \quad (5.69)$$

The former leads to the unstable intermediate which then rearranges at a measurable rate, but the latter proceeds directly to the thermodynamically stable product. Both observations are consistent with remote attack.

Haim and Sutin[498a, b] showed that isomerisation of the green complex $CrSCN^{2+}$ is catalysed by chromium(II), implying transfer across a three-atom chain, i.e.

$$Cr\underline{S}CN^{2+} + Cr^{2+} \rightarrow Cr^{2+} + SC\underline{N}Cr^{2+} \quad (5.70)$$

Two forms of adjacent attack are possible, and in addition, the reaction has a base-catalysed pathway for which it has been suggested that the transition state contains a six-membered ring structure. A summary of all the known and proposed reactions of the system $(Cr^{3+} + Cr^{2+} + NCS^-)$ is shown in *Figure 5.1*. A number of binuclear complexes with bridging thiocyanate ion have been prepared[298], including the linkage isomers $(H_3N)_5CoNCSCo(CN)_5$ and $(H_3N)_5CoSCNCo(CN)_5$[172].

With azide ion as bridge remote attack has generally been assumed[403, 960, 961, 1095] but there is no proof of this. In the reaction cis-$Cr(H_2O)_4(N_3)_2^+ + Cr^{2+}$ both azide ions bridge simultaneously. Snellgrove and King[960] have suggested the transition state III.

$$\left[\begin{array}{c} \diagup | \diagdown \\ \diagup Cr \diagdown \\ \diagup | \diagdown \end{array} \begin{array}{c} N\!\!=\!\!N\!\!=\!\!N \\ \\ N\!\!=\!\!N\!\!=\!\!N \end{array} \begin{array}{c} \diagup | \diagdown \\ \diagup Cr \diagdown \\ \diagup | \diagdown \end{array} \right]^{3+}$$

III

but the alternative structure

$$\left[\begin{array}{c} \diagup | \diagdown \\ \diagup Cr \diagdown \\ \diagup | \diagdown \end{array} \begin{array}{c} N\!\!\equiv\!\!N \\ \| \\ N \\ \| \\ N \\ \| \\ N\!\!\equiv\!\!N \end{array} \begin{array}{c} \diagup | \diagdown \\ \diagup Cr \diagdown \\ \diagup | \diagdown \end{array} \right]^{3+}$$

IV

is possible. From x-ray structural studies, examples of both types of bridging, M–NNN–M and M_2NNN, are known[20, 1123].

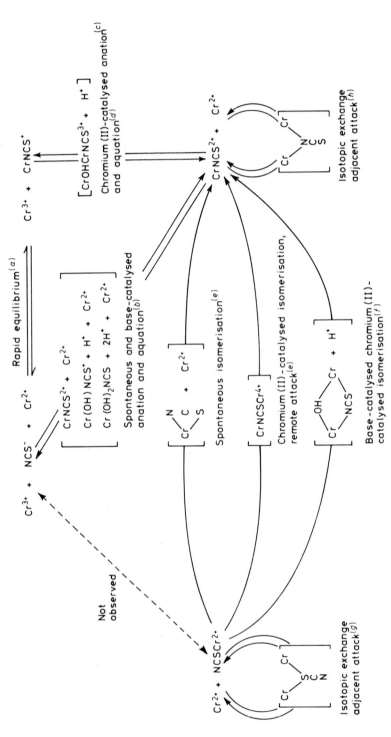

Figure 5.1 Observed and proposed reaction pathways in the system $Cr^{3+} + SCN^- + Cr^{2+}$. (a) Reference 1115, mechanism unknown; (b) Reference 854; (c) Reference 564, semi-quantitative observations; (d) not reported, but proposed by analogy with Cr^{2+}-catalysed aquations of CrX^{2+}, $X = N_3$ (Reference 342), Cl, Br, I (References 841, 14, 12, respectively); (e) References 498, 499; (f) Reference 499; (g) not reported but proposed by analogy with the $Co(NH_3)_5SCN^{2+} + Cr^{2+}$ reaction[943]; (h) Reference 499—but against this, see Reference 1095

(b) Organic systems

The results of a comprehensive study by Nordmeyer and Taube[810, 811] of the chromium(II) reduction of some nicotinamide and iso-nicotinamide complexes are summarised in the reactions below ($Ro = Co(NH_3)_5$):

$$Ro-N\overset{\frown}{\underset{\smile}{}}\text{(V)} + Cr^{2+} \longrightarrow Co^{2+} \quad (5.71)$$

$$\text{VI} \xrightarrow{} \text{VII} + Cr^{2+} \longrightarrow Co^{2+} + \text{[HN-py-CONH}_2\text{]} + Cr^{3+} \quad (5.72)$$

$$\text{VIII} \longrightarrow \text{IX, X} \quad (5.73)$$

Reduction of the nicotinamide complex VI by chromium(II) yielded two products identifiable as $Cr(H_2O)_6^{3+}$ and a new complex, VII, capable of being isolated by ion exchange methods, which in turn dissociated slowly into $Cr(H_2O)_6^{3+}$ and free nicotinamide. Visible and UV spectroscopy established that complex VII is a 1:1 nicotinamide–chromium(III) complex: and two further lines of evidence established the point of attachment of the chromium as the amide oxygen rather than the ring nitrogen: the changes in the infra-red spectrum of the carbonyl group, and the overall charge greater than +3 (from the ion-exchange behaviour) consistent with protonation as shown.

Experiments with the iso-nicotinamide complex VIII gave evidence of two successive reactions by remote attack. Reduction by chromium(II) yielded two primary products, complexes IX and X in proportions dependent on the initial chromium(II) concentration, and on the acid concentration. By working with high chromium(II) and high acid concentrations, complex IX could be isolated. Visible infra-red spectra, and ion-exchange behaviour all confirmed the amide-bonded structure, analogous to complex VI. The other product, complex X, favoured at low acid concentration, had the infra-red spectrum characteristic of a free amide group, and a visible spectrum consistent with the $Cr^{III}O_5N$ chromophore. In separate experiments the reversible isomerisation reactions between complexes IX and X were shown to be chromium(II)-catalysed (equation 5.73). Evidently both steps VIII → IX and IX → X proceed by remote attack, the latter being a bridged chromium(II)–chromium(III) exchange reaction, detectable only because of the asymmetry of the ligand.

The pyridine complex V was studied for comparison with these results. The chromium(III) product was exclusively $Cr(H_2O)_6^{3+}$ and a careful spectrophotometric study gave no evidence of, for example, $Cr(H_2O)_5py^{2+}$ as product or intermediate. This confirms the expected outer-sphere mechanism; by analogy with this, the direct production of $Cr(H_2O)_6^{3+}$ from the nicotinamide complex VI is presumed to be outer-sphere. Presumably an outer-sphere path operates also in the the reduction of complex VIII; no $Cr(H_2O)_6^{3+}$ could be detected, but this may be explained by the much faster rate of reduction of complex VIII than of complex VI.

5.3.3 Primary products inferred from decomposition products

The chromium(II) reduction of complex XII is believed[5,6,7] to proceed in two successive remote attack steps, i.e.

$$\text{Co(NH}_3)_5\text{OOC}-\text{CH=CH}-\text{C(OCH}_3)=\text{O} \xrightarrow{Cr^{2+}} \text{HOOC}-\text{CH=CH}-\text{C(OCH}_3)=\text{OCr}^{3+} \xrightarrow{Cr^{2+}} \text{CrOOC}-\text{CH=CH}-\text{C(OCH}_3)=\text{O}$$

XI   XII   XIII

$$\downarrow \text{(slow)}$$

$$\text{HOOC}-\text{CH=CH}-\text{COOCH}_3 + Cr^{3+}$$

XIV   (5.74)

The structure of the complex XIII is assigned on the basis of the rate of the slow aquation process XIII → XIV, which is characteristic of carboxylato chromium(III) complexes. The presence of the intermediate complex XII, is implied by product analysis studies. When complex XI reacts with chromium(II) a fraction of the product consists of $Cr(H_2O)_6^{3+}$ and a free half-ester, XIV. This fraction is too great to be accounted for by the aquation of complex XIII and it varies with chromium(II) concentration. When chromium(II) is added slowly to complex XI and each portion is allowed to react before the addition of the next, the yield of free ester is greater than when all the chromium(II) is added at once. This is consistent with the above mechanism, in which the spontaneous aquation XII → XIV competes with the chromium(II)-catalysed isomerisation XII → XIII.

A similar mechanism has been proposed for the reduction of the amidoester complex XV. (This compound is made by the reaction of methylamidofumarate and $Co(NH_3)_5OH_2^{3+}$; the evidence for the structure shown is that the rate of aquation is characteristic of an amide-bound rather than an ester-bound cobalt complex.) Reduction by chromium(II) yields $Cr(H_2O)_6^{3+}$ as the only detectable chromium(III) product, yet the rate is greatly in excess of that expected for an outer-sphere reaction. Adjacent attack at the amide group is also ruled out, since this would lead to an amide-bound chromium(III) complex, which would not aquate to $Cr(H_2O)_6^{3+}$ within the time-scale of the experiments. Remote attack, however, leads to the ester-bound chromium(III) complex XVII which is expected to be labile:

$$\text{XV} \xrightarrow{Cr^{2+}} \text{XVI} \longrightarrow H_2NCO\text{-CH=CH-}COOCH_3 + Cr^{3+} \quad (5.75)$$

These and other examples of remote attack are summarised in *Table 5.6*.

5.3.4 Other evidence

Several systems have been reported in which there is no mechanistic evidence other than the composition of the transition state, yet only the remote mechanism seems feasible. The N-bonded isomer of glycinato pentaammine cobalt(III) reacts with chromium(II) by the rate law[998].

$$\text{Rate} = k[Co(NH_3)_5NH_2CH_2COOH^{3+}][Cr^{2+}][H^+]^{-1} \quad (5.76)$$

The chromium(III) product is chelated, but a chelated transition state seems impossible: presumably electron transfer is followed by relatively rapid ring closure. The rate is greater than that expected for an outer-sphere reaction and the most reasonable transition state is therefore $Co(NH_3)_5NH_2CH_2COOCr^{4+}$.

5.4 Outer-sphere mechanisms

When an outer-sphere mechanism is identified by one of the tests described in section 5.2, the most that can be said with certainty is that during the reaction event, the *inner* coordination spheres of the reactants remain undisturbed. This still leaves open the possibility of structural variations in the outer coordination spheres, and in fact several different varieties of outer-sphere mechanism have been proposed. These are summarised in *Table 5.7*. In all cases the argument turns only on a comparison, or implied comparison, of rates of reaction.

Table 5.6 Examples of remote attack (organic ligands)

Reaction (showing point of attack by labile reactant)	Evidence	References
Co(NH$_3$)$_5$OOC–CH=CH–C(OCH$_3$)=O → Cr^{2+}	Indirect (see text, p. 160). Primary product aquates to give Cr(H$_2$O)$_6^{3+}$	567
Co(NH$_3$)$_5$O–C(OCH$_3$)=CH–CH=C(NH$_2$)–O → Cr^{2+}	Indirect (see text, p. 161). Primary product aquates to give Cr(H$_2$O)$_6^{3+}$	567
Co(NH$_3$)$_5$–OOC–CH=CH–C(NH$_2$)=O → Cr^{2+}	No direct evidence. Observed product is (CrOOC–CH=CH–CONH$_2$)$^{2+}$. This could arise by two successive remote attacks	567
Co(NH$_3$)$_5$OOC–C$_6$H$_4$–CHO → Cr^{2+}	Indirect. Primary product assumed, aquates to give Cr(H$_2$O)$_6^{3+}$	1119
Co(NH$_3$)$_5$OOC–C$_6$H$_4$–CHO → Cr^{2+}	Primary product observed	1119
Co(NH$_3$)$_5$OOC–C$_6$H$_4$–CH=CH–CHO → Cr^{2+}	Primary product observed	479
Co(NH$_3$)$_5$OOC–C$_6$H$_4$–COC$_6$H$_5$ → Cr^{2+}	Indirect. Primary product rapidly aquates to give Cr(H$_2$O)$_6^{3+}$	480

Structure	Notes	Ref.
Co(NH$_3$)$_5$OOC—⟨py⟩—N ← Cr^{2+}	Rate comparisons. Primary product not observed: presumed to undergo a second remote attack	480
[CrN⟨py⟩—COO]$^{2+}$ ← Cr^{2+}		
COOCo(NH$_3$)$_5$ on quinoline N ← Cr^{2+}	N-bonded chromium(III) product	475
Co(NH$_3$)$_5$—N⟨py⟩—C(NH$_2$)=O ← Cr^{2+}	Primary product observed (see text, p. 159)	810, 811
Co(NH$_3$)$_5$—N⟨py⟩—C(NH$_2$)=O ← Cr^{2+} (para)	Primary product observed (see text, p. 159)	810, 811
Cr^{III}—N⟨py⟩—C(R)(=O)—C(NH$_2$)=O ← Cr^{2+}	Primary product observed (see text, p. 159)	810, 811
Co(NH$_3$)$_5$—N⟨py⟩—C(R)=O ← Cr^{2+}	R is CH$_3$ or C$_6$H$_5$. Indirect (rapid rate). Primary product aquates to Cr^{3+}	477
Co(NH$_3$)$_5$—N⟨py⟩—C(C$_6$H$_5$)=O ← Cr^{2+}	Indirect (rapid rate). Primary product aquates to Cr^{3+}	477
Co(NH$_3$)$_5$NC—⟨C$_6$H$_4$⟩—C(CH$_3$)=O ← Cr^{2+}	Primary product observed	68

Table 5.6 continued

Reaction (showing point of attack by labile reactant)	Evidence	References
Co(NH$_3$)$_5$OOC–[pyridine]–Cr^{2+} (4-position)	Indirect (rate comparisons)	480
Co(NH$_3$)$_5$OOC–[pyridine]–Cr^{2+} (3-position)	Indirect (rate comparisons)	480
Co^{III}(tetrasulphoporphinate)	Chromium(III)–cobalt(II) binuclear product identified	417
[Co(NH$_3$)$_3$(HO)$_2$ and Co(NH$_3$)$_3$ bridged benzoate], COO→Cr^{2+}	Final chromium(III) product is binuclear	1036
[Co(NH$_3$)$_4$(H$_2$N) and Co(NH$_3$)$_4$ bridged], COO→Cr^{2+}	Final chromium(III) product is binuclear	580
[Co(NH$_3$)$_4$(H$_2$N) and Co(NH$_3$)$_4$ bridged], COO$^-$→Cr^{2+}	Final chromium(III) product is binuclear	580
[Co(NH$_3$)$_3$O(HO)$_2$ and Co(NH$_3$)$_3$O bridged benzaldehyde], CHO→Cr^{2+}	Indirect (rapid rate). Primary product aquates to Cr^{3+}	581

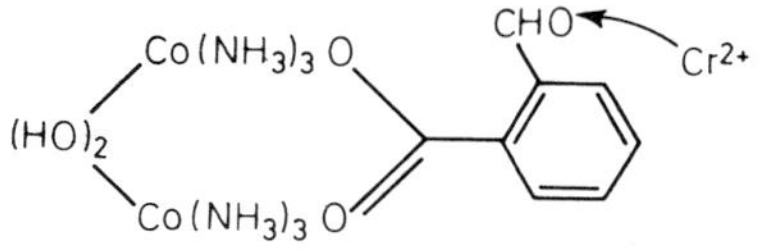	Indirect (rapid rate). Primary product equates to Cr^{3+}	581
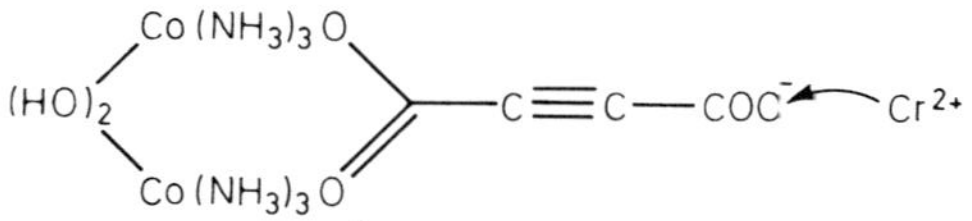	Final chromium(III) product is binuclear	540
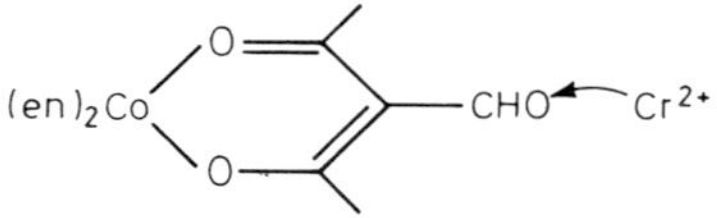	Indirect (rapid rate). Primary product undergoes ring closure	67
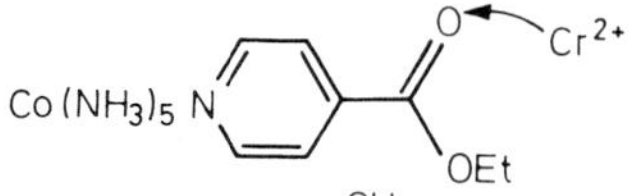	Indirect (rapid rate compared with Eu^{2+}-reduction)	318
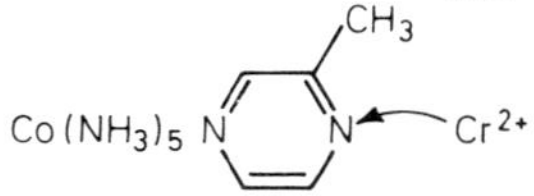	Indirect (rapid rate). Chromium(III) product is N-bonded	318
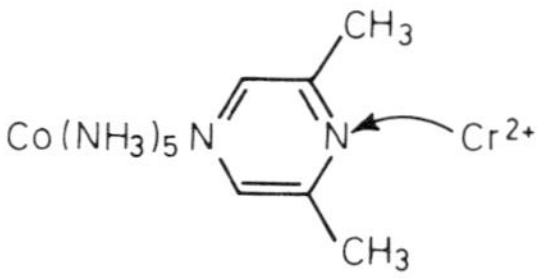	Indirect (rapid rate). Chromium(III) product is N-bonded	318
$Co(NH_3)_5NC_5H_4COOH^{2+}$ + V^{2+}	Indirect. (Kinetic evidence of inner-spere mechanism)	813
+ Eu^{2+}		
$Co(NH_3)_5NH_2CH_2COO^-$ + Cr^{2+}	Indirect (see text p. 161)	998

Table 5.7 Subdivisions of the outer-sphere electron transfer mechanism, with proposed examples

(For references and discussion, see text)

Mechanism		Examples
Outer-sphere bridging	(A⁺) (X) (B)	MnO_4^-, $Cs^+\ MnO_4^{2-}$
Specific orientation	(A⁺.X) (B)	$Co(NH_3)_5OOC\text{-}C_6H_4\text{-}NO_2 + CO_2^-$
Facial attack	A⁺.X ···▶ B	$O_2 + Ru(NH_3)_5py^{2+}$

5.4.1 Outer-sphere bridging

Some outer-sphere reactions show strong catalysis by a third substance which itself is also confined to the outer coordination sphere. Representative examples are shown in *Table 5.8*. The reaction $MnO_4^- + MnO_4^{2-}$ is so sensitive to the nature of the cations present that it is questionable whether a specific rate can be assigned to the uncatalysed reaction at all. Rate constants defined by Rate = $k[MnO_4^-][MnO_4^{2-}]$ (*see* equation 5.78) increase in the order $Li^+ \simeq Na^+ < K^+ < Cs^+ \ll Co(NH_3)_6^{3+}$. The effects of salts on interionic reactions in general have been extensively studied, and have been distinguished as 'primary' effects (anions increasing cation–cation reactivity, or cations increasing anion–anion reactivity), and 'secondary' effects (cations decreasing cation–cation reactivity or anions decreasing anion–anion reactivity)[445]. The effects referred to here come into the former category, and when secondary effects are controlled or eliminated they can be interpreted in terms of transition states of definite composition:

$$A^+ + X + B = [A^+.X.B]^\ddagger \rightarrow A + X + B^+ \qquad (5.77)$$

$$\text{Rate} = k[A^+][B][X] \qquad (5.78)$$

Thus, in the reaction $MnO_4^- + MnO_4^{2-}$, with varying concentration of Cs^+ ion, and with Na^+ to maintain the ionic strength, there is a third-order rate term $k[MnO_4^-][MnO_4^{2-}][Cs^+]$[465]. Similarly, in the reaction $Fe(CN)_6^{3-} + Fe(CN)_6^{4-}$, the rate law requires transition states containing one and two cations per reactant pair. On the other hand, in the reaction $Co(en)_3^{3+} + Co(en)_3^{2+}$, secondary but not primary salt effects have been observed: substitution of Ba^{2+} ion for Na^+, at constant Cl^- concentration, decreased the reaction rate, but substitution of ClO_3^- for Cl^-, at constant cation concentration, had no effect[667].

Table 5.8 Catalytic effects in outer-sphere electron transfer reactions

Oxidant	Reductant	Catalyst	References
MnO_4^-	MnO_4^{2-}	$K^+ < Cs^+ < Co(NH_3)_6^{3+}$	946
MnO_4^-	MnO_4^{2-}	Cs^+	465
$Fe(CN)_6^{3-}$	$Fe(CN)_6^{4-}$	$(n\text{-}C_5H_{11})_4N^+ < (n\text{-}C_4H_9)_4N^+ <$ $(C_6H_5)_4As^+ < (n\text{-}C_3H_7)_4N^+ < K^+ <$ $(C_2H_5)_4N^+ < Co(C_5H_5)_2^+ < (CH_3)_4N^+$	187
$Fe(CN)_6^{3-}$	$Fe(CN)_6^{4-}$	$H^+ \ll Li^+ < Na^+ < K^+ < Rb^+ < NH_4^+$ $Mg^{2+} < Ca^{2+} < Sr^{2+}$	948
$Co(OOCCH_2COO)_3^{3-}$	$Fe(CN)_6^{4-}$	Na^+, K^+	976
L_2PtX_4	$Fe(CN)_6^{4-}$	K^+	78[a]
(carbazole-type ligand structure)			1099
$Fe(phen)_3^{3+}$	Fe^{2+}	$Br^- < Cl^- \ll I^- < N_3^- \lesssim SCN^-$	990
$Fe(bipy)_3^{3+}$	Fe^{2+}	$Cl^- \ll SCN^-$	990
$Co(phen)_3^{3+}$	Fe^{2+}, Cr^{2+}, V^{2+}	$Cl^- < SCN^-$	864
$Co(NH_3)_6^{3+}, Co(en)_3^{3+}$	Cr^{2+}, V^{2+}	$Cl^- < SCN^-$	864
$Co(NH_3)_6^{3+}$	Cr^{2+}	$Cl^- > Br^-$	719
Cytochrome-c^{III}	Cr^{2+}	$Cl^- < I^- < N_3^- \simeq SCN^-$	1114
Fe^{3+}	Tl^{2+}	2- and 4-aminopyridine	1046[b]
Fe^{3+}	Tl^{2+}	2-hydroxypyridine	597[b]
$Co(en)_3^{3+}$	$Cr(bipy)_3^{2+}$	4,4'-bipyridyl	1048[c]

(a) X is Cl, Br; L is NH_3 or primary amine
(b) Tl^{2+} generated as intermediate in the reaction $Tl^{3+} + 2Fe^{2+} = Tl^+ + 2Fe^{3+}$
(c) Also other cobalt(III) oxidants

It is suggested that the counter-ions play a definite structural role. Electrostatically the most favourable configuration would have the three centres in a straight line, but it is always possible that a triangular arrangement, with the A^+ and B centres closer together would be more favourable for electron transfer [h].

The reaction $Fe(phen)_3^{3+} + Fe^{2+}$ is catalysed by various anions[990]. In the case of thiocyanate ion at least, it is known that the rate process is $Fe(phen)_3^{3+} + FeNCS^+$, since the immediate iron(III) product is $FeNCS^{2+}$ [i]. The order of effectiveness of catalysts is $Cl^- \ll I^- < N_3^- \simeq SCN^-$, and it is argued that the stronger catalysts act as bridging groups, either by nucleophilic attack on a carbon atom of a phenanthroline ring, or by a less specific attachment to the π-electron system. In a further series of comparisons of the same type, Przystas and Sutin[864] have invoked the orbital symmetries of the oxidant, reductant and catalyst, an argument which again requires that the catalyst acts as a bridge. Second-order rate constant k_0 were measured for reactions $A^+ + B$, where A^+ is $Co(phen)_3^{3+}$, $Co(en)_3^{3+}$ or $Co(NH_3)_6^{2+}$ and B is Fe^{2+}, Cr^{2+} or V^{2+}, and third-order rate constants for the rate terms $k_X[A^+][B][X^-]$, where X^- is Cl^- or NCS^-. The acceptor orbital of $Co(phen)_3^{3+}$ is presumed to be a ligand orbital of π symmetry, while the acceptor orbitals of $Co(en)_3^{3+}$, $Co(NH_3)_6^{2+}$ are metal e_g orbitals of σ symmetry; the donor orbitals of V^{2+} and Cr^{2+} are $t_{2g}(\pi)$ and $e_g(\sigma)$ respectively. The thiocyanate ion is considered to be a good π-bonding ligand while Cl^- is a good σ-bonding ligand. The maximum value of the ratio k_{NCS}/k_0 occurs in $Co(phen)_3^{3+}.SCN^-.V^{2+}$ where the orbital system runs $\pi-\pi-\pi$; the maximum value of the ratio k_{Cl}/k_0 is for $Co(NH_3)_6^{3+}.Cl^-.Cr^{2+}$, where the orbital system runs $\sigma-\sigma-\sigma$ [864].

Recently, Ulstrup[597, 1046] has shown that certain uncharged organic molecules effectively catalyse the reaction $Fe^{3+} + Tl^{2+}$, and he has attributed this to an outer-sphere bridged mechanism in which the electron is carried through the orbitals of the bridging group. Specific rate terms of the form of equation (5.78) are observed and it is found that only the more reducible molecules such as 2- and 4-aminopyridine, and not the 3-isomer, are effective. It should be noted, however, that an alternative mechanism, in which the X group is not simultaneously associated with oxidant and reductant, is also consistent with these data:

$$A^+ + X \underset{-1}{\overset{1}{\rightleftharpoons}} A + X^+ \quad (5.79)$$

(h) Kharkats[614, 615, 620] has considered these reactions theoretically using an adaptation of Marcus' theory (chapter 6). Taking all three reactant molecules as conducting spheres, he calculates the induced charge distribution and hence the reorganisation energy, for different ionic radii and relative positions[615]. The optimum condition for electron transfer is thus found to be the linear configuration A^+-X-B, with reactants in close contact

(i) The $FeNCS^{2+}$ complex referred to here has the N-bonded structure. This has been confirmed by the recent characterisation of the alternative S-bonded form $FeSCN^{2+}$ [852]

$$X^+ + B \xrightarrow{2} X + B^+ \tag{5.80}$$

giving $k = k_1 k_2/k_{-1}$. The two mechanisms are distinguishable in principle. According to equation (5.77) the oxidised form X of the catalyst need not appear in the reaction mixture, but according to equations (5.79) and (5.80) it must build up to a steady state concentration $[X^+] = k_1[A^+][X]/(k_{-1}[A] + k_2[B])$. In suitable cases this might be detectable, either directly or by its reactions with other substances.

5.4.2 Specific orientation of reactants

Since no complex has true spherical symmetry, there is the question whether a particular relative orientation of the reactant is preferred at the moment of electron transfer (this orientation might or might not be the most stable arrangement in the outer-sphere association complex). Again there is no conclusive evidence, but some plausible cases have been put forward. In the reaction $Co(NH_3)_5OOCC_6H_4NO_2^{2+} + CO_2^-$, it has been suggested (see p. 240) that the electron from the CO_2^- ion is accepted first into the nitro group; presumably this requires a close approach of CO_2^- to the nitro group. The molecule cytochrome-c contains an iron-porphyrin unit partially buried in a roughly spheroidal protein molecule. Sutin[988] has argued that electron transfer takes place at a restricted area where the edge of the porphyrin ring is exposed. The argument is based on a quantitative comparison, as follows. Since the presumed active site amounts to $\simeq$ 3 per cent of the total surface area of the molecule, the specific rate of the cytochrome-c^{III} + cytochrome-c^{II} exchange reaction contains a steric factor of $0.03 \times 0.03 \cong 10^{-3}$. It is further supposed that in the absence of this steric factor, the specific rate would have been the same as for the $Fe(phen)_3^{3+} + Fe(phen)_3^{2+}$ reaction; in other words, the heme and phenanthroline ring system have the same electron-mediating properties. The actual rate constants are $5 \times 10^4 \, M^{-1} s^{-1}$ [631] and $\simeq 10^8 \, M^{-1} s^{-1}$ [988]; a ratio of 5×10^{-3}, in good agreement with the model.

The reaction cytochrome-c^{III} + Cr^{2+} is catalysed by anions, in the order of effectiveness $Cl^- < I^- < N_3^- \simeq SCN^-$ [1114]. This is similar to the order of catalysis of the $Fe(phen)_3^{3+} + Fe^{2+}$ reaction and suggests again that the effective site of transfer is the edge of the prophyrin ring system.

A related question is whether, other things being equal, the rate of transfer between optically active centres might be stereoselective. The rates of exchange between the radical ion 1-(α-naphthyl)-1-phenyl-ethane and the corresponding neutral molecule, have been measured by the esr method[171]. The activation energy, E_a, for the mixture containing only the d-enantiomer was found to be specifically higher than for the racemic mixture. The difference is small: in DME as solvent $E_a = 1.6 \pm 0.5$ kcal mol^{-1} for the d–d mixture and 0.6 ± 0.5 kcal mol^{-1} for the racemic mixture. The specific rates differ by a factor of two in DME solvent but in THF they are the same. A report[991] of stereoselectivity in metal–metal systems (e.g. that $[(+)_{D^-} - Co(phen)_3]^{3+}$ reacted at different rates with the two isomers of $Cr(phen)_3^{2+}$) has been shown to be in error[607]; but the possibility of such effects remain intriguing and merits further study. La Mar and Van Hecke[641,642] have argued that stereoselectivity could result when the electron to be transferred is initially delocalised in the π-electron system of the ligands, rather than on the central metal ion, and they have presented evidence of delocalisation in $Cr(phen)_3^{2+}$ and related complexes.

5.4.3 Facial attack

The reaction $Cr(OH)OAc^{2+} + V^{2+}$ appears to be of the outer-sphere type since the specific rate $k = 2.2 \times 10^2 \, M^{-1}$ exceeds the normal limit for substitution in $V(H_2O)_6^{2+}$. Since this reaction represents a base-catalysed pathway for the reaction $CrOAc^{2+} + V^{2+}$, the question arises as to why such a path should be significant, since as noted on p. 149 base catalysis is normally characteristic of inner-sphere pathways. Deutsch and Taube[310] have suggested a transition state in which the OH group is presented to one of the faces of the $V(OH_2)_6^{2+}$ octahedron. Whether such a reaction should be classed as outer or inner-sphere seems a moot point. The authors call it 'an inner-sphere activated complex which does not make use of a normal coordination position in V^{2+}'. This geometry is perhaps related to the symmetry of the donor orbitals $t_{2g}(V^{2+})$ and $e_g(Cr^{2+})$. Facial attack, or expansion of coordination number, has also been postulated in a number of reactions with $Ru(NH_3)_6^{2+}$ as reductant, where again the transferring electron is of the t_{2g} type[420,656a]. In some cases at least, the attacking group becomes attached to the ruthenium(III) ion, and the mechanism can best be described as inner-sphere, or as a bimolecular electrophilic substitution[38]:

$$Ru(NH_3)_6^{2+} + NO^+ \rightarrow Ru(NH_3)_5 NO^{3+} + NH_3 \tag{5.81}$$

Other reactions such as $Ru(NH_3)_6^{2+} + O_2$[849], $Ru(NH_3)_5py^{3+} + O_2$[423] give ruthenium(III) products with the inner-sphere ligand undisturbed, but still 'one is faced ... with the kinetically indistinguishable possibility of some inner-sphere type of interaction involving the electron-deficient O_2 and the electron-rich face of the Ru^{II} octahedron'[420].

Some other peculiar features of ruthenium(II) chemistry may also be related to its ability to accept a seventh coordinated ligand. The aquation $Ru(NH_3)_6^{2+} \rightarrow Ru(NH_3)_5OH_2^{2+} + NH_3$ is acid-catalysed, in contrast to the corresponding reactions of other transition metal ammines such as $NiNH_3^{2+}$ and $Cr(NH_3)_6^{3+}$[421]. Interaction of the proton with the d-electron shell of the metal ion has been suggested. This could be described by an electron transfer model, with resonance between the $Ru^{2+} \ldots H^+$ and $Ru^{3+} \ldots H^0$ configurations. A similar, if not identical, transition state has been postulated for the reaction $Ru(NH_3)_6^{3+} + H \rightarrow Ru(NH_3)_6^{2+} + H^+$[786]. The electron transfer reaction $Co(NH_3)_5F^{2+} + Ru(NH_3)_6^{2+}$[832], is acid-catalysed, in contrast to

$Co(NH_3)_5F^{2+} + V^{2+}$ [680]. Linck[678] suggests that a transition state with expanded coordination of the ruthenium centre may be favoured by protonation of the coordinated fluoride—though the details of this proposal are not entirely clear.

Part 3
Energetics of electron transfer

Chapter 6
Theory of electron transfer

6.1 Models of the electron transfer process

6.1.1 Early work

Prior to about 1950, it was widely felt that 'pure' – electron transfer reactions, such as the $Fe^{3+} + Fe^{2+}$ exchange in aqueous solution, must necessarily be very rapid[a, b]. Platzman and Franck[851] wrote in 1952:

> In the early years of work in this field, it was commonly thought that this reaction could proceed very rapidly, because an electron could 'jump' from the ferrous ion to the ferric ion, with no change in energy, across great distances. It is now widely appreciated, though it may bear repetition, that such an electronic transition is quite impossible . . . .

Actually this seems unjust, at any rate towards the leading workers in the field, for although one can well imagine that such misapprehensions might have existed, they are nowhere clearly stated in the literature. It is true that the weight of the earlier evidence seemed to favour rapid electron transfer as a general rule, but for this very reason several groups began systematically searching for examples of a symmetrical isotope exchange reaction that would be measurably slow, and by 1948 two successes were reported. Lewis and

---

(a) 'Because many, perhaps most, ionic reactions in solution are immeasurably rapid it has seemed to some logical to infer that ions in general are in an activated state and require for reaction little or no added energy of activation . . . . . And as far as I know no basis has been suggested for predicting whether a given reaction will be immeasurably rapid or very slow.' (P. A. Shaffer (1936)[941])

(b) The following are typical of remarks made later in recollection:
(i) 'In 1946 it was thought that most exchange reactions between ions of the same structure in different valence states were rapid . . . . Our attentions were then turned to amine complexes of cobaltous ions and cobaltic ion, where the addendum group might be thought of as non-conducting, and exchange might consequently be slow . . .' (Lewis and Coryell[666])
(ii) 'Oxidation–reduction reactions have been thought until recently to proceed very rapidly, and the term "instantaneous" has frequently been used to characterise their rates.' (R. H. Betts[118])

Coryell[666] succeeded in measuring rates of exchange in cobalt (III)–cobalt (II) systems such as

$$Co(en)_3^{3+} + Co(en)_3^{2+} \rightleftharpoons Co(en)_3^{2+} + Co(en)_3^{3+} \qquad (6.1)$$

Explaining the basis of their choice, they argued that the saturated organic ligand would be effectively an insulator, preventing the flow of electrons. Harbottle and Dodson[516] reported measurements on various thallium(III)/thallium(I) systems. They argued that a two-electron transfer process would be less probable than a one-electron transfer, and the reaction accordingly would be slower.

The distinction between redox reactions which involved the making and breaking of bonds, and electron transfer pure and simple, came out particularly clearly in the data on actinide ions, which by the end of the 1940s was already extensive. For the three elements U, Np, Pu, oxidation states V and VI had been characterised as the dioxoions MO_2^+ and MO_2^{2+}, but valencies III and IV were aquo ions M^{3+}.aq and M^{4+}.aq, though liable to extensive hydrolysis and complexation. In 1951, Huizenga and Magnusson[563] offered a concise summary of rates of reactions of these ions, namely that reactions of the types $MO_2^{2+} + MO_2^+$ and $M^{4+} + M^{3+}$ were much faster than reactions between a dioxo and an aquo ion. They suggested that in the latter case 'the rate of electron transfer is dependent on the mechanism for the addition or removal of oxygen'.

As more systems were explored and techniques of rate measurements were refined, it became apparent that even the supposedly simple exchange processes spanned a very wide range of rates. Several exchange systems involving 'simple' ions, such as $Fe^{3+} + Fe^{2+}$, $Eu^{3+} + Eu^{2+}$, $Ce^{4+} + Ce^{3+}$, while not as slow as the cobalt(III)–cobalt(II) ammine systems, were found to be measurable by the available techniques, while most systems involving 'complex' ions, such as $Os(bipy)_3^{3+} + Os(bipy)_3^{2+}$, $MnO_4^- + MnO_4^{2-}$, and $Fe(CN)_6^{3-} + Fe(CN)_6^{4-}$, were faster by several orders of magnitude. Early explanations in terms of the electronic structure of the ions are still of interest, but the first generally useful theory was that of W. F. Libby announced in 1949[671] and finally published in 1952[672]. According to Libby, these 'contradictory and amazing results' could be rationalised by considering the role of the solvent molecules[c]. For example, in the iron(III)–iron(II) system it could be argued that direct electron transfer would require the expenditure of a large

(c) According to Libby, it was J. Franck himself who first pointed out the decisive role of the solvent in homogeneous thermal electron transfer reactions, though not apparently in any published writings. In the abstract to reference 672 Libby writes, 'The bearing of the Franck–Condon principle on electron exchange reactions as set forth recently by Franck is described ...', and later in the same paper (p. 864), 'Franck has clearly enunciated the principle that electron transfer in aqueous solution should be inhibited by the relatively longer times required for movement of the heavy water molecules constituting the hydration atmospheres as compared to the transit time of the electron.' See also reference 851 as quoted in the text, and reference 427

amount of energy since, in the time taken for the electron to move from Fe^{2+} to Fe^{3+}, the water molecules of hydration could not move appreciably. Hence, the new Fe^{3+} ion would be generated with the hydration configuration appropriate to Fe^{2+}. This is the Franck–Condon principle of molecular spectroscopy, and the process just described is in fact an optical and not a thermal one—an intervalence charge transfer of the type discussed in Chapter 8. However, this description accounts qualitatively for the fact that an activation energy exists, and it explains the much smaller activation barriers for systems such as $Fe(CN)_6^{3-} + Fe(CN)_6^{4-}$, since on the assumption that the Fe–C and C–N bond distances are similar in the two complexes, the hydration energy barrier is smaller owing to the larger radii. On this basis there was proposed a further classification of electron transfer reagents according to the extent of molecular rearrangement involved: Class I with nearly identical geometry in both oxidised and reduced forms, and sufficiently large radii for the near-neighbouring solvent molecules to be only slightly affected, e.g. $Fe(CN)_6^{3-}/Fe(CN)_6^{4-}$; Class II with small but significant changes in bond length, e.g. $Fe(H_2O)_6^{3+}/Fe(H_2O)_6^{2+}$; and Class III with more substantial changes in geometry, e.g. $Cr(H_2O)_6^{3+}/Cr^{2+}$.aq (the latter presumably regarded as four-coordinate $Cr(H_2O)_4^{2+}$). Later work would have necessitated extension to cover cases like $VO^{2+}/V(H_2O)_6^{3+}$. It remains broadly true that for symmetrical exchange processes, rates diminish in the order of the classes, I, II, III.

Both Libby[672] and Weiss[1082] made semi-quantitative estimates of the activation energies, using the ionic model to estimate the hydration energies of the ground and excited states. Weiss' treatment in particular is of interest for its recognition of the distinction between atomic and electronic polarisation, using the Landau[647] expression ($\varepsilon_{op}^{-1} - \varepsilon_s^{-1}$). This, however, is more conveniently discussed below (section 6.3.2(d)).

Weiss[1082] also considered in some detail the energetics of unsymmetrical electron transfer processes, using the formalism previously developed by Bates and Massey for reactions between unlike atoms in the gas phase. It will be recalled (see section 1.2), that for such a reaction there is in general a critical internuclear distance R_c, at which the energy curves of the reactants and products intersect, and this is the point of maximum probability of electron transfer. For a non-adiabatic process, the actual probability and hence the reaction cross section are given by the Landau–Zener formula (equations (1.45)–(1.47)). The distance R_c is the value of R at which

$$\Delta U^0(R) = U_f^0(R) - U_i^0(R) = 0 \qquad (6.2)$$

Where $U_i^0(R)$ and $U_f^0(R)$ are the interaction energies for the systems $(A^+ + B)$ and $(A + B^+)$ respectively, at separation distance R. In the gas phase system, the energies $U_f(R)$ and $U_i(R)$ can be represented by electrostatic expressions (see equations 1.39, 1.40), and Weiss assumed that the same was true in solution. On this basis he wrote down expressions for the activation energies of electron transfer reactions of several different charge types. The details of his argument are not clear, however, and the physical principles involved are again

6.1.2 The two-state description[662] [(d)]

In the remainder of this chapter we shall be concerned mainly with one-electron intramolecular reactions of the type

$$(A^+ \ldots B) \text{ (env)} \xrightarrow{k} (A \ldots B^+) \text{ (env)} \qquad (6.3)$$
$$\quad\;\; p \qquad\qquad\qquad\qquad\;\; s$$

where A and B are the central ions (usually metal ions) and the positive sign denotes the higher of the two valencies, regardless of the actual ionic changes. The state symbol (env) denotes the bridging ligands (if present), the non-bridging ligands, and the solvent; k is the first-order rate constant.

The states p and s (precursor and successor) differ in electronic configuration. In those theories which we shall call two-state theories, the difference in configuration is ascribed solely to the transferring electron, which is presumed to move independently of all the other electrons in the system. Thus we may write two one-electron wavefunctions ϕ_p and ϕ_s corresponding to localisation of the electron around the nuclei B and A respectively. To each of these wavefunctions there correspond electronic energies given by

$$E_p = \int \phi_p H \phi_p^* \, d\tau \qquad (6.4)$$

$$E_s = \int \phi_s H \phi_s^* \, d\tau \qquad (6.5)$$

where the Hamiltonian H is a function of the coordinates of all nuclei and other electrons in the system, and the integration is carried out over all space. We may also define the total energies U_p and U_s, comprising the energies of electron plus ions and solvent molecules, as functions of the coordinates $q_1, q_2 \ldots$ etc. of all atoms. The functions $U_p(q_1, \ldots)$ and $U_s(q_1, \ldots)$ may be represented as multidimensional surfaces with minima U_p^0 and U_s^0, and intersecting on a multidimensional line when $U_p = U_s$. A reaction pathway is a line connecting the two minima, and a projection of such a line into two dimensions gives the diagram of *Figure 6.1*, where the energy surfaces have become curves, the nuclear coordinates have been replaced by the generalised reaction coordinate q and the surface crossing is the point at $q = q^\ddagger$. These energy curves are analogous to those discussed in Chapter 1 for the reaction

---

(d) General reviews of the theoretical literature are given by Marcus[727] and Ruff[915]; the distinctive contributions of the Russian theoretical workers are reviewed in references 327, 331, 662. For unified accounts of electron transfer covering the topics mentioned in section 6.3, see especially references 662 and 230

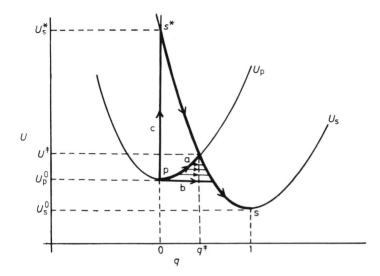

Figure 6.1 Representation of an electron transfer reaction p → s in terms of intersecting energy curves. The lettered arrows refer to the reaction pathways specified in the text.

between atoms in the gas phase, and the mechanism of electron transfer can be described qualitatively in the same way.

From a strict quantum-mechanical viewpoint, the problem of calculating the rate constant k involves calculating the probability W that a reacting system represented initially by a point in the vicinity of p will be transformed in unit time to one represented by a point in the vicinity of s, and of averaging this probability over all relevant positions and velocities of nuclei. In practice it is usual to distinguish limiting mechanisms as follows.

In the *thermal* mechanism (*Figure 6.1*, pathway a) the representative point moves along the U_p curve to the crossing then transfers to the U_s curve. If the resonance energy is sufficiently small, electron transfer is virtually restricted to a small range of the coordinate q, in the vicinity of the crossing. The activation energy is $(U^{\ddagger} - U_p^0)$. In the *tunnel* mechanism, the representative point moves horizontally from a point on the U_p curve to a point on the U_s curve (paths shown by horizontal arrows). The total potential energy U remains constant through the transition, but the electronic energy E and the kinetic energy of the nuclei do not. A tunnel transition involves simultaneous movement of the nuclei and a change in the character of the electronic wavefunctions. The total reaction probability thus consists of contributions from the thermal pathway, and for various tunnelling pathways as shown. These, however, become rapidly less probable as the extent of nuclear motion increases. Hence, normally, only two limiting cases are considered: the high-temperature limit when the energy required to reach the crossing point is readily available and only the thermal mechanism need be considered, and the low-temperature limit, when the reactants are virtually confined to the energy level U_p^0 and tunnelling becomes the relatively favoured mechanism (path b). In the latter case, the activation energy for the forward reaction is zero, and for the backward reaction is $(U_p^0 - U_s^0)$. In either direction the overall rate is of course much less than it would be at any higher temperature.

The vertical transition (path c) corresponds to the absorption of a photon. In accordance with the Franck–Condon principle, the nuclear positions remain unchanged. The required energy $hv_{CT} = U_s^* - U_p^0$ is also equal to the difference in electronic energies $E_s(0)$–$E_p(0)$ where $E_s(0)$ and $E_p(0)$ are given by equations of the same type as equations (6.4) and (6.5) with the Hamiltonian appropriate to the coordinate $q = 0$. In his pioneer discussions of the rates of thermal electron transfer process, Libby[672] assumed that path c was the normal reaction path, with activation energy $U_s^* - U_p^0$. It was later recognised that path a is more economical of energy and therefore preferred. Some authors[81] have argued that path c is forbidden thermodynamically, but this is incorrect: path c could be followed even in the absence of a photon, if the necessary energy could be supplied rapidly from some other source, such as sudden large energy fluctuation elsewhere in the medium, but this is highly improbable. The energy $U_s^* - U_p^0$ is of the order of 100 $k_B T$, hence vibrational excitation would require the simultaneous absorption of about 100 phonons, or of a phonon with energy 100 times the mean energy.

Electron transfer reactions have also been discussed in terms of 'electron tunnelling'[673, 732, 1131]. This, however, is really another way of expressing the probability of change in wavefunction at the transition point. The relationship between the two is best expressed by considering explicitly the potential field within which the transferring electron moves. In the precursor state p, the electron is localised at atom B. The potential energy V of the transferring electron, as a function of space coordinates, has wells located at each atomic nucleus and the lowest stationary energy level is at nucleus B (*Figure 6.2(a)*). Optical excitation would take the electron to atom A. The thermal reaction p → s, however, involves progressive changes in the coordinates of the atoms, and hence in the shape of the potential energy surface, from *Figure 6.2(a)* to *Figure 6.2(c)*. At the transition state, the energy levels for $A^+ \ldots B$ and $A \ldots B^+$ are equal (*Figure 6.2(b)*). This is the resonance condition. The electron initially located at B has a 50% probability, over time, of being located at A. The electron is said to have 'tunnelled' from A to B, since its potential energy did not pass through the intermediate value corresponding to point C.

6.2 The thermal mechanism

6.2.1 Energy curves

In section 6.3 we shall discuss specific models of the reacting complex which lead to quantitative expressions for the function U_p and U_s plotted in *Figure 6.1*, and hence to quantitative predictions of rate constants. Some of the most

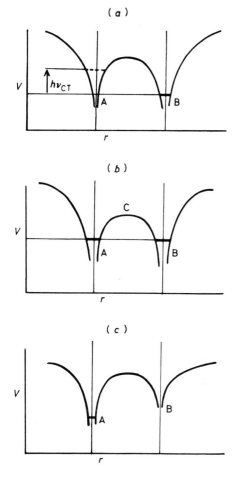

Figure 6.2 Potential wells of the transferring electron at different stages of the reaction $A^+ + B = A + B^+$; (a) precursor state; (b) transition state; (c) successor state

important features of electron transfer reactions can, however, be discussed on a more general basis, and we may conveniently take these first.

The simplest form of a curve with single minimum is the parabola, and the models to be discussed below suggest that for electron transfer reactions in general, and especially for outer-sphere reactions, the parabolic energy curve is a good approximation. We may, therefore, write

$$U_p = A_p x^2 \tag{6.6}$$

$$U_s = A_s(1-x)^2 + \Delta U_0 \tag{6.7}$$

where x is a dimensionless reaction coordinate varying from 0 to 1 as the reaction proceeds and A_p and A_s are constants characteristic of the reaction system. For symmetrical exchange reactions these constants are identical. For the sake of simplicity we assume that for other, unsymmetrical reactions we have, also,

$$A_p = A_s = A' \tag{6.8}$$

It is by no means intuitively obvious that equation (6.8) should hold, but in terms of the two principal reaction models (discussed in section 6.3) it is possible to specify conditions when it does hold.

With this assumption, the reaction coordinate $x^{\ddagger}$ at the crossing and the activation energy $\Delta U^{\ddagger}$ relative to the precursor complex are given by

$$x^{\ddagger} = \tfrac{1}{2}(1 + \Delta U_0/A') \tag{6.9}$$

$$\Delta U^{\ddagger} = \tfrac{1}{4}A'(1 + \Delta U_0/A')^2 \tag{6.10}$$

where $\Delta U_0 = U_s - U_p$. For a symmetrical reaction, $\Delta U_0 = 0$ giving

$$U^{\ddagger} = \tfrac{1}{4}A' \tag{6.11}$$

The energy A' will later be called the reorganisation energy or 'Franck–Condon energy' ΔU_{FC} (see Chapter 8). The quantity $\tfrac{1}{4}A'$ is sometimes called the *intrinsic* energy barrier for the reaction[e].

Equations of the type of equations (6.9), (6.10), were apparently first developed by Kubo and Toyozawa[636] in connection with the theory of non-radiative electronic transitions. In the field of reaction kinetics, these equations, or rather their analogues, for free energies, are now associated with the name of Marcus, who deduced them in a different way from the dielectric continuum model. Later the simplicity of the derivation from energy curves was re-emphasised by Newton[795]. Similar equations have been deduced in the same way for various classes of reaction not involving electron transfer[236, 730], notably proton transfer reactions[183, 600].

For completeness, we note here the analogous results when equation (6.8) is not satisfied:

$$\Delta U^{\ddagger} = A_p(A_p - A_s)^{-2}\{A_s - [A_p A_s - \Delta U_0(A_p - A_s)]^{1/2}\}^2 \tag{6.12}$$

When the second term in the square root is small compared with the first this may be expanded as

$$\Delta U^{\ddagger} = \frac{A_p A_s}{(A_p^{1/2} + A_s^{1/2})^2} + \frac{\Delta U_0 A_p^{1/2}}{(A_p^{1/2} + A_s^{1/2})} + \frac{(\Delta U_0)^2}{4 A_p^{1/2} A_s^{1/2}} \tag{6.13}$$

(see also references 627, 795).

6.2.2 Transition probability

A convenient model for the reaction system is that of a harmonic oscillator with a characteristic frequency ω, so that the representative point P moves according to the equations

(e) The prime is used to distinguish A' from the analogous free energy A (equation 6.74)

$$x = \alpha \sin \omega t \tag{6.14}$$

$$\dot{x} = dx/dt = \alpha\omega \cos \omega t \tag{6.15}$$

where x is the reaction coordinate. For a given oscillation the amplitude α is related to the total energy ε (potential energy plus kinetic energy) by

$$\alpha = (\varepsilon/A')^{1/2} \tag{6.16}$$

Hence, for a collection of oscillators of different energies obeying the Boltzman distribution law, the distribution of amplitudes, of x coordinates and of $\dot{x}$ can be calculated. Thus, it is easy to show that at any instant in time, the fraction d^2W of representative points falling in the ranges x to $(x+dx)$ and $\dot{x}$ to $(\dot{x}+d\dot{x})$ is given by

$$d^2W = \frac{A'}{\pi\omega k_B T} \exp\left(-\frac{A'}{k_B T}(x^2 + \omega^{-2}\dot{x}^2)\right) dx\, d\dot{x} \tag{6.17}$$

This gives the distribution of velocities among the reacting systems passing through the crossing point x. The fraction of representative points which, in time dt, cover the distance from $x^{\ddagger}$ to $(x^{\ddagger} + \dot{x}\,dt)$ is

$$\left(\frac{d^2 W}{d\dot{x}\,dx}\right)_{x=x^{\ddagger}} d\dot{x}\,\dot{x}\,dt = W_s d\dot{x}\,dt \tag{6.18}$$

where

$$W_s = \left(\frac{A'}{\pi\omega k_B T}\right) \exp(-\Delta U^{\ddagger}/k_B T)\exp(-A\omega^{-2}\dot{x}^2/k_B T)\dot{x} \tag{6.19}$$

The probability W_e of a p $\to$ s transition at the crossing point is a function of $\dot{x}$, hence the reaction probability W, integrated over all velocities, is

$$W = \int_0^\infty W_s W_e\, d\dot{x} \tag{6.20}$$

The value of W_e is given by Zener's equation (chapter 1, equation 1.45) as

$$W_e = 1 - \exp[4\pi^2\beta^2/h\dot{x}S^{\ddagger}] \tag{6.21}$$

where β is the resonance integral and $S^{\ddagger}$ is the difference in slope between the two energy curves, both evaluated at the crossing point

$$\beta = \int \phi_p H \phi_s^* \, d\tau \tag{6.22}$$

$$S^{\ddagger} = \left|\left(\frac{d}{dx}(U_p - U_s)\right)_{x=x^{\ddagger}}\right| \tag{6.23}$$

Three special cases of equation (6.20) are usually distinguished. In the *weak-overlap* case, when β is very small, W_e is small, and the rate of reaction is low.

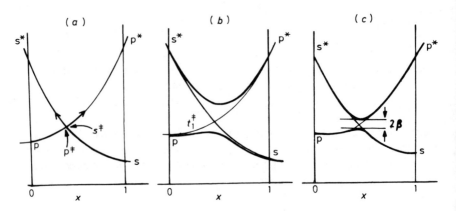

Figure 6.3 Energy curves for electron-transfer reactions: (a) non-adiabatic case; (b) strongly adiabatic case; (c) the Marcus–Hush approximation

The activation energy is given by $\Delta U^{\ddagger} = \tfrac{1}{4}A'$ (*Figure 6.3(a)*). In the *strong-overlap* case, the resonance integral is large so that the activation energy is appreciably lower, $\Delta U^{\ddagger} = \tfrac{1}{4}A' - \beta$ (*Figure 6.3(b)*). The transition probability W_e approaches unity, but the harmonic oscillator model presumably cannot be used to calculate the probability of reaction. In the *medium-overlap* case (the Marcus–Hush approximation), the resonance energy is assumed large enough so that $W_e \simeq 1$, but still small enough so that $\Delta U^{\ddagger} \simeq \tfrac{1}{4}A'$ (*Figure 6.3(c)*). This is the most convenient assumption from the theoretical point of view, and several authors have argued that it does in fact have wide application.

A rough calculation based on the harmonic oscillator model is instructive. If we take $W_e = 0.99$ as being acceptably close to unity, we have, from equation (6.21),

$$\beta^2 = (2\ln 10/\pi^2)h\dot{x}S^{\ddagger} \qquad (6.24)$$

For intersecting parabolae, $S = d(U_p - U_s)/dx = 2A'$. For the velocity $\dot{x}$ we take the root mean square value, $\langle \dot{x} \rangle = \omega\sqrt{2}$, giving

$$(\beta/\Delta U^{\ddagger})^2 = (4\ln 10\sqrt{2}/\pi^2)h\omega/\Delta U^{\ddagger}) \qquad (6.25)$$

Taking ω as the frequency of a typical infra-red vibration, 10^{12} s^{-1} and taking the molar activation energy $L\Delta U^{\ddagger}$ as typically 10 kcal mol^{-1} we have finally $(\beta/\Delta U^{\ddagger}) \simeq 0.1$; a molar resonance energy of $L\beta \cong 1$ kcal mol^{-1}. Evidently the range of validity of the medium overlap approximation is very limited: a decrease in W_e from 0.99 to 0.9, or a decrease of 1 kcal in $\Delta U^{\ddagger}$, are both experimentally measurable. A theory based on this model may, however, be more successful in predicting rates of reaction than activation parameters. The effects of variation of β on the overall reaction, via W_e and $\Delta U^{\ddagger}$, work in opposite directions and to that extent the model is self-compensating.

Marcus[720] estimated the actual resonance energy for an outer-sphere electron transfer by the following argument. Within the transition state, the electronic configurations p and s are of equal energy in order to fulfil the resonance condition. There is, however, a limit to the precision with which that condition may be specified, due to the Uncertainty Principle. Taking the lifetime of the transition state as $\tau \simeq 10^{-13}$ s, and identifying the resonance energy as the extent of 'broadening' of the energy due to this uncertainty, Marcus obtains $2\beta \simeq 0.15$ kcal mol^{-1} for an outer-sphere activated complex. On the other hand for a bonded inner-sphere activated complex, with strong electronic interaction, τ might be as low as 10^{-15} s and $2\beta \simeq 15$ kcal mol^{-1}. Hush[570] made rough estimates of the interaction energy by referring to the order of magnitude of the interaction energy of metal ions in compounds like cupric acetate, i.e. $\beta \simeq 0.15$ kcal mol^{-1}. Thus, it seems likely that for outer-sphere electron transfer the resonance energy will be small enough to fulfil the medium overlap approximation. The possibility that the overlap might be too small cannot be so easily dismissed. From the Zener formula, Hush calculated that interaction energies as large as in the copper acetate case were enough to ensure adiabatic behaviour. Stearn and Eyring[965], who discussed this problem from a similar point of view, concluded that the transmission coefficient approaches unity when the resonance energy is of the order of $\frac{1}{2}k_b T$ per molecule, i.e. 0.3 kcal mol^{-1}.

Assuming medium overlap the oscillator model discussed above gives the total reaction probability as

$$W = \int_0^\infty W_s \, d\dot{x}$$
$$= (\omega/2\pi)\exp(-\Delta U^{\ddagger}/k_B T) \quad (6.26)$$

The more familiar derivation from transition state theory[466] gives

$$W = (Lh/RT)\exp(-\Delta U^{\ddagger}/k_B T) \quad (6.27)$$

and on averaging over all reacting systems,

$$k = (Lh/RT)\exp(-\Delta G^{\ddagger}/RT)$$
$$= (Lh/RT)\exp(-\Delta H^{\ddagger}/RT)\exp(\Delta S^{\ddagger}/R) \quad (6.28)$$

In the weak overlap case, equation (6.28) is modified by introducing the transmission coefficient κ

$$k = \kappa(Lh/RT)\exp(-\Delta H^{\ddagger}/RT)(\Delta S^{\ddagger}/R) \quad (6.29)$$

and in the strong overlap case, the resonance energy is presumed to affect $\Delta H^{\ddagger}$, and thereby $\Delta G^{\ddagger}$ but not $\Delta S^{\ddagger}$

$$k = (Lh/RT)\exp(-(\Delta H^{\ddagger} - L\beta)/RT)\exp(\Delta S^{\ddagger}/R) \quad (6.30)$$

6.3 The activation process

6.3.1 Metal–ligand bond stretching

Consider the symmetrical exchange reaction

$$A^+(\text{env}) + A(\text{env}) \rightarrow [A^+(\text{env})^\ddagger + A(\text{env})^\ddagger] \rightarrow A(\text{env}) + A^+(\text{env}) \quad (6.31)$$

$$\underbrace{(1) \quad\quad (2)}_{p} \quad \underbrace{(1) \quad\quad (2)}_{t} \quad \underbrace{(1) \quad\quad (2)}_{s}$$

where the reactant ions are held at fixed distance so that the overall rate constant k is first order. The state symbol env denotes the total ligand–solvent environment, but in the present case we shall assume that the metal solvent interaction energy is dominated by the inner-sphere metal–ligand bond energies. We assume further that the ions A^+ and A have the same coordination geometry and that for each ion the bond-stretching or compression is characterised by a single coordinate with an appropriate force constant. In quantitative calculations, the activation process is usually taken to be the symmetrical 'breathing' mode of vibration. Labelling the atoms 1, 2 as shown in equation (6.31), the energy U_p of the precursor state can then be written

$$U_p = U^0 + \tfrac{1}{2} N f_A^+ (r_1 - r_A^+)^2 + \tfrac{1}{2} N f_A (r_2 - r_A)^2 \quad (6.32)$$

where r_1 and r_2 are the bond distances in complexes 1 and 2, r_A and r_A^+ are ground state bond distances appropriate to the oxidation states A and A^+; f_A, f_A^+ are the corresponding force constants and N is the coordination number. Similarly for the successor state

$$U_s = U^0 + \tfrac{1}{2} N f_A (r_1 - r_A)^2 + N f_A^+ (r_2 - r_A^+)^2 \quad (6.33)$$

The terms U_p, U_s as functions of the independent variables r_1, r_2, are three-dimensional paraboloids. They may be plotted by means of contours in a two-dimensional graph (*Figure 6.4(a)*). The surfaces cross where $U_p = U_s$, and by symmetry this occurs when $r_1 = r_2$. The transition state occurs at the point T where the crossing energy is lowest. (This is also the point where the contour ellipses are tangential to the line $r_1 = r_2$.) It may easily be shown that at this point, the bond distance r is

$$r^\ddagger = r_1 = r_2 = \frac{f_A r_A + f_A^+ r_A^+}{f_A + f_A^+} \quad (6.34)$$

and the activation energy $\Delta U^\ddagger$ is

$$\Delta U^\ddagger = \frac{L a f_A^+}{2(f_A + f_A^+)} (r_A - r_A^+)^2 \quad (6.35)$$

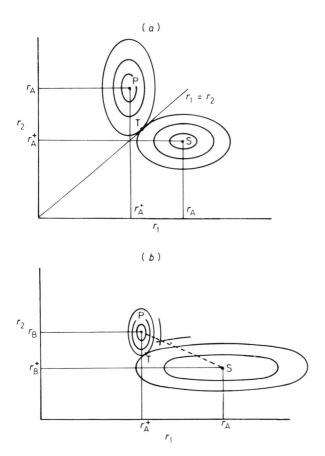

Figure 6.4 Potential energy surfaces for outer-sphere electron transfer reactions, with energy determined only by metal–ligand bond length: (a) symmetrical reaction $A^+ + A = A + A^+$; (b) unsymmetrical reaction $A^+ + B = A + B^+$. See text and equations (6.32), (6.33)

Equation (6.35) was first derived by Sutin[985(f)]. It is noteworthy that the 'compromise' bond distance $r^{\ddagger}$ is not in general the arithmetic mean of r_A and r_A^+; though in some cases it is probably close to the mean since for most ions $r_A > r_A^+$ while $f_A^+ > f_A$.

The case of an unsymmetrical reaction

$$A^+(\text{env}) + B(\text{env}) \rightarrow A(\text{env}) + B^+(\text{env}) \qquad (6.36)$$
$$\underbrace{(1) \qquad\quad (2)}_{p} \quad \underbrace{(1) \qquad\quad (2)}_{s}$$

(f) Previously Orgel[824] had obtained the equation $\Delta U^{\ddagger} = (3/4)(f_A + f_A^+)(r_A - r_A^+)^2$

is shown in *Figure 6.4(b)*. Here we have

$$U_p = U_p^0 + f_A^+(r_1 - r_A^+)^2 + f_B(r_2 - r_B)^2 \tag{6.37}$$
$$U_s = U_s^0 + f_A(r_1 - r_A)^2 + f_B^+(r_2 - r_B^+)^2 \tag{6.38}$$

with force constants f_B, f_B^+, and bond distances r_B, r_B^+ appropriate to ions B and B^+. The crossing of the surfaces, $U_p = U_s$, now occurs on a curve defined by an equation of the second degree in both r_1 and r_2. (The curve may be shown to be a hyperbola.) Locating the lowest-energy point T on this curve is a tedious mathematical exercise, but it can be simplified by assuming that $\alpha\beta = \alpha^+\beta^+$, where $\alpha = f_A(r_A - r_A^+)^2$, $\beta = f_B(r_B - r_B^+)^2$, $\alpha^+ = f_A^+(r_A - r_A^+)^2$, $\beta^+ = f_B^+(r_B - r_B^+)^2$. In that case the point T lies on the straight line joining points P and S, and the energy surfaces can be replaced by their cross-sections in the plane PTS, which are two parabolas of different curvature. The energies U_P and U_S may then be expressed as

$$U_P = U_P^0 + px^2 \tag{6.39}$$
$$U_S = U_S^0 + s(1-x)^2 \tag{6.40}$$

where $x = (r_1 - r_A^+)/(r_A - r_A^+) = (r_B - r_2)/(r_B - r_B^+)$, $p = \alpha^+ + \beta$, and $s = \alpha + \beta^+$. The activation energy $\Delta U^\ddagger$ is given by equation (6.12), with $A_p = p$, $A_s = s$. In the limit where $p = s$, the activation energy is given by equation (6.10), with $A' = p = s$. (See also reference 457.)

6.3.2 The solvent as dielectric continuum

(a) The ionic model

It is well known that the hydration energies of simple ions in aqueous solution can be calculated by treating the ion as a charged sphere and the solvent as a continuous dielectric medium. Thus, Born[142] obtained the expression

$$\Delta G_{hyd} = \left(\frac{L}{4\pi\varepsilon_0}\right)\left(\frac{z^2e^2}{2a}\right)\left(1 - \frac{1}{\varepsilon_s}\right) \tag{6.41}$$

where ε is the dielectric constant, a is the radius of the ion, and z is the ionic charge number. When Pauling's radii are used the results are in error by up to 100%; but Latimer, Pitzer and Slansky[653] showed that better agreement with experimental data could be obtained by using 'effective radii', $a + c$, where c is a constant, $\simeq 0.8\text{Å}$ for all cations and $\simeq 0.1\text{Å}$ for all anions. The failure of the original equation can be ascribed to neglect of the detailed structure of the solvent in the near vicinity of the ion. The bulk dielectric constant ε measures the response of the solvent to an applied field. This response entails partial reorientation of the solvent molecules. In the inner-coordination sphere of a cation the solvent molecules are almost rigidly held and do not reorientate with variation in ionic charge. In terms of the continuum model, the solvent is dielectrically saturated and the local dielectric constant is less than in the bulk phase. Some authors have sought to improve the continuum theory of solvation by allowing for dielectric saturation[826]. Laidler[644] has calculated effective dielectric constants as a function of distance from a central ion. Others have used the bond model for the inner-sphere and the continuum model for the outer-sphere reactions[81, 82, 845]. In this section we briefly review the continuum theory of solvation but we express the results in a general form, applicable to molecules of complex structure and therefore to the binuclear complex or electron transfer transition state.

The process of ionisation of an atom or molecule A in the gas phase and in solution[816] may be defined by the following equations:

$$A(g) \rightarrow A^+(g) + e^-(g) \tag{6.42}$$

$$A(aq) \rightarrow A^+(aq) + \{e^-\}_{aq} \tag{6.43}$$

where the state symbol (g) denotes a free species *in vacuo*, with zero translational kinetic energy; (aq) denotes the total solvent environment at thermal equilibrium with the solvated species, and the symbol $\{e^-\}_{aq}$ denotes an electron delocalised through the bulk of the solvent (and hence not specifically interacting with any of the species present) but having zero kinetic energy[(g)]. Such an electron has been called 'quasifree'[268, 305], or 'dry'[324]. In quantum-mechanical terms, it is at the lowest continuum energy level of the system.

If the charged species in equation (6.42) are deemed to be at zero potential, then it may be supposed that those in equation (6.43) are at some other potential, such as the average potential over the interior of the bulk solvent phase, but as Guggenheim has pointed out[489], there is no way of defining unambiguously the inner potential, and in any case it cancels out at a later stage of the calculation.

In the electrostatic theory of solvation [142], the ionisation processes are compared with the classical process of reversible charging. To obtain the electrostatic analogue of equation (6.42), we start with a body A' having the same boundary surface as the ion A^+, then remove electric charge by infinitesimal stages to yield finally the charge distribution characteristic of A^+. This process may be written:

$$A'(g) \rightarrow A'^+(g) + [-]_0 \tag{6.44}$$

(g) The notation $\{e^-\}_{aq}$, $\{-\}_{aq}$, is based on that of Noyes, whose paper[816] should be consulted for a clear discussion of the problems of defining transfer energies of charged species

where the state symbols (g) have the same meaning as before, and $[-]_0$ denotes (see footnote $^{(g)}$ p. 189) one mole of 'electricity', infinitely dispersed and at zero potential[489]. For the analogue of equation (6.43), the solvent is considered as a uniform dielectric, the body A' is a cavity, and the electric field is applied reversibly as before to yield the final charge distribution characteristic of A^+ (aq). This process can be written:

$$A'(aq) \rightarrow A'^+(aq) + [-]_{aq} \tag{6.45}$$

where A' and A'^+ denote cavities as described, (aq) denotes the dielectric medium and $[-]_{aq}$ denotes one mole of 'electricity' infinitely dispersed and at the potential prevailing in the medium. This potential can be defined conceptually[972] but not operationally[489]; but again this is of no concern since it cancels out at a later stage.

All of the above can be restated, with appropriate verbal changes, for the formation of a negative species A^- (aq), or of an uncharged species with dipole or higher multipole fields.

The work W_0 done in equation (6.44) and the work W done in equation (6.45) are taken to be free energies $^{(h)}$ and the assumption of the ionic model is that the difference in work between equation (6.44) and (6.45) is the same as between equations (6.42) and (6.43), so that for the transfer process

$$A^+(g) + e^-(g) \rightarrow A^+(aq) + [e^-]_{aq} \tag{6.46}$$

representing the solvation of ion A^+, the free energy change is given by

$$\Delta G_{solv} = L(W_0 - W) \tag{6.47}$$

Strictly speaking, an additional term ΔG_{cav} should be added to the right-hand side of equation (6.47) to account for the work done in excavating a cavity in the solvent to accommodate the A^+ ion[366], but in all later applications of equation (6.47) this term will cancel, hence it may be neglected.

(h) W_0 and W are identified as free energies on the grounds that both equations (6.44) and (6.45) are reversible. The question remains as to whether we should consider the Gibbs (constant pressure) or Helmholtz (constant volume) free energy. For rigid dielectrics (the only ones normally considered in the standard texts on electrostatics) there is no difference. For liquids under constant external pressure, $(W_0 - W)$ is commonly identified as Gibbs free energy but this is not strictly correct, since it fails to account for the work of compressing the dielectric under the influence of the applied field. For a rigorous discussion, see reference 429. For uniform applied fields it would be possible to define two dielectric constants, one for charging at constant pressure and one for charging at constant volume. Here, however, we are dealing with a non-uniform field. The problem is partly avoided by dividing the solvent into inner- and outer-sphere regions, as noted below

Following Noyes[816] we include the electron on either side of the equation so that no net charge is transferred between phases, and the question of inner potentials does not arise.

(b) Electrostatic considerations

The reversible work W_0 done in equation (6.44) is given by

$$W_0 = \tfrac{1}{2}\varepsilon_0 \int_V \boldsymbol{E}_c \cdot \boldsymbol{E}_c \, dV \tag{6.48}$$

where the vector $\boldsymbol{E}_c$ is the field set up in the vacuum by the ion A^+, and ε_0 is the permittivity of free space. The integration is with respect to volume V over all space[(i)]. The work W_0 is sometimes called the 'self energy' of the charge distribution. The proof of this equation is a standard theorem in electrostatics[690].

In the gradual process (6.45) charges are built up which in the absence of dielectric would produce the field $\boldsymbol{E}_c$. At the same time, the material of the dielectric suffers a polarisation which can be expressed by saying that every volume element dV acquires a dipole moment $d\boldsymbol{p}$. The dipole moment per unit volume is the *polarisation* $\boldsymbol{P}$, and associated with this is the field $\boldsymbol{E}$. If the medium is isotropic, and if the field is not too large, the polarisation is always proportional to the field, thus

$$\boldsymbol{P} = \varepsilon_0 \alpha \boldsymbol{E} \tag{6.49}$$

where α is the *polarisability*[(j)]. In the classical literature, another vector, the *displacement* is defined by

$$\boldsymbol{D} = \varepsilon_0 \boldsymbol{E} + \boldsymbol{P} \tag{6.50}$$

$$= \varepsilon_0 (1 + \alpha) \boldsymbol{E} \tag{6.51}$$

$$= \varepsilon_0 \varepsilon \boldsymbol{E} \tag{6.52}$$

where ε is the relative permittivity or 'dielectric constant'. The work done in charging the system, i.e. in the gradual process (6.45), is then given by[690]

$$W = \tfrac{1}{2} \int_V \boldsymbol{E} \cdot \boldsymbol{D} \, dV \tag{6.53}$$

(i) This statement needs qualification when the charge distribution of A^+ involves point charges, but the conclusions are not affected. For a rigorous discussion of this point, see reference 903
(j) The problem of dielectric saturation at high field strength is dealt with by Padova[825].

where again the integral is taken over all space[k]; hence the solvation energy is

$$\mu = W - W_0$$

$$= \tfrac{1}{2} \int_V (\boldsymbol{E} \cdot \boldsymbol{D} - \varepsilon_0 \boldsymbol{E}_c \cdot \boldsymbol{E}_c) \mathrm{d}V \tag{6.54}$$

where

$$\mu = \Delta G_{\text{solv}}/L.$$

To evaluate this integral we need to know $\boldsymbol{E}$ and $\boldsymbol{D}$ as functions of position in space, and these are often difficult to find, since they depend not only on $\boldsymbol{E}_c$ and the dielectric constant of the solvent, but also on the shape of the boundary of the molecule and the value assigned to the dielectric constant in the interior. Pictorially this is expressed by saying that the lines of force, which show the direction of $\boldsymbol{D}$ at any point are refracted at the boundary (*Figure 6.5*). In some cases in which the functional form of the boundary is known, the problem has been solved by a procedure of successive approximations known as the method of images[1]. In some other cases exact integrals have been given[623, 624, 1087]. When image effects are neglected, the integral (6.53) can be represented approximately by

$$W = (\varepsilon_0/2\varepsilon) \int_V \boldsymbol{E}_c \cdot \boldsymbol{E}_c \mathrm{d}V \tag{6.55}$$

giving

$$\mu = -(\varepsilon_0/2)\left(1 - \frac{1}{\varepsilon}\right) \int_V \boldsymbol{E}_c \cdot \boldsymbol{E}_c \mathrm{d}V \tag{6.56}$$

Following Hush[571], we shall use this approximate expression in the general discussion, sections (c) and (d) below, but when we consider specific cases in section (e) we shall revert to more exact forms.

(k) Although equations 6.53, 6.54 etc. are commonly quoted they are not strictly applicable in cases where the charge distribution of the ion involves point charges, point dipoles, etc., and at these points both $\boldsymbol{E}_c$ and $\boldsymbol{E}$ are infinite. An alternative equation which is not open to this objection is

$$\mu = \tfrac{1}{2} \sum_i q_i \phi_i^R$$

where q_i is the value of the ith point charge and ϕ_i^R is the potential at that point due to the polarisation only. An example in which this equation leads to the correct result but equation 6.54 does not, is the case where the molecule is represented by a sphere with a point dipole at the centre (reference 623, note 10)

(l) For the case of two, spherical, equally charged ions in contact, Marcus has estimated that image effects contribute about 8% of the total interaction energy, and that they contribute negligibly to $\Delta G^{\ddagger}$ (reference 728, note 29). For a worked example including image effects, see reference 187, note 38.

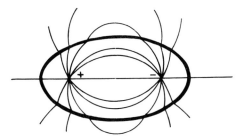

Figure 6.5 Electric field of a dipolar, ellipsoid molecule in a continuous medium. (Kirkwood–Westheimer model), with the interior of lower dielectric constant than the exterior. Lines of force are refracted at the boundary and the field in the medium depends on the dielectric constant of the interior of the molecule.

(c) *Non-equilibrium polarisation*

A polarisation field can exist independently of any electrostatic charges. This can occur if, for example, a dielectric liquid is subjected to an external field, then cooled to a rigid glass, and removed from the field. Alternatively, more generally, if a field is applied and then rapidly altered, and the material of the dielectric fails to respond at the same rate, then after the alteration the polarisation will contain an 'inert' part preserved from the earlier state, and a 'labile' part, in equilibrium with and so jointly determined by the new applied field and the inert polarisation. Fluid dielectrics in general show different dielectric 'constants' depending on the frequency of oscillation of the applied field. In the low-frequency limit the 'static' dielectric constant ε_s reflects the motion of ions and molecules under the influence of the field; in the frequency range of visible radiation, the 'optical' dielectric constant ε_{op} reflects the response of electron clouds to the rapidly oscillating field, the atomic nuclei remaining fixed. This is the classical equivalent of the Franck–Condon principle. By standard arguments of electromagnetic theory, ε_{op} is identified as n^2, the square of the refractive index of the medium, for light of the appropriate frequency[955]. Tables of dielectric constants[830, 1052] and refractive indices[648, 827] as functions of frequency, are available for a wide range of materials.

The relevance of these considerations to electron transfer reactions lies in the fact that the electrons in the reacting system move at much faster rates than the atoms, and, hence, can be said to interact with different parts of the dielectric polarisation. The same principle applies to any other reacting system in which charged particles move at very different rates, for example proton transfer reactions[639].

Marcus[720] has calculated the self-energy W of a system containing a charge distribution, an arbitrary inert polarisation P_u and a labile polarisation P_e in equilibrium jointly with E_c and P_u. Two polarisabilities are defined so that

$$\varepsilon_{op} = 1 + \alpha_{op} \tag{6.57}$$

$$\varepsilon_s = 1 + \alpha_{op} + \alpha_s \tag{6.58}$$

and the final result, again neglecting image effects, is

$$W = \int \left(\frac{\varepsilon_0}{2} \mathbf{E}_c^2 + \frac{1}{2\varepsilon_0 \alpha_s} \mathbf{P}_u^2 - \tfrac{1}{2} \mathbf{P} \cdot \mathbf{E}_c - \tfrac{1}{2} \mathbf{P}_u \cdot \mathbf{E} \right) dV \qquad (6.59)$$

where $\mathbf{P}$ and $\mathbf{E}$ are the total polarisation and field, i.e.

$$\mathbf{P} = \mathbf{P}_u + \mathbf{P}_c = \mathbf{P}_u + \varepsilon_0 \alpha_{op} \mathbf{E} \qquad (6.60)$$

$$\mathbf{E} = \mathbf{E}_c - \varepsilon_0^{-1} \mathbf{P} \qquad (6.61)$$

but for computation purposes it is more convenient to eliminate $\mathbf{E}$ and $\mathbf{P}$, giving

$$W = \frac{1}{\varepsilon_{op}} \int \left(\frac{\varepsilon_0}{2} \mathbf{E}_c^2 + \frac{\varepsilon_s}{2\varepsilon_0 \alpha_s} \mathbf{P}_u^2 - \mathbf{E}_c \cdot \mathbf{P}_u \right) dV \qquad (6.62)$$

It may be verified that when $\mathbf{P}_u$ is replaced by its equilibrium value, $\mathbf{P}_u = \varepsilon_0 \alpha_s \mathbf{E} = \varepsilon_0 (\alpha_s/\varepsilon_s) \mathbf{E}_c$, equation (6.62) reduces to equation (6.53).

(d) Derivation of ΔG^*

Marcus' treatment[720]   The argument used by Marcus can be expressed in the thermodynamic cycle, *Figure 6.6(a)*. States p and s are shown in the gas phase and in solution, μ_p and μ_s being the solvation energies; states p‡ and s‡ are shown only in solution. States p(g), p(aq), p‡(aq) have the same charge distribution and vacuum field E_c^p; states s(g), s(aq), s‡(aq) have the vacuum field E_c^s. States p and s have the respective atomic and electronic polarisation in equilibrium with E_c^p and E_c^s. States p‡ and s‡ have different vacuum fields and therefore different electronic polarisations, but they have the same atomic polarisation, $\mathbf{P}_u^{\ddagger}$, different from those of either state p or state s, and not in equilibrium with either of the four charge distributions. Marcus calculates this polarisation, and thus obtains the free energy of the transition state.

By the resonance condition[m]

$$\Delta U_e = U_s^{\ddagger} - U_p^{\ddagger} = 0 \qquad (6.63)$$

Usually ΔG_e, ΔH_e and ΔS_e will also be zero. If the p‡ → s‡ transition involves a change in electronic degeneracy, there will be a finite entropy change

$$\Delta S_e = R \ln(\Omega_s/\Omega_p) \qquad (6.64)$$

where Ω_s and Ω_p are the degeneracies of the two configurations. Neglecting this, the cycle then gives

$$\Delta G^* = L(-\mu_p + \mu_p^{\ddagger} - \mu_s^{\ddagger} + \mu_s) \qquad (6.65)$$

[m] Marcus[720] gives a quantum-mechanical proof of the equality of energies, which amounts to a formal justification of the application of the Franck–Condon principle to this situation

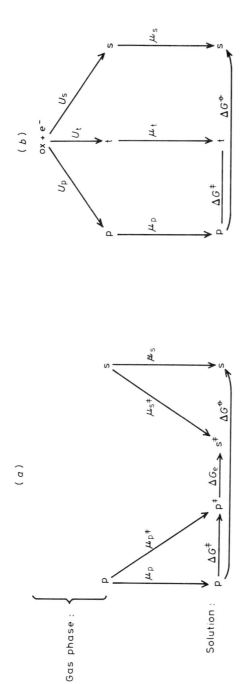

Figure 6.6 Thermodynamic relations between states in an electron-transfer reaction; (a) illustrating Marcus' treatment; (b) illustrating Hush's treatment. Terms labelled U are electron binding energies in the gas phase, terms labelled μ are solvation energies. The observables are the overall standard free energy charge $\Delta G^\ominus$ and the free energy of activation $\Delta G^\ddagger$. The symbols ox, p. etc. are defined in the text

The solvation energies of the equilibrium states are given directly by equation (6.56):

$$\mu_p = -\frac{\varepsilon_0}{2}\left(1 - \frac{1}{\varepsilon_s}\right)\int_V \boldsymbol{E}_c^p \cdot \boldsymbol{E}_c^p \, dV \tag{6.66}$$

$$\mu_s = -\frac{\varepsilon_0}{2}\left(1 - \frac{1}{\varepsilon_s}\right)\int_V \boldsymbol{E}_c^s \cdot \boldsymbol{E}_c^s \, dV \tag{6.67}$$

The solvation terms for the non-equilibrium states are given by equation (6.62):

$$\mu_p^{\ddagger} = \frac{\varepsilon_0}{2}\left(1 - \frac{1}{\varepsilon_{op}}\right)\int_V \boldsymbol{E}_c^p \cdot \boldsymbol{E}_c^p \, dV - \frac{\varepsilon_s}{2\varepsilon_0 \varepsilon_{op} \alpha_s}\int_V \boldsymbol{P}_u^{\ddagger} \cdot \boldsymbol{P}_u^{\ddagger} \, dV +$$

$$+ \frac{1}{\varepsilon_{op}}\int_V \boldsymbol{P}_u^{\ddagger} \cdot \boldsymbol{E}_c^p \, dV \tag{6.68}$$

$$\mu_s^{\ddagger} = \frac{\varepsilon_0}{2}\left(1 - \frac{1}{\varepsilon_{op}}\right)\int_V \boldsymbol{E}_c^s \cdot \boldsymbol{E}_c^s \, dV - \frac{\varepsilon_s}{2\varepsilon_0 \varepsilon_{op} \alpha_s}\int_V \boldsymbol{P}_u^{\ddagger} \cdot \boldsymbol{P}_u^{\ddagger} \, dV +$$

$$+ \frac{1}{\varepsilon_{op}}\int_V \boldsymbol{P}_u^{\ddagger} \cdot \boldsymbol{E}_c^s \, dV \tag{6.69}$$

To identify $\boldsymbol{P}_u^{\ddagger}$ Marcus argues as follows. Different fields $\boldsymbol{P}_u^{\ddagger}$ would produce different total energies for the transition state, and these would represent different reaction paths over the 'saddle' point in the energy surface. However, the usual assumption of transition state theory is that the bulk of the reaction passes over the barrier at its lowest point, hence the energy of the transition state is a minimum with respect to $\boldsymbol{P}_u^{\ddagger}$. Applying this condition to equations (6.68) and (6.69) yields, after some manipulation, a family of solutions defined by taking

$$\boldsymbol{P}_u^{\ddagger} = \frac{\alpha_s}{\varepsilon_s}\left((1+m)\boldsymbol{E}_c^p - m\boldsymbol{E}_c^s\right) \tag{6.70}$$

where m can have any value. The physical significance of this parameter is that the atomic polarisation in the transition state is the same as that which would exist at equilibrium in a medium of polarisability α_u, under the influence of a field $\boldsymbol{E}^{\ddagger} = \boldsymbol{E}_c^{\ddagger}/\varepsilon_s$ where

$$\boldsymbol{E}_c^{\ddagger} = (1+m)\boldsymbol{E}_c^p - m\boldsymbol{E}_m^s \tag{6.71}$$

which is a weighted average of the fields of the precursor and successor complexes. If $m = 0$, the transition state has the atomic configuration of the precursor complex; if $m = -1$ it has the atomic configuration of the successor complex.

Substituting in equations (6.68), (6.69) and using equations (6.66), (6.67) we obtain

$$\mu_p^\ddagger - \mu_p = m^2 A/L \tag{6.72}$$
$$\mu_s^\ddagger - \mu_s = (m+1)^2 A/L \tag{6.73}$$

where

$$A = \frac{L\varepsilon_0}{2}\left(\frac{1}{\varepsilon_{op}} - \frac{1}{\varepsilon_s}\right) \int_V (E_c^p - E_c^s)^2 \, dV \tag{6.74}$$

Introducing this into equation (6.65) we obtain the value of m

$$-m = \tfrac{1}{2}(1 + \Delta G^\circ/A) \tag{6.75}$$

whence

$$\Delta G^\ddagger = \tfrac{1}{4}A(1 + \Delta G^\circ/A)^2 \tag{6.76}$$

These may be compared with equations (6.9), (6.10) above. The free energy A will later be called the reorganisation free energy or 'Franck–Condon' free energy, ΔG_{FC} (see p. 204 and Chapter 8). The quantity $\tfrac{1}{4}A$ may be considered as an intrinsic *free* energy barrier, analogous to $\tfrac{1}{4}A'$.

Hush's treatment[570] The arguments of Hush are represented in the cycle in *Figure 6.6(b)*. States p and s are defined as before, together with a single transition state t, in both gas and solution phases. It is also convenient to define a state 'ox' in the gas phase which is the species $A^+ \cdot B^+$, i.e. with the transferring electron ionised to infinity, but atoms A^+ and B^+ in the positions characteristic of s and p.

The energies U_p and U_s defined in *Figure 6.6(b)* are then the one-electron binding energies defined by equations (6.4), (6.5) above. The wavefunction of the transferring electron in the transition state is assumed to be the linear combination

$$\phi_t = c_1\phi_s + c_2\phi_p \tag{6.77}$$

where c_1 and c_2 are constants and the binding energy U_t is thus

$$U_t = c_1^2 \int \phi_s H \phi_s^* \, d\tau + 2c_1 c_2 \int \phi_s H \phi_p^* \, d\tau + c_2^2 \int \phi_p H \phi_p^* \, d\tau \tag{6.78}$$

where the Hamiltonian H in each case is determined by the atomic coordinates in the transition state. Using the small overlap approximation to eliminate the second term in this equation gives

$$U_t = \lambda U_s + (1-\lambda)U_p \tag{6.79}$$

where

$$\lambda = c_1^2, \quad 1 - \lambda = c_2^2$$

From the two closed cycles we have

$$\Delta G^\circ = (U_s - U_p) + L(\mu_s - \mu_p) \tag{6.80}$$

$$\Delta G^\ddagger = (U_t - U_p) + L(\mu_t - \mu_p) \tag{6.81}$$

whence

$$\Delta G^\ddagger = \lambda \Delta G^\circ - \lambda L(\mu_s - \mu_p) + L(\mu_t - \mu_p) \tag{6.82}$$

Expressions for μ_p and μ_s have been given above. To obtain μ_t, Hush argues as follows[n]. For any ion or complex A^+ the total solvation energy μ is given by equation (6.56).

$$\mu = \frac{\varepsilon_0}{2}\left(1 - \frac{1}{\varepsilon_s}\right)\int_V \mathbf{E}_c \cdot \mathbf{E}_c \, dV \tag{6.83}$$

This can be written

$$\mu = \mu_1 + \mu_2 \tag{6.84}$$

where

$$\mu_1 = \frac{\varepsilon_0}{2}(1 - \varepsilon_{op}^{-1})\int_V \mathbf{E}_c \cdot \mathbf{E}_c \, dV \tag{6.85}$$

$$\mu_2 = \frac{\varepsilon_0}{2}(\varepsilon_{op}^{-1} - \varepsilon_s^{-1})\int_V \mathbf{E}_c \cdot \mathbf{E}_c \, dV \tag{6.86}$$

The terms μ_1 and μ_2 correspond to distinguishable physical processes. If the complex is introduced into the solvent 'instantaneously'—i.e. faster than the time-scale of electron motion (though still reversibly in the thermodynamic sense!) the electron clouds of the solvent molecules will adjust to the charge but the atoms will remain in their original positions. This gives the non-equilibrium state which may be denoted $A^+(aq^*)$

$$A^+(g) + e^-(g) \to A^+(aq^*) + [e^-]_{aq} \tag{6.87}$$

Then on a longer time-scale the molecules reorientate and solvate the complex, the electron clouds meanwhile undergoing further readjustment

$$A^+(aq^*) \to A^+(aq) \tag{6.88}$$

By the argument already given, the free energy change for equation (6.87) is given by equation (6.85); therefore the free energy change for equation (6.88) is given by equation (6.86). Equation (6.86) can also be deduced from the modified Marcus equation (6.62) by noting that the state $A^+(aq^*)$ has the vacuum field $\mathbf{E}_c$, but the non-equilibrium polarisation $\mathbf{P}_u = 0$.

[n] Hush's argument is couched in terms of a specific ion geometry, i.e. the sphere with central charge. In the present text, the physical ideas have been retained but the calculations are done in terms of a general charge distribution in order to emphasise the agreement with Marcus' method

Hush considers the solvation of the transition state t(g) in two steps, but from the definition of the transition state he argues that the charge distribution, and hence the vacuum field, must be evaluated differently for the two stages. The transition state is defined as a species in which the electron has a probability λ of being found in the configuration of s and probability $(1-\lambda)$ of being found in configuration p; but the time of transit of the electron is of the same order of magnitude as the time of motion of the solvent electrons. Hence, when the species t is transferred rapidly from gas to solvent phase, the effect is polarisation of the electron cloud by two complexes, a fraction of $(1-\lambda)$ with the same charge distribution as that on the precursor and λ with the same charge distribution as that on the successors. By equation (6.85) the solvation energy at this stage is therefore

$$\mu_{t1} = -\frac{\varepsilon_0}{2}(1-\varepsilon_{op}^{-1})\left((1-\lambda)\int_V \mathbf{E}_c^p \cdot \mathbf{E}_c^p \, dV + \lambda \int_V \mathbf{E}_c^s \cdot \mathbf{E}_c^s \, dV\right) \qquad (6.89)$$

In the second stage the motion of the solvent molecules is so much slower than the transit time that the effect is one of polarisation by a group of similar molecules, each exerting a field $\mathbf{E}_c^\ddagger$, the mean of the fields of p and s, weighted in accordance with the probabilities:

$$\mathbf{E}_c^\ddagger = (1-\lambda)\mathbf{E}_c^p + \lambda \mathbf{E}_c^s \qquad (6.90)$$

(compare equation 6.71) Thus, by equation (6.86)

$$\mu_{t2} = -\frac{\varepsilon_0}{2}(\varepsilon_{op}^{-1} - \varepsilon_s^{-1})\int_V [(1-\lambda)\mathbf{E}_c^p + \lambda \mathbf{E}_c^s]^2 \, dV \qquad (6.91)$$

The total solvation energy is then given by $\mu_t = \mu_{t1} + \mu_{t2}$. The expressions for μ_p and μ_s are the same as before and substitution into equation (6.82) gives

$$\Delta G^\ddagger = \lambda \Delta G^\circ + \lambda(1-\lambda)A \qquad (6.92)$$

where A is as defined in equation (6.74). Recalling that λ is a reaction coordinate, we observe that the value of λ at the transition state must be such as to make $\Delta G^\ddagger$ a maximum; differentiating equation (6.92) we find the appropriate value $\lambda^\ddagger$ as

$$\lambda^\ddagger = \tfrac{1}{2}(1 + \Delta G^\circ/A) \qquad (6.93)$$

whence, as before

$$\Delta G^\ddagger = \tfrac{1}{4}A(1 + \Delta G^\circ/A)^2 \qquad (6.94)$$

Thus Hush's reaction coordinate λ at the transition state, coincides with Marcus' parameter $-m$.

Interpretation of parameters In a later review[727], Marcus made a detailed comparison of the two treatments. In the accounts given here some of the differences have been eliminated by the use of more general forms for the electrostatic expressions. Also, some corrections which were considered at different stages and in different ways by the two writers, are dealt with later on. The important remaining difference, which goes back to the assumptions of the two models, is in the interpretation of the parameter $\lambda^{\ddagger}$, or $-m$. For Hush, λ describes the actual distribution of charge (or electron probability density) in the $(A-B)^+$ complex as the reaction event progresses. It functions as a reaction coordinate and the value of $\lambda^{\ddagger}$ describes the distribution in the transition state. For Marcus, the distribution of charge is not a possible reaction coordinate, since it does not of itself significantly control the energy of the transition state. (In his model the distribution switches sharply from state $p^{\ddagger}$ to state $s^{\ddagger}$ without affecting the energy.) The parameter m, which Marcus defines only at the transition state, but which clearly could be used as a reaction coordinate, characterises the state of the non-equilibrium polarisation and is some function of the coordinates of all the solvent molecules. This polarisation $P_u^{\ddagger}$, as equation (6.70) shows, has the form of a *possible* equilibrium polarisation, namely the polarisation which would be produced by the reacting pair, if it had the charge distribution postulated by Hush, at equilibrium with the medium. The terms 'virtual charge density' and 'virtual probability distribution' have been proposed to describe the physical condition defined in this way[727]. Marcus concludes that this polarisation 'need not describe (the) actual electronic distribution'. Which of these two models is more appropriate will depend on the magnitude of the interaction energy β.

Independently of these arguments, equations (6.70) and (6.90) embody what has elsewhere been called the 'principle of similitude'[356], since it implies that, other things being equal, when the free energies of the transition state and reactants are similar ($\Delta G^{\ddagger} \simeq 0$) so also are their structures; and likewise when the free energies of the transition state and products are similar ($\Delta G^{\ddagger} \simeq \Delta G^{\ominus}$). These conditions in turn correspond to highly favourable or unfavourable overall free energy changes. This follows directly from any model based on crossing of energy surfaces and similar conclusions have been reached for many other classes of reaction, notably proton transfer and non-metal substitution reactions[356].

(e) Explicit formulae for the intrinsic energy barrier

To evaluate the energy barrier in any given case it is necessary to integrate equation (6.74). This requires consideration of the boundary conditions at the 'surface' of the molecule or reacting pair, and in most cases this is a complicated problem. Mathematically, however, it is exactly the same problem as was already discussed in connection with equilibrium hydration energies. This is because the vector $(\boldsymbol{E}_c^p - \boldsymbol{E}_c^s)$ in equation (6.74) is itself a field, and can be regarded as having arisen from an appropriate hypothetical charge distribution. The quantity A may, therefore, be interpreted as the free energy of transfer of a body having the boundary surface of complex p and bearing the appropriate charge distribution, from a medium of dielectric constant ε_s to a medium of dielectric constant ε_{op}. Some examples of fields $(\boldsymbol{E}_c^p - \boldsymbol{E}_c^s)$ are shown in *Figure 6.5* and the expression for A are given below.

Long-range transfer between spherical ions For transfer between localised centres, the field is that of a finite dipole composed of charges $\pm e$. If the ions are spherical, with radii a_1, a_2, and the distance between them is infinitely large, the Born equation can be applied to the separate ions to yield

$$A = \left(\frac{1}{4\pi\varepsilon_0}\right) Le^2 \left(\frac{1}{\varepsilon_{op}} - \frac{1}{\varepsilon_s}\right)\left(\frac{1}{2a_1} + \frac{1}{2a_2}\right) \quad (6.95)$$

Medium-range transfer between spherical ions This is the model always considered so far for outer-sphere electron transfer reactions. The approximation used by Marcus and Hush assumes that the ions are far enough apart for the fields around each one to be spherically symmetrical, but not so far apart that the Coulombic forces can be neglected. This gives

$$A = \left(\frac{1}{4\pi\varepsilon_0}\right) Le^2 \left(\frac{1}{\varepsilon_{op}} - \frac{1}{\varepsilon_s}\right)\left(\frac{1}{2a_1} + \frac{1}{2a_2} - \frac{1}{R}\right) \quad (6.96)$$

where R is the internuclear distance. Kharkats[617] has also given formulae for the analogous case of medium-range transfer between ellipsoidal (oblate or prolate) ions.

Short-range transfer between spherical ions Most authors have assumed that outer-sphere electron transfer takes place predominantly between ions at close distance, possibly in actual contact, and for this case both Marcus and Hush have calculated A from equation (6.96), using $R = a_1 + a_2$. Although Platzman and Franck emphasised the incorrectness of this procedure[850, 851] it must be admitted that exact calculations are excessively difficult; certainly there is no possible solution in closed form[746].

Recently Kharkats[615] has proposed the equation

$$a = \left(\frac{1}{4\pi\varepsilon_0}\right) Le^2 \left(\frac{1}{\varepsilon_{op}} - \frac{1}{\varepsilon_s}\right)\left(\frac{1}{2a_1} + \frac{1}{2a_2} - \frac{1}{R} - f(R, a_1) - f(R, a_2)\right) \quad (6.97)$$

where the functions $f(R, a_1)$ and $f(R, a_2)$ are given by

$$f(R, r) = \tfrac{1}{4}\frac{R}{R^2 - r^2}\left[\frac{r}{R} - \tfrac{1}{2}\left(1 - \frac{r^2}{R^2}\right)\ln\frac{R+r}{R-r}\right] \quad (6.98)$$

but this is still an approximation. It is obtained by allowing for the volume of the ions but it still involves taking the field around each ion as spherically symmetrical, hence the boundary conditions are not observed. For the case of a contact ion pair ($R = a_1 + a_2$), the additional terms account for $\simeq 6\%$ of the total when $a_1 = a_2$, or 11% when $a = \tfrac{1}{3}a_2$.

In a more sophisticated treatment[616], Kharkats has attempted to allow for the mutual influence of the fields of the two ions, by assuming the ions to be 'completely polarisable' i.e. metallic conductors, and calculating as a first approximation the dipole induced on each ion by the charge of the other. The resulting equations are extremely complicated, but from published graphs it can be seen that the modification is significant: when $R = a_1 + a_2$, and $a_1 = a_2$ the additional terms constitute a 25% increase in A. A complete calculation, however, would consider not only the dipole, quadrupoles, etc. induced at sphere A by the charge on sphere B, but also the new dipoles etc. induced at B by the induced dipoles etc. of A: and so *ad infinitum*. For conducting spheres the net effect must be that, as the ions are allowed to approach towards contact, the effective centres of their charges move to the point of contact, in which case the expression for A diverges to infinity[199].

All these formulae neglect the change in ionic radius as a result of the reaction. Kharkats[619] has proposed more elaborate formulae allowing for this effect.

Ellipsoidal complex A more realistic model has been proposed for electron transfer over medium distance with particular reference to bridged binuclear complexes[198]. The boundary surface is a prolate ellipsoid of revolution and the field ($E_c^p - E_c^s$) is taken to be that of a dipole consisting of equal and opposite charges placed at the foci. If the boundary ellipsoid, when chosen to fit the actual molecule under discussion, has foci which do not coincide with the metal ion centres, the charges at the foci are chosen so that the dipole moment change p is equal to eR, where R is the actual distance between centres. The problem of the necessary integration has been thoroughly treated by Westheimer and Kirkwood[624, 1087]. Adapting their results we obtain

$$A = \left(\frac{1}{4\pi\varepsilon_0}\right) \frac{Lp^2}{2a^2b} \left(\frac{1}{\varepsilon_{op}} - \frac{1}{\varepsilon_s}\right) S(\lambda_0) \tag{6.99}$$

where a and b are the semi-major and semi-minor axes of the ellipsoid, λ_0 is the coordinate of the boundary expressed in confocal coordinates ($\lambda_0^2 = a^2/(a^2 - b^2)$) and $S(\lambda_0)$ is a 'shape' function given by

$$S(\lambda_0) = \sum_{n=0}^{\infty} \tfrac{1}{2}[1 - (-1)^n](2n+1)\lambda_0(\lambda_0^2 - 1)Q_n(\lambda_0)/P_n(\lambda_0) \tag{6.100}$$

where $P_n(\lambda_0)$ and $Q_n(\lambda_0)$ are Legendre polynominals of the first and second kinds of degree n. $S(\lambda_0)$ varies with eccentricity of the ellipsoid from $S \simeq 0.65$ when $\lambda_0 = 1$ to $S = 1$ when $\lambda_0 \to \infty$ [198].

Spherical ion, point dipole From equation (6.99), on proceeding to the limit $\lambda_0 \to \infty$, as the radii A, B become equal, the ellipsoid becomes a sphere and the foci converge to the centre. If the dipole moment p is held constant this leads to

$$A = \left(\frac{1}{4\pi\varepsilon_0}\right) \frac{Lp^2}{2a^3} \left(\frac{1}{\varepsilon_{op}} - \frac{1}{\varepsilon_s}\right) \tag{6.101}$$

where a is the radius of the sphere. The same expression can be deduced from an earlier calculation by Kirkwood[623].

6.3.3 The phonon model

Levich, Dogonadze and others have approached the problem of electron transfer from a different standpoint based on the theory of radiationless transitions in ionic crystals[143, 636]. Mathematically, their treatments are less straightforward than those of Marcus and Hush, but the physical principles are closely analogous, and when suitable approximations are introduced to obtain useable quantitative predictions, the final results are in close agreement. We shall mention here only some of the most distinctive features of these theories. For further details, reviews by Levich[662], and by Dogonadze and co-workers[327, 331] may be consulted, as well as the original papers, most of which have appeared in English translation[321, 322, 323, 325, 663, 664].

In a crystal, every atom may be considered as an oscillator, and the vibrations of the whole lattice may be expressed in terms of normal modes, so that for N atoms, there are N modes with normal coordinates q_k ($k = 1, 2 \ldots N$), all mutually orthogonal in an N-dimensional space. The modes may be classified into two types, according to whether neighbouring nuclei, on average, move in-phase or out-of-phase. In-phase vibrations (acoustic) lead to sound waves with variations in density through the body of the crystal. Out-of-phase vibrations (optical) lead to polarisation waves with variations in dipole moments through the body of the crystal. It is the optical branch which affects the probability of electron transfer between localised sites.

In a liquid, a similar description applies, except that diffusion jumps must be added to the list of possible motions. The optical modes, which are still the relevant ones, include, in descending order of frequency, displacement of the electron clouds relative to the nuclei, bond stretching and deformation, vibration of solvent molecules within the solvent cage (libration), and rotational motion of molecules. All these can also be expressed in terms of a continuum model. The set of time-dependent dipole orientations which constitute one phonon is replaced by a continuous function, the specific polarisation P_k with which is associated a polarisability α_k. The total energy is then

$$U = \int \sum_k \alpha_k (P_k^2 + \omega_k^{-2} P_k'^2) \mathrm{d}V \qquad (6.102)$$

(compare the exponent of equation (6.17) above). The frequencies of the polarisation waves may be grouped into the two main regions of ultraviolet (electronic motions) and infra-red (nuclear motions), but in both regions the wavelengths range from values similar to the interatomic distance, to infinity. Physically, long waves correspond to the polarisation or movement of relatively large numbers of solvent molecules. In the theory the wavelength is allowed to tend to infinity, in which case all the frequencies ω_k tend to a single limiting value, ω_0, with a corresponding polarisability $\alpha(\omega_0)$. This is equivalent to the assumption in the Marcus and Hush theories that all molecules are identical, thus neglecting the difference between inner- and outer-sphere coordination, and between bond vibration frequencies in reactants and products. The distinction between inert and labile polarisation is then introduced, and by an argument due to Frohlich[441, 442] (a version of which was also used by Hush as described above, p.198), the polarisability is identified as

$$\alpha(\omega_0) = 2\pi \left(\frac{1}{\varepsilon_{op}} - \frac{1}{\varepsilon_s}\right)^{-1} \tag{6.103}$$

With the assumption of a single frequency, the Hamiltonian of the pure solvent is shown to take the simple form

$$H = \sum_k \frac{\hbar\omega_0}{4\pi}\left(q_k^2 - \frac{\partial^2}{\partial q_k^2}\right) \tag{6.104}$$

where the terms q are dimensionless normal coordinates.

In a quantum-mechanical treatment, Hamiltonians are obtained for the systems (solvent + reactants) and (solvent + products) and the probability of transition from one to the other is obtained by the use of time-dependent perturbation theory. This is physically equivalent to the two-state treatment with medium to weak interaction, a simpler version of which has been given in section 6.1.2. It is also shown that the $2N$ normal coordinates q_{kp}, q_{ks} ($k = 1, \ldots N$) associated with reactants and products can be replaced by different sets of which only one need be considered as a reaction coordinate. Hence the energy terms are represented as two-dimensional parabolas. With this simplification, which also depends on the assumption of a single frequency, the theory takes the form outlined in section 6.3.1, and the expression for the intrinsic energy barrier A is the same as that of Marcus and Hush.

More recently, theories have been developed which avoid the assumption of a single frequency, using instead the range and distribution of frequencies defined by the vibrational spectrum of the solvent[326, 329, 332, 921, 922].

6.4 Inner- and outer-sphere energy terms

In general, hydration and Frank–Condon free energies (*see* p. 197) can be subdivided into inner- and outer-sphere contributions

$$\Delta G_{FC}^* = \Delta G_{FC}^{in} + \Delta G_{FC}^{out} \tag{6.105}$$

and the continuum theory applies only to the latter term. In principle this gives a method of distinguishing the two terms experimentally. Assuming that ΔG_{FC}^{in} will be independent of the nature of the solvent, while ΔG_{FC}^{out} will be proportional to $(\varepsilon_{op}^{-1} - \varepsilon_s^{-1})$, a plot of ΔG_{FC} against $(\varepsilon_{op}^{-1} - \varepsilon_s^{-1})$ will give a straight line which may be extrapolated to yield ΔG_{FC}^{in} as an intercept. This procedure has been used to analyse optical charge transfer data (Chapter 8) but attempts to verify the expected solvent dependence for the activation energies of thermal reactions, although begun as long ago as 1955[238] have yielded no simple conclusions. These experiments have involved varying the dielectric constant by using solvent mixtures of varying composition. In some cases, no significant variation of rate has been observed[238], in others, not only the rate but also the order of reaction with respect to one reactant, or both, has been found to vary in a complicated manner[237, 584, 599, 742]. One possible reason for this is the fact that the measured activation parameters include the work terms (see pp. 206–210), which must be highly dependent on the solvent. Another is the fact that in mixed solvents, the mole ratio of solvent molecules in the immediate vicinity of the reactant may be quite different from the average over the bulk of the solution [754]. The stability constants for aquo complexes in organic solvents have been measured for several metal ions: some are above, and some below the statistical values.[950].

Several authors have calculated activation energies including both inner- and outer-sphere contributions[570, 724, 726, 795, 985]. (Some earlier calculations by Marcus, using the outer-sphere, continuum model only, gave activation free energies significantly smaller than the experimental values.[722]) When the reaction is unsymmetrical ($\Delta G^\ominus \neq 0$) so that $\alpha = \frac{1}{2}$, a problem arises in that the value of λ (in Hush's terminology) which minimises ΔG_{FC}^{in} may not be the same as the value which minimises ΔG_{FC}^{out}. Moreover, the overall radii of the complex ions, which determine the outer-sphere hydration energies, are themselves functions of λ. Hence the two contributions are interdependent. Reynolds and Lumry[886] have described an iterative procedure to overcome this problem, which they have applied to equation (6.150) below, but they have not published the results in detail. The problem disappears if it can be shown that the dependence of ΔG_{FC}^{in} on λ is of the same form as that of ΔG_{FC}^{out}. Hush argued that this would be a satisfactory approximation for inner-sphere energies determined by ion–dipole interaction; but in considering a reaction of transition metal ions with incomplete d shells, he found it necessary to use successive approximations. German et al.[457] have given an explicit general formula for ΔG_{FC}, expressed in terms of the phonon model of solvent polarisation. Non-adiabatic transition probabilities have been calculated from this model also, with allowance for differences between inner- and outer-coordination layers[333, 613, 921, 923].

6.5 Comparison with experiment [1]: rate comparisons and correlations

6.5.1 The Marcus cross-relation[727]

A particularly elegant deduction from the above theories is the relation between the rate constants of two symmetrical exchange reactions and the corresponding *cross-reaction*

$$A^+ \ldots A \rightarrow A \ldots A^+ \tag{6.106}$$

$$B^+ \ldots B \rightarrow B \ldots B^+ \tag{6.107}$$

$$A^+ \ldots B \rightarrow A \ldots B^+ \tag{6.108}$$

Again the reactions are written as first-order processes, p → s. Applying the Marcus equation to each we have

$$\Delta G^{\ddagger}_{11} = \tfrac{1}{4} A_{11} \tag{6.109}$$

$$\Delta G^{\ddagger}_{22} = \tfrac{1}{4} A_{22} \tag{6.110}$$

$$\Delta G^{\ddagger}_{12} = \tfrac{1}{4} A_{12}(1+\Delta G^{\ominus}_{12}/A_{12})^2 \tag{6.111}$$

where the A's are the intrinsic barriers. Marcus assumes that for sufficiently closely related reactions, A_{12} is given by

$$A_{12} = \tfrac{1}{2}(A_{11}+A_{22}) \tag{6.112}$$

whence

$$\Delta G^{\ddagger}_{12} = \tfrac{1}{2}(\Delta G^{\ddagger}_{11}+\Delta G^{\ddagger}_{22})\left(1+\frac{\Delta G^{\ominus}_{12}}{2(\Delta G^{\ddagger}_{11}+\Delta G^{\ddagger}_{22})}\right) \tag{6.113}$$

Introducing the corresponding rate and equilibrium constants, $k_{12} = (RT/Lh)\exp(-\Delta G^{\ddagger}_{12}/RT)$ etc., $K_{12} = \exp(-\Delta G^{\ominus}_{12}/RT)$, this gives

$$\log k_{12} = \tfrac{1}{2}\log(k_{11}k_{22}) + \tfrac{1}{2}\log K_{12} + \tfrac{1}{2}\log f \tag{6.114}$$

i.e.,

$$k_{12} = (k_{11}k_{22}K_{12}f)^{1/2} \tag{6.115}$$

where

$$\log f = (\log K_{12})^2/4\log[k_{11}k_{12}/RT/Lh)^2] \tag{6.116}$$

When $\Delta G^{\ominus}_{12}$ is fairly close to zero, $f \simeq 1$ and the cross relation becomes

$$k_{12} = (k_{11}k_{22}K_{12})^{1/2} \tag{6.117}$$

This condition is in fact satisfied by the majority of known reactions.

When second-order reactions are considered it is necessary to take account of the stability constants of the precursor and successor complexes, often referred to as the work terms:

$$A^+ + A \xrightleftharpoons{K^{ip}_{11}} A^+ \ldots A \xrightarrow{k_{11}} A \ldots A^+ \tag{6.118}$$

$$B^+ + B \xrightleftharpoons{K^{ip}_{22}} B^+ \ldots B \xrightarrow{k_{22}} B \ldots B^+ \tag{6.119}$$

$$A^+ + B \xrightleftharpoons{K^{ip}_{22}} A^+ \ldots B \xrightarrow{k_{12}} A \ldots B^+ \xrightleftharpoons{K^{sf}_{12}} A + B^+ \tag{6.120}$$

For the second-order rate constants $k'_{11} = K^{ip}_{11}k_{11}$, etc., and the overall equilibrium constant $K'_{12} = K^{ip}_{12}K_{12}K^{sf}_{12}$, the cross relation analogous to equation (6.117) is

$$k'_{12} = (k'_{11}k'_{22}K'_{12})^{1/2}(K^{ip}_{12}/K^{ip}_{11}K^{ip}_{22}K^{sf}_{12})^{1/2} \tag{6.121}$$

If, however, the reagents A^+ and B^+ are of similar charge types, we have, approximately

$$K_{12}^{ip} = (K_{11}^{ip} K_{22}^{ip})^{1/2} \tag{6.122}$$

and if, moreover, the charge type is unchanged by the reaction (as for example in $Fe^{3+} + Cr^{2+} \rightarrow Fe^{2+} + Cr^{3+}$ but not $Cu^{2+} + Cr^{2+} \rightarrow Cu^+ + Cr^{3+}$), then $K_{12}^{ip} \simeq 1/K_{12}^{sf}$ also. With these conditions equation (6.121) simplifies to

$$k'_{12} = (k'_{11} k'_{22} K'_{12})^{1/2} \tag{6.123}$$

and, with somewhat less justification, equation (6.115) may be replaced by

$$k'_{12} = (k'_{11} k'_{22} K'_{12} f')^{1/2} \tag{6.124}$$

where

$$\log f' = (\log K'_{12})^2 \log (k'_{11} k'_{22}/k_0^2) \tag{6.125}$$

and k_0 is the diffusion controlled rate constant (previously written Z, e.g. in equations (4.10) and (4.12). These last equations are the forms originally derived by Marcus and usually quoted.

The important assumption is equation (6.112). It does not follow directly from either of the two models (bond-stretching and dielectric polarisation) discussed in section 6.3. It can, however, be obtained as a first approximation from both these models, if the differences in solvation characteristics (force constants and radii) between the ions A^+ and B^+, and likewise A and B, are sufficiently small. Since symmetrical crossing of the energy surfaces is also only an approximation in the case of the bond-stretching model, we may conclude that the assumption in equation (6.112) is intuitively reasonable. Certainly it seems to be justified by the results.

The usefulness of the cross relation has been verified many times. Provided that the reagents in the two symmetrical exchange reactions are similar in charge type and in the nature of the inner sphere ligands, equation (6.124) generally predicts rates constant within a power of ten or better[916] and similarly for organic systems[723,881,1101]. The equation is now regularly used to calculate rates of reactions which are otherwise difficult to measure[227,847,1040]. It has also been used in reverse, to calculate $\log K_{12}$ from rate data and hence to obtain otherwise unknown redox potentials[109,1020]. In cases where the cross relation fails, some authors[594] have recommended replacing observed exchange rates by a self-consistent set of 'apparent' rates which correctly reproduce the cross reaction rates. More generally, however, conspicuous failures of the cross relations are taken as *prima facie* evidence of some special effect, as shown, for example, in the next section.

6.5.2 Effects of work terms

Table 6.1 lists some examples of reactions which apparently fail to obey the cross relation. In the first two examples, the experimental rate is lower than the calculated value even though consideration of charge type alone would

Table 6.1 Calculated and observed specific rates of some outer-sphere electron transfer reactions[a]

$$A^+ + B \rightarrow A + B^+$$

Reactants[b]		k_{11}/ $M^{-1}s^{-1}$	k_{22}/ $M^{-1}s^{-1}$	K_{12}	$k_{12}(calc)$/ $M^{-1}s^{-1}$ [d]	$k_{12}(exp)$/ $M^{-1}s^{-1}$	$k_{12}(exp)/$ $k_{12}(calc)$	References and notes[c]
A^+	B							
$Fe(phen)_3^{3+}$	$Fe(H_2O)_6^{2+}$	1×10^6	4.0	2.5×10^5	5.6×10^5	3.7×10^4	0.066	189
$Co(phen)_3^{3+}$	$V(H_2O)_6^{2+}$	5.0	1.0×10^{-2}	4×10^{10}	2.3×10^4	3.8×10^3	0.16	864
$Co(phen)_3^{3+}$	$Fe(CN)_6^{4-}$	4.4×10	1.9×10^4	5.0	2.1×10^3	6.0×10^6	3×10^3	502
$IrCl_6^{2-}$	$Fe(DMP)^{2+}$	2.3×10^5	3×10^8	0.30	4.6×10^6	1.1×10^9	0.24×10^3	507
$Mo(CN)_8^{3-}$	$Os(bip)_3^{2+}$	2.0×10^4	$3 \times 10^{8\,(e)}$	0.50	2×10^6	2.0×10^9	1×10^3	188[f]
$HMnO_4$	$Mo(CN)_8^{4-}$	1.1×10^3	3×10^4	1.2×10^{-4}	2.4×10^7	1.9×10^7	0.8	1025
MnO_4^-	$Fe(CN)_6^{4-}$	$4.0 \times 10^{3\,(g)}$	1.9×10^4	$2 \times 10^{2\,(h)}$	1.1×10^5	1.3×10^4	1.2	
AmO_2^{2+}	NpO_2^+	2.4	9.6×10^1	6.4×10^7	5.0×10^4	2.5×10^4	0.5	1106

(a) 25 °C, ionic strengths various (0.1 to 1.0 M)
(b) phen is 9:10-phenanthroline; DMP is 4, 7-dimethyl-1:10-phenanthroline
(c) Unless otherwise stated, this reference contains the measurement of $k_{12}(exp)$ and the calculation of $k_{12}(calc)$ together with references to the original data used
(d) Calculated using $k_{12} = (k_{11}k_{12}K_{12}f)^{1/2}$ where $\ln f = (\ln K_{12})^2/4\ln(k_{11}k_{12}/Z^2)$ with $Z = 10^{11}\,M^{-1}\,s^{-1}$
(e) Assumed same as for $Fe(DMP)^{2+} + Fe(DMP)^{3+}$
(f) 10 °C
(g) Reference 946
(h) Reference 470

support the cross relation. This can be explained in terms of short-range interactions. Marcus suggested[726] that non-Coulombic effects would contribute to the work terms. Presumably the association of $Fe(H_2O)_6^{2+}$ and $Fe(H_2O)_6^{3+}$ is favoured by dipole–dipole interactions on hydrogen bonding, while the association of $Fe(phen)_3^{2+}$ with $Fe(phen)_3^{3+}$ is favoured by van der Waals forces between the polarisable organic ligands. In the cross reaction both these factors are lacking.

In the examples where the reactants are oppositely charged, the Coulombic interactions favour the cross reaction but disfavour the exchange reactions. In these cases the experimental rate constants are greater than the calculated values.

A most elegant application of these arguments to the inner-sphere reaction series has been advanced by Fay and Sutin[403]. The following three reactions have all been shown to proceed by the inner-sphere mechanism with transfer of the bridging group.

$$CrSCN^{2+} + Cr^{2+} \xrightarrow{k^{SCN}} Cr^{2+} + SCNCr^{2+} \qquad (6.126)$$

$$CrNNN^{2+} + Cr^{2+} \xrightarrow{k^{NNN}} Cr^{2+} + NNNCr^{2+} \qquad (6.127)$$

$$CrNCS^{2+} + Cr^{2+} \xrightarrow{k^{NCS}} Cr^{2+} + NCSCr^{2+} \qquad (6.128)$$

The rate constants have been determined, and for the equilibrium constants we have $K^{SCN} = k^{SCN}/k^{NCS}$; $K^{NNN} = 1$, and $K^{NCS} = 1/K^{SCN}$. Fay and Sutin assume further that the intrinsic energy barriers A for the thiocyanate and azide bridged systems are the same. With this assumption, noting that $\log K$ is small or zero in all three reactions, and assuming that the work terms cancel, equation (6.123) predicts

$$k^{NNN} = (k^{SCN}k^{NCS})^{1/2} \qquad (6.129)$$

In fact this is not the case; the azide-bridged reaction is faster than predicted by a factor of $\simeq 80$, but on introducing the precursor complex formation constants K_1^{NNN}, K_1^{SCN}, K_1^{NCS}, the prediction becomes

$$\frac{k^{NNN}}{(k^{SCN})^{1/2}(k^{NCS})^{1/2}} = \frac{K^{NNN}}{(K_1^{SCN})^{1/2}(K_1^{NCS})^{1/2}} \qquad (6.130)$$

Thus the deviation of the three rate constants from the pattern suggested by equation (6.123) is attributed to a corresponding variation in the stabilities of the precursor complexes. Making the reasonable assumption that $K_1^{NNN} \simeq K_1^{SCN}$, since both relate to formation of a Cr^{II}–N bond, Fay and Sutin conclude that $K_1^{NNN}/K_1^{NCS} \simeq 7 \times 10^3$. This may be compared with the corresponding ratio of chromium(III) complex stability constants, $K(Cr^{3+} + NCS^-)/K(Cr^{3+} + SCN^-) = K^{NCS} = 3 \times 10^5$. Evidently both chromium(II) and chromium(III) show the preference for nitrogen–over sulphur-bonding ligands which characterises the 'class (a)' group of metals: but the ion of lower valency shows this preference less strongly, which is also typical[21].

6.5.3 Linear free energy relations

The parameter $\lambda^{\ddagger}$ in equation (6.93) is not, strictly speaking, an experimental quantity, since it is not possible to vary ΔG° without changing other conditions as well. For practical purposes, however, it is assumed that we can vary ΔG° by considering a series of reactions involving different oxidant/reductant pairs, and assuming that the intrinsic barriers and work terms are the same throughout the series. The extent of validity of these assumptions has not been fully explored, though most authors using this approach have discussed the point and justified their treatment of the particular set of data under review. In general, it seems that differences in work terms are relatively unimportant, provided that the reactions concerned are all of the same mechanism, i.e. inner- or outer-sphere, and of the same or similar charge-type. Differences in A are usually assumed to be negligible if the central metal ions are the same throughout the series, so that changes in ΔG° are effected by changes in the ligand environment only. (Changes in outer-sphere ligands or in solvent would presumably have even less effect on A but these have apparently not been exploited). According to Marcus' assumption, equation (6.112), the barrier A_{ij} for a reaction between reagents i and j is given by $A_{ij} = \frac{1}{2}(A_{ii} + A_{jj})$, hence the only requirement for constant A_{ij} is that the reagents which change along the series should have the same barrier.

For a series of reduction B_i with a common oxidant A^+ (or vice-versa), taking A as a constant through the series, the dependences of $\Delta G^{\ddagger}$ on ΔG°, and of $\log k$ on $\log K$, are given by

$$\Delta G_i^{\ddagger} = \tfrac{1}{4}A(1 + \Delta G_i^{\circ}/A)^2 \qquad (6.131)$$

$$\log k_i = \log k_A + \tfrac{1}{2}\log K_i + (\log K_i)^2/16\log(k_A/Z) \qquad (6.132)$$

where $K_A = \exp(-A/4RT)$.

The derivation of these equations is usually illustrated by means of intersecting free energy curves at various relative heights (*Figure 6.7*). Thus the plots of $\Delta G_i^{\ddagger}$ against ΔG_i°, or $\log k_i$ against $\log K_i$, are parabolae (*Figure 6.8*) with slope α at any given value of ΔG° or $\log K$ given by

$$\alpha = d(\Delta G^{\ddagger})/d(\Delta G^{\circ}) = \tfrac{1}{2}(1 + \Delta G^{\circ}/A) \qquad (6.133)$$

$$\alpha = d(\log k)/d(\log K) = \tfrac{1}{2}[1 + \log K/4\log(k_A/Z)] \qquad (6.134)$$

By analogy with electrochemical theory[107], the slope α is sometimes called the *transfer coefficient*. If the range of variation of ΔG_i° is not too great, the dependence approximates to a straight line, which we may take to be a tangent to the overall parabola. The analytical equation of the tangent is

$$(\Delta G_i^{\ddagger} - \Delta G_0^{\ddagger}) = \alpha_0(\Delta G_i^{\circ} - \Delta G_0^{\circ}) \qquad (6.135)$$

where $\Delta G_0^\ominus$ and $\Delta G_0^\ddagger$ are the coordinates of the point of contact and α_0 is the slope at that point. A convenient value of $\Delta G_0^\ominus$ would be the mean value for the set of reactions being considered. Using equations (6.131), (6.133) to eliminate $\Delta G_0^\ominus$ and $\Delta G_0^\ddagger$ this becomes

$$\Delta G_i^\ddagger = \alpha_0(1-\alpha_0)A + \alpha_0 \Delta G_i^\ominus \tag{6.136}$$

$$\log k_i = 4\alpha_0(1-\alpha_0)\log k_A + (2\alpha_0 - 1)^2 \log Z + \alpha_0 \log K_i \tag{6.137}$$

When $\Delta G_0^\ominus = 0$, so that $\alpha_0 = \frac{1}{2}$, this reduces to

$$\Delta G_i^\ddagger = \tfrac{1}{4}A + \tfrac{1}{2}\Delta G_i^\ominus \tag{6.138}$$

Even if the energy curves are not parabolic, a slope $\alpha = \frac{1}{2}$ is thus expected provided only that the curves have equal and opposite gradients at the crossing point. Detailed calculations of the effects of anharmonicity (using the multiphonon model) have been made by Søndergaard, Ulstrup and Jortner[962].

An alternative form of linear free energy relationship is obtained by comparing the reactions of a series of reductants B_i with two oxidants A_1^+ and A_2^+ (or vice versa). Again, provided the range of oxidants is not too long, both sets of activation energies obey the linear equations

$$\Delta G_{i1}^\ddagger = \alpha_{01}(1-\alpha_{01})A_1 + \alpha_{01}\Delta G_{i1}^\ominus \tag{6.139}$$

$$\Delta G_{i2}^\ddagger = \alpha_{02}(1-\alpha_{02})A_2 + \alpha_{02}\Delta G_{i2}^\ominus \tag{6.140}$$

The variation of $\Delta G_{i1}^\ddagger$ in terms of $\Delta G_{i2}^\ddagger$ is

$$\frac{\Delta G_{i1}^\ddagger}{\alpha_{01}} - \frac{\Delta G_{i2}^\ddagger}{\alpha_{02}} = \tfrac{1}{2}(A_1 - A_2) + \tfrac{1}{2}\Delta G_{12} \tag{6.141}$$

where ΔG_{12} is the standard free energy change of the reaction $A_1^+ + A_2 = A_1 + A_2^+$; or more conveniently

$$\Delta G_{i1}^\ddagger = a\Delta G_{i2}^\ddagger + b \tag{6.142}$$

where $a = \alpha_{01}/\alpha_{02}$, $b = (\alpha_{01}/2)(A_1 - A_2 + \Delta G_{12})$. Thus under certain conditions, the intrinsic barrier may be estimated by plotting $\Delta G_{i1}^\ddagger$ against $\Delta G_{i2}^\ddagger$. For example, if A_1 and A_2 are assumed to be the same, or similar enough to be replaced by a mean value A, the calculation simplifies to $A = (\Delta G_{12})^2/2b(1-a^{-1})$. Alternatively, if it is found by experiment that $a = 1$ (as is the case for outer-sphere reactions of Cr^{2+} and V^{2+}) then $(A_1 - A_2) = b - \Delta G_{12} - (b^2 - 2b\Delta G_{12})^{1/2}$.

Equations (6.138) has been generally confirmed, the slope being close to 0.5 when $\Delta G^\ominus$ is close to zero (*Table 6.2*). In view of the derivation from Marcus' continuum theory, these equations were at first assumed to be valid only for outer-sphere reactions, but there are now equally well attested linear

Table 6.2 Correlation of $\log k$ with $\log K$, for electron transfer reactions (see also Chapter 5, Table 5.4)

System[a]	Slope, α	References
$FeL_3^{3+} + Fe^{2+}$	0.5	189
$Ce^{IV} + FeL_3^{2+}$	0.5	189
$Mn^{III} + FeL_3^{2+}$	0.5	189
$Co^{III} + FeL_3^{2+}$	0.27 ($\Delta G^\circ \ll 0$)	189
$Cr^{III} + Cr^{II}$ (various ligands)	0.5	353
$Fe^{3+} + Fe(CPD)_2$	0.44	848
Co^{III}(various) $+ Fe^{2+}$	—	820
$M^{n+} + M'^{m+}$		
$MO_2^{2+} + M'^{m+}$	0.2 ($\Delta G^\circ \ll 0$)	640a
$MO_2^+ + MO_2'^+ \to M^{4+} + MO_2'^{2+}$		
$MO_2^+ + M'^{m+} \to M^{4+} + M'^{(m+1)+}$	0.13 ($\Delta G^\circ \ll 0$)	640a
Co^{III}(various) $+ Ru(NH_3)_6^{2+}$	0.58	893
Co^{III}(various) $+ V^{2+}$	0.56	893
$M^{n+} + e^-.aq$	$\log k$ not solely dependent on $\log K$ (all k close to diffusion limit)	30, 490, 1024
Organic dyes + free radicals	Curved plots consistent with a diffusion limit	873
FeL_3^{3+} & $RuL_3^{3+} + Co(TIM)^{2+}$	0.32	892
FeL_3^{3+} & $RuL_3^{3+} + Co(trans-1:4-diene)^{2+}$	0.34	892

(a) L is 1, 10-phenanthroline, 2, 2-bipyridyl, or various substituted derivatives; CPD is $C_5H_5^-$ and derivatives; *Trans*-1:4-diene is the ligand **I** shown on page 128; TIM is 2, 3, 9, 10-tetramethyl-1, 4, 8, 11-tetraaza-*cyclo*tetra-4, 11-diene.

free energy relations for inner-sphere reactions. For these, the usual assumption is that A is constant for a series of reactions involving the same pair of metal ions and the same bridging group; but even this has been broadened in some cases, as in the comparison between the N_3^-- and NCS^--bridged reactions described above, p. 209. The variation in slope α with ΔG^* has also been verified in a number of instances. For example, reactions of Co^{3+} with a series of substituted derivatives of $Fe(phen)_3^{2+}$ having ΔG° in the range -14 to -19 kcal mol^{-1} gave $\alpha = 0.27$. In such cases the equation

$$\Delta G^\ddagger = \Delta G^{ip} + \tfrac{1}{4}A + \tfrac{1}{2}\Delta G^\circ - \tfrac{1}{2}RT\ln f \tag{6.143}$$

may be used. The 'curvature term' f as defined in equation (6.125) is calculated for each pair of reactants using the appropriate value of K'_{12} but taking $\ln(k'_{11}k'_{22})$ to be constant for the series. Plots of $(\Delta G^\ddagger + \tfrac{1}{2}\ln f)$ against ΔG° are then expected to show slopes of 0.5 and intercepts $\tfrac{1}{4}A_{12} = \tfrac{1}{2}(\Delta G^\ddagger_{11} + \Delta G^\ddagger_{22})$.

Although linear free energy relationships have been studied in some detail, most discussion has centred on the rationalisation of differing slopes. Relatively little detailed work has been done in testing the variation of A_{12} with $\Delta G_{11}^{\ddagger}$ and $\Delta G_{22}^{\ddagger}$. Endicott et al.[892, 893], however, have concluded that for reactions involving certain cobalt(III)/cobalt(II) couples, the dependence of A_{12} on the self-exchange rate parameters is considerably weaker than the Marcus equation requires. One possibility[893] is that the equation $A_{12} = 2(\Delta G_{11}^{\ddagger} + \Delta G_{22}^{\ddagger})$ holds only when $\Delta G_{11}^{\ddagger}$ and $\Delta G_{22}^{\ddagger}$ are fairly similar in magnitude.

6.5.4 Non-linear free energy relations

Although the variation in α with $\Delta G^{\ominus}$ from one series to another is strong evidence of the validity of the quadratic equation (6.131), it is natural to seek a further test by studying a single series of reactions with enough variation in $\Delta G^{\ominus}$ to show the curvature in a single plot. It seems unlikely that this will be feasible by merely varying the non-bridging ligands; instead it is necessary to further relax the restrictions on the constancy of the intrinsic barrier A by considering a series of different reducing metals with a common oxidant or vice versa.

Hyde, Davies and Sykes[578] studied the reactions of Co^{3+} with the powerful reductants Cr^{2+}, V^{2+} and Eu^{2+}; and showed that a plot of $\Delta G^{\ddagger}$ against $\Delta G^{\ominus}$, for reductants ranging from Co^{2+} to Cr^{2+} is indeed curved in the expected sense. More recently, Ekstrom, Maclaren and Smythe[365] added an additional point to this plot, for the reaction $Co^{3+} + U^{3+}$, which fits well on to the same curve; but they pointed out that data for a large number of cation–cation reactions involving other oxidants also cluster well round the same curve. Falcinella, Felgate and Lawrence[399] have produced an analogous plot for the reactions of Tl^{2+} acting alternatively as oxidant and reductant. Inspection of these plots in comparison to *Figure 6.7* strongly suggests that the fastest reactions are close to the theoretical maximum limit of $k = Z$. There has been some discussion of this point since at first it was felt[578] that the limiting value of $\Delta G^{\ddagger} \simeq 9$ kcal was rather too high to be accounted for simply by work terms. One explanation considered for the series of Co^{3+} reactions was a rapid equilibrium between the two spin states of cobalt(III), $(t_{2g}^6 \rightleftharpoons t_{2g}^5 e_g)$ prior to electron transfer[578]. The data from other oxidants have weakened, if not altogether removed, the necessity for this postulate. It has in any case been argued that for hexa-aquo cobalt(III) the spin-change free energy difference is likely to be small[133].

It is accordingly now argued that the limiting rate does represent the diffusion-controlled value. Using the Debye equation (p. 100), Ekstrom, Maclaren and Smythe[365] calculate for the reaction $Co^{3+} + U^{3+}$ the collision rate $Z = 1.7 \times 10^4 \, M^{-1} s^{-1}$ (at 25 °C, zero ionic strength) to be compared with the measured electron transfer rate, $7.1 \times 10^3 \, M^{-1} s^{-1}$. In theory there should be differing limiting rates depending on the ionic charges. It remains somewhat

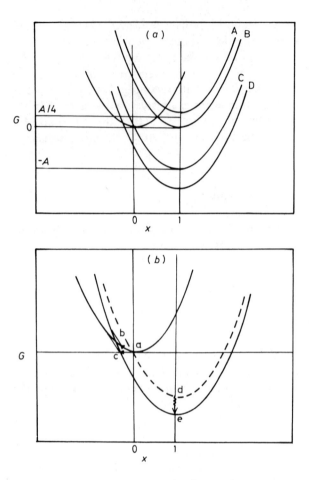

Figure 6.7 Free energy curves for the reacting system p→s. (a) Showing various relative energies of reactants and products: A, $\Delta G^\circ > 0$; B, $\Delta G^\circ = 0$, $\Delta G^{\ddagger} = \frac{1}{4}A$; C, $\Delta G^\circ = -A$, $\Delta G^{\ddagger} = 0$; D, $\Delta G^\circ < -A$. (b) Showing alternative reactions paths a–b–c–e, a–c–e, a–d–e in the 'anomalous region'

puzzling therefore that reactions as diverse as $Co^{3+} + NpO_2^+$, $UO_2^{2+} + Eu^{2+}$ and $Fe^{3+} + U^{3+}$ all fall near the same curve.

Even if the limiting rate were not reached the work term could still be evaluated by fitting the data to equation (6.143). This has not been done for any electron transfer system, but Kreevoy and Konasevich[262] have used this method to estimate work terms in a series of proton transfer reactions. Other

(p) Kuznetsov and Letnikov[640a] have published non-linear $\Delta G^{\ddagger}/\Delta G^\circ$ plots for two sets of reactions covering a wide variety of transition metal and actinide ions, but they correlate them with two straight lines of differing slope, according to whether $\Delta G^{\ddagger}$ is positive or negative. Since they include every reaction twice (forward and reverse) their correlation is equivalent to a linear correlation with slope $\alpha < \frac{1}{2}$ when $\Delta G^\circ \ll 0$, or with $\alpha > \frac{1}{2}$ when $\Delta G^\circ \gg 0$. (See Table 6.2)

6.5.5 The 'anomalous region'

A remarkable prediction of equation (6.131) is that when ΔG° becomes more negative than $-A$, a further decrease in ΔG° leads to an *increase* in $\Delta G^\ddagger$. This follows straightforwardly from *Figure 6.3(a)*: when $\Delta G^\circ = -A$, the products' curve intersects the reactants' curve at $x = 0$, (curve B) while when $\Delta G^\circ < -A$ the intersection point moves to the left of the diagram, with negative reaction coordinate(curve D). In terms of the mechanism discussed above, this would mean for example that the metal–ligand bonds in the complex of higher oxidation state would be compressed, and those in the complex of lower oxidation state would be stretched, until the energies of the products' and reactants' electronic configurations become equal. The electron distribution coordinate of Hush's model would clearly be inapplicable here since the transferring electron cannot be located on the reductant with a probability greater than 1; Marcus' equivalent polarisation could, however, be defined so as to give $m < 0$.

It has, however, since been argued that this somewhat bizarre prediction is not likely to be fulfilled. Marcus and Sutin[731] point out that electron transfer in the abnormal region requires a non-adiabatic transition (*see Figure 6.8(b)*) presumably of lower probability than the adiabatic transitions in the normal region. Other processes such as 'horizontal' nuclear tunnelling may, therefore, compete effectively (*Figure 6.7(b)*, path a–c–e). (Even so, some lowering of rate might then be expected since nuclear tunnelling is generally considered to be relatively improbable, as pointed out above (p. 179)). Another theory[1050] based on the non-adiabatic description leads to a linear dependence of ΔG° on $\Delta G^\ddagger$.

In one system, the deviation from the Marcus equation has been clearly established. In a series of reactions between organic molecules and radical ions, Rehm and Weller[873] obtained a parabolic plot of $\Delta G^\ddagger$ against ΔG° in the normal region (ΔG° ranging from -5 to $+5$ kcal mol^{-1}); but in the range $\Delta G^\circ = -5$ to -62 kcal mol^{-1}, $\Delta G^\ddagger$ remained constant.

The constancy of $\Delta G^\ddagger$ can also be explained without recourse to non-adiabatic processes. It has been suggested[357] that when the energy curve of the ground state products passes the minimum on the reactant curve, other curves for higher energies levels may become available, as shown in *Figure 6.7(b)*. Hence the reaction may still proceed with zero thermal activation energy along the route a → d → e. It was originally suggested that the relevant higher levels might be electronically excited levels, but Efrima and Bixon[352] have argued that vibrational levels are more likely. The high density of vibrational states then ensures that a convenient level will be available for a wide range of product energies, so that $\Delta G^\ddagger$ can remain constant over a correspondingly wide range of ΔG°. More sophisticated quantum mechanical models have also been discussed[332,920,1047].

6.5.6 Activation parameters

From equation (6.94) the activation parameters may be obtained as[731]:

$$\Delta H^{\ddagger} = -\partial(\Delta G^{\ddagger}/T)/\partial(1/T) = \Delta H_A(1-4\beta^2) + \tfrac{1}{2}\Delta H^{\ominus}(1+2\beta) \quad (6.144)$$

$$\Delta S^{\ddagger} = -\partial(\Delta G^{\ddagger})/\partial T = \Delta S_A(1-4\beta^2) + \tfrac{1}{2}\Delta S^{\ominus}(1+2\beta) \quad (6.145)$$

where ΔH_A and ΔS_A are activation parameters corresponding to the intrinsic barrier A, and $\beta = 2\Delta G^{\ominus}/A = 2(2\alpha - 1)$. This may be positive or negative, but is usually small: hence if $\Delta H^{\ominus}$ is sufficiently negative, $\Delta H^{\ddagger}$ may be negative even when ΔH_A is positive. Complete calculations to verify these equations require values of ΔH_A and ΔS_A from the appropriate exchange reactions, as well as of $\Delta H^{\ominus}$ and $\Delta S^{\ominus}$. For the reactions $Fe(bipy)_3^{3+} + Fe^{2+}$ and $Ru(bipy)_3^{3+} + Fe^{2+}$, theory agrees with experiment at least as regards the negative sign i.e. $\Delta H^{\ddagger} = -3.2$ and -2.9 kcal mol^{-1} (calculated)[731], -0.8 and -0.3 kcal mol^{-1}

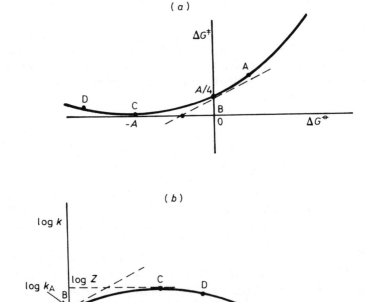

Figure 6.8 Predicted dependence of $\Delta G^{\ddagger}$ upon $\Delta G^{\ominus}$, and of $\log k$ upon $\log K$, for a series of electron transfer reactions with constant intrinsic free energy barrier. (See text and equation (6.13), (6.132)). The broken diagonal lines have gradient 0.5. Points lettered A to D correspond to the curve-crossing situations shown in Figure 6.7

by experiment[153, 261]. More recent measurements on a number of related systems also agree with equation (6.144): a plot of $(\Delta H^{\ddagger} - \frac{1}{2}\Delta H_A^{\ddagger})$ against $\frac{1}{2}\Delta H^{\circ}(1+2\beta)$ conforms to the expected straight line of unit slope[152]. The activation entropies of exchange and cross reactions have also been discussed by Jacks and Bennett[594].

6.6 Comparison with experiment [2]: absolute calculations

In attempting to calculate rate constants direct from theory, all authors have used the outer-sphere contact model for the transition state, and have calculated the outer-sphere contribution to $\Delta G^{\ddagger}$ via equation (6.96). They have assumed that the inner-sphere contribution can be regarded as the sum of separate contributions from the two reacting metal ions. To calculate these contributions, the central metal ions are considered as point charges and the coordinated solvent molecules are represented as point dipoles aligned towards the centre.

For the inner-sphere energy, Hush[570] used equations consisting of an electrostatic ion–dipole term for the attractive potential and an arbitrary $(-m^{\text{th}})$ power term for the repulsive potential:

$$U(z) = -\frac{L}{4\pi\varepsilon_0}\left(\frac{zep}{r^2} - \frac{NC}{r^m}\right) \qquad (6.146)$$

where ze is the ionic charge, p the dipole moment of the solvent molecule, r the metal-ligand centre-to-centre distance, N the coordination number and C is a constant. After minimising this expression with respect to r, to find the equilibrium distance r_e, Hush obtained an expression for r_e as a function of charge z and so estimated the coefficient m by comparing the radii of ions of different charge. For ions with incomplete d-shells he included an additional term for the splitting of t_{2g} and e_g levels using the crystal field model and the same dipole moments p. Sutin and others[81, 918, 985] replaced equation (6.146) by a more elaborate expression with allowance for the additional dipole produced on each water molecule by the central charge. For a symmetrical octahedral complex this gives

$$U = -\frac{L}{4\pi\varepsilon_0}\left\{-\frac{6ze}{r^3}(p_0 + p_i) + \frac{6 \times 1.19}{r^3}(p_0 + p_i)^2 + \frac{6p_i^2}{2\alpha} + \frac{6C}{r^m}\right\} \qquad (6.147)$$

where p_0 and p_i are the permanent and induced dipole moments, and α is the polarisability. Following Basolo and Pearson[81], the value of $m = 9$ was assumed. The value of p_i was obtained by solving the simultaneous equations

$$p_i = \alpha E \qquad (6.148)$$

$$E = \left(\frac{L}{4\pi\varepsilon_0}\right)\left(\frac{ze}{r^2} - \frac{2 \times 1.19(p_0 + p_i)}{r^3}\right) \qquad (6.149)$$

where E is the magnitude of the component of field, at the centre, directed along the axis of $(p_0 + p_i)$.

In a different set of calculations, Sutin[985] used equations (6.35) and (6.147) above, but since the force constants f of the breathing modes of vibration were not available, these were estimated using the definition $f = d^2U/dr^2$ where $r = r_e$. Later, Bockris, Khan and Matthews[132] used experimental force constants, or in some cases force constants estimated by analogy from other experimental values. German et al.[457] based their calculations on the phonon model, assuming the same vibrational wavenumber $\Omega = 450$ cm^{-1} for all trivalent ions and $\Omega = 360$ cm^{-1} for all divalent ions.

Since all the experimental data consist of second-order rate constants, the work terms have also to be estimated in each case. Different formulae have been used (chapter 4, p. 101); but the variation between them is not great. The greatest uncertainty is in the choice of internuclear distance R. Hush and Sutin estimated R as the sum of reactant ion radii, while German et al.[457] chose a larger constant value $R = 7.0 \times 10^{-10}$ m. Comparisons of calculated and experimental activation free energies are shown in *Figure 6.9*. When the same system has been considered by different authors there are some wide discrepancies; but for any given set of calculations the correlation between $\Delta G^{\ddagger}_{\text{calc}}$ and $\Delta G^{\ddagger}_{\text{obs}}$ is satisfactory: indeed it may be considered remarkable in view of the difficulties which have been encountered in calculating absolute rates of other types of reaction. However, the agreement may be more apparent than real. Reynolds and Lumry[886] performed a detailed calculation on the reaction

$$\text{Fe(phen)}_3^{3+} + \text{Fe(H}_2\text{O)}_6^{2+} \rightarrow \text{Fe(phen)}_3^{2+} + \text{Fe(H}_2\text{O)}_6^{3+} \tag{6.150}$$

using the iterative procedure referred to above, p. 205. They found that the calculated and observed activation free energies agreed within about 10%, but on calculating the contributions of the enthalpy and entropy terms, they found that $\Delta H^{\ddagger}$ made a major contribution to $\Delta G^{\ddagger}$, whereas experimentally for this reaction $\Delta H^{\ddagger}$ is almost zero.

All the calculations mentioned here have been based on the 'medium overlap' model, assuming a transmission coefficient $\kappa = 1$. Where experimental rates fall significantly below calculated rates, one obvious possibility is that the rate is limited by a low transition probability at the crossing point. On this basis, Levich[662] estimated $\kappa \simeq 10^{-3}$ and 10^{-4} for several of the reactions represented in *Figure 6.9*, but he also pointed out that in view of the approximations involved in the calculation, this conclusion was far from certain.

6.7 Other theories of electron transfer

In this book we have considered only the more elementary branches of electron transfer theory, closely related to the familiar transition-state theory. The most important assumption of the theory is the separability of atomic and

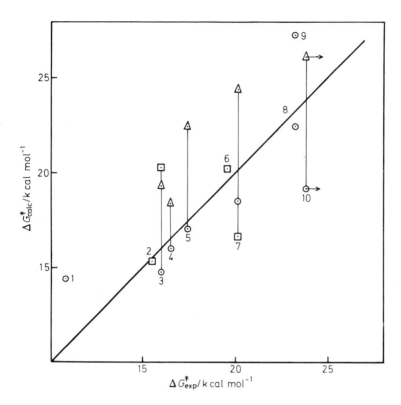

Figure 6.9 Comparison of calculated and experimental activation parameters for symmetrical one-electron transfer reactions $A^+ + A = A + A^+$; *Sources of calculations:* □ *Hush*[570], ⊙ *Bockris, Khan and Matthews*[132], △ *German et al.*[457]. *The reacting systems, shown by oxidant* A^+, *and source of data, are as follows (all at 25°C unless otherwise stated). 1,* $Fe(CN)_6^{3-}$, *reference 688; 2,* NpO_2^{2+}, *reference 239; 3,* $Fe(H_2O)_6^{3+}$, *reference 951; 4,* $Co(H_2O)_6^{3+}$, *reference 139; 5,* $Mn(H_2O)_6^{3+}$, *reference 314 ($\Delta G^{\ddagger}$ estimated from some other reactions using the cross relation, equation (6.124)); 6,* Ce^{IV} *in 6M* HNO_3, *reference 486; 7,* $V(H_2O)_6^{3+}$, *reference 633; 8,* $Co(en)_3^{3+}$, *reference 667; 9,* $Co(NH_3)_6^{3+}$, *at 64.5°C, reference 127; 10,* $Cr(H_2O)_6^{3+}$, *reference 31*

electronic motions. This leads to the use of potential energy surfaces, and to the distinction between adiabatic and non-adiabatic transfer. It also leads to the concept of nuclear tunnelling. More general theories have also been developed, from which the results quoted above can be deduced as limiting cases. Christov has given a more general form of the transition-state theory over a continuous temperature range, from which the adiabatic, non-adiabatic and nuclear tunnelling cases can be deduced under specified conditions[228, 229] and has applied this to electron transfer and proton transfer reactions[230, 231]. Other general approaches have been developed and reviewed by Schmidt[924-8], but the conclusions of this latter work have not yet been cast in a form convenient for experimental testing.

6.8 Connection with electrochemical kinetics

There is a close analogy between a homogeneous electron exchange reaction

$$A^+(aq) + A(aq) \rightleftharpoons A(aq) + A^+(aq) \tag{6.151}$$

and an electrochemical reaction

$$A^+(aq) + e^- \underset{k_{-el}}{\overset{k_{el}}{\rightleftharpoons}} A(aq) \tag{6.152}$$

where e^- denotes the electron in the electrode. In the latter case, the net rate of reaction is measured by the current, given by

$$i = (k_{el}[A^+] - k_{el}[A])\mathcal{A}F \tag{6.153}$$

where $\mathcal{A}$ is the area of the electrode and F is the Faraday constant. k_{el} and k_{-el} are heterogeneous first-order rate constants (dimensions [concentration]$^{-1}$ [area]$^{-1}$ [time]$^{-1}$; usual units, cm s^{-1}). The standard free energy change associated with reaction (6.152) is

$$\Delta G^\circ = F(\varepsilon - \varepsilon^\circ) \tag{6.154}$$

where ε is the applied potential and ε° is the standard potential of the A^+/A couple. Mechanisms of electrochemical transfer exhibit many of the same complications as those of homogeneous transfer, reviewed in this book: for example, catalysis, and rate-determining preliminary steps, and bridged electron transfer when the reacting complexes are adsorbed at the electrode. The theory of electrochemical transfers runs closely parallel to that of homogeneous reactions, and has been developed by many of the same authors, though some, notably Christov, have tended to concentrate on electrochemical aspects. Some of the key concepts were derived first for electrochemical systems and then applied to homogeneous systems; with others the reverse is true. Occasionally, comparisons with electrochemical systems have been used to illustrate points in connection with homogeneous mechanisms[373].

Electrochemical reactions have been described in terms of intersecting energy surfaces, and as before the probability of an individual reaction event can be related to the energy of reorganisation of the solvent and the probability of an adiabatic crossing between energy surfaces. We shall not discuss these points: good reviews of both theoretical[107, 328, 662, 727, 929, 938] and experimental aspects[725, 1055] are available. Here we mention the 'medium overlap' case, where the electronic interaction between the reactants and the electrode is small, but the transmission coefficient is also close to unity. In this case (neglecting work terms) the electrochemical rate constant is given by[728, 729] the equation

$$k_{el} = k_{el}^0 \exp(-\Delta G_{el}^\ddagger/RT) \tag{6.155}$$

where Z_{el} is an appropriate specific collision rate (see chapter 4, equation (4.12), and Marcus obtained

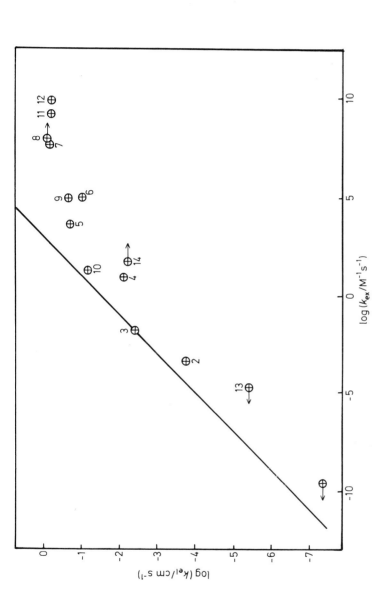

Figure 6.10 Correlation of electrochemical rate constants $k_{el}(A^+ + e^- \to A)$ and exchange rate constants $k_{ex}(A^+ + A \to A + A^+)$ for different redox couples in aqueous media. The straight line is calculated from Marcus' equation (6.159), with $k_{ex}^0 = 10^{11} M^{-1} s^{-1}$, $k_{el}^0 = 10^4 cm s^{-1}$. Oxidants A^+: 1, $Co(NH_3)_6^{3+}$; 2, Eu^{3+}; 3, V^{3+}; 4, Fe^{3+}; 5, MnO_4^-; 6, 9, $Fe(CN)_6^{3-}$; 7, $Fe(bipy)(CN)_4^-$; 8, $Fe(bipy)_3^{3+}$; 10, $Co(phen)_3^{3+}$; 11, $Cr(bipy)_3^{2+}$; 12, perylene; 13, Cr^{3+}; 14, UO_2^{2+}. Sources: 1–6, reference 726; 7–12, reference 1078; 13, reference 919; 14, references 472, 640

$$\Delta G_{el}^{\ddagger} = \tfrac{1}{4}A_{el}(1 + \Delta G_{el}^{\circ}/A_{el})^2 \qquad (6.156)$$

where A_{el} is an intrinsic free energy barrier. Thus a plot of $\Delta G_{el}^{\ddagger}$ against ΔG_{el}° is expected to be a parabola, but over a sufficiently small range of ΔG_{el}°, near zero ($\varepsilon \simeq \varepsilon^{\circ}$) this reduces to a straight line with the equation

$$\Delta G_{el}^{\ddagger} = \tfrac{1}{4}A_{el} + \tfrac{1}{2}\Delta G_{el}^{\circ} \qquad (6.157)$$

Linear correlations between $\Delta G_{el}^{\ddagger}$ and ΔG_{el}° (more commonly in the experimental literature, between $\log k_{el}$ and ε or between $\log i$ and ε), are well known and have been much discussed[107]. Attempts to verify the predicted parabolic dependence, however, have led to some controversy through difficulties of interpretation of the electrochemical data. For the most recent discussion, see reference 1078.

By applying the bond-stretching and continuum models, Marcus further shows that[728, 729]

$$A_{el} \simeq \tfrac{1}{2}A \qquad (6.158)$$

where A is the intrinsic free energy barrier of the homogeneous reaction equation (6.151)[(q)]. Combining equations (6.138) and (6.157) with equation (6.158), leads to the following relationship between k_{el}, the electrochemical rate constant, and k_{ex}, the homogenous exchange rate constant:

$$(k_{ex}/k_{ex}^0)^{1/2} = (k_{el}/k_{el}^0) \qquad (6.159)$$

For collisions between neutral reactants, $k_{ex}^0 \simeq 10^{11}\,M^{-1}\,s^{-1}$ and $k_{el}^0 \simeq 10^4\,cm\,s^{-1}$ [(r)]. For charged reactants different values apply, but these differences can be neglected if the work terms in the respective reactions are small, or constant along a reactions series. The actual correlation so far observed is shown in *Figure 6.10*. Although the Marcus equation is fairly well obeyed at lower rates, there are strong deviations at higher rates. Of the various explanations which have been considered, the most likely suggestion is that the electrochemical rate constants reach a diffusion-controlled limit significantly below the expected value, i.e. at $\simeq 1\,cm^{-1}\,s^{-1}$ [919].

(q) According to Saveant[919a], earlier theories of Hush[569, 570] lead to $A_{el} = A$. See also reference 572, footnote on p. 1008

(r) Marcus[728] quotes theoretical expressions derived from the kinetic theory of gases, i.e. $k_{ex}^0 = (8\pi kT/m)^{1/2}R^2$ and $k_{el}^0 = (kT/2\pi m)^{1/2}$, for ions of mass m and collision diameters R. Using Chapman's equation for the diffusion coefficient, (reference 111, p. 184), these give $k_{ex}^0 = 8\pi DRL$ (see chapter 4, equation 4.8) and $k_{el}^0 = 2D/\pi R$. Taking $D = 2 \times 10^5\,cm^2\,s^{-1}$ and $R = 4 \times 10^{-10}$ m, $k_{ex}^0 = 1.2 \times 10^{10}\,M^{-1}\,s^{-1}$ and $k_{el}^0 = 3.2 \times 10^2\,cm^{-1}\,s$ for reactions in aqueous solution at 25 °C

Chapter 7
Bridged electron transfer: theory and experiment

The discovery of inner-sphere electron transfer reactions (Chapter 5) immediately showed that rates of electron transfer could depend greatly on the nature of the bridging group between oxidant and reductant centres. Most work has been done on inner-sphere reactions—to such an extent that the terms 'inner-sphere' and 'bridged' electron transfer have often been treated as synonymous. There is, however, some evidence for bridging in outer-sphere reactions, as noted in Chapter 5, and again in section 7.1.3 below. The function of the bridging group has been discussed theoretically but, in comparison with non-bridged reactions, it is fair to say that the theory of bridged reactions is still comparatively undeveloped.

7.1 Theoretical models

7.1.1 The polarisation model

One suggestion, by analogy with electron conduction and magnetic exchange in the solid state, was that the polarisability of the bridging group was a determining factor; thus an iodide ion would function more effectively than a fluoride ion, and an unsaturated or aromatic organic system more effectively than an aliphatic system. In particular, the idea that a conjugated bond system can conduct electrons had an obvious appeal. In the field of optical spectroscopy, considerable success has resulted from the use of 'free-electron' models, in which the π-electrons of an aromatic ring or conjugated polyene system are assumed to move freely, as non-interacting particles in a one- or two-dimensional 'box'[595, 782]. Direct evidence of electron mobility in aromatic systems comes from measurements of ring-currents induced by a magnetic field[256]. However, the analogy between electron mobility in this sense and the transfer of an electron between localised reducing and oxidising centres is not exact. This can be seen by translating the descriptions of the two processes into a common language, i.e. that of molecular orbitals. When a time-dependent magnetic field is applied to the aromatic ring electrons, the energy levels can be said to split into two, corresponding to clockwise and anti-clockwise currents, and there will be a net ring current when the population in one level is greater than that in the other. When electron transfer between two atoms is facilitated by bridging groups, this means that the molecular orbital centred on the two atoms is lowered in energy by mixing with another, empty orbital of the bridging group. Effective transfer depends on the presence of such an empty orbital, of appropriate symmetry and sufficiently low in energy, and this need not necessarily be the same orbital as is required for other forms of polarisation.

This argument leads to the idea that if electron transfer between two centres is facilitated by a bridging group, the process occurring on the bridge is analogous to chemical oxidation or reduction. As will be seen later, this idea has proved fruitful both theoretically and experimentally.

7.1.2 The three-state model

In the same way as for the two-state model of outer-sphere electron transfer (Chapter 6), we may define energy terms U_p, U_x, U_s for the localisation of the transferring electron on atoms B, X or A respectively. These energies would correspond to three zero-order states, $A^+.X.B$ (p), $A^+.X^-.B^+$ (x) and $A.X.B^+$ (s). Alternatively, the state x may be $A.X^+.B$. A plot of any of these terms against all relevant atomic coordinates defines a multidimensional energy surface. In the harmonic approximation each surface is a paraboloid, and a suitable cross section gives a two-dimensional parabola corresponding to motion along a normal coordinate. *Figure 7.1* shows projections of such parabolae on to a single plane. Three possibilities are distinguished, according to the energy of the middle term, U_x. In case (a), (*Figure 7.1(a)*), the lowest intersections of U_x with U_p and U_s are higher than the intersection of U_p and U_s. The point P representing the reaction p → s will therefore traverse the U_p and U_s curves with a single transition at point T. If the U_x term is sufficiently high it may be ignored altogether and the two-state description may be used, as reviewed in the previous chapter. Otherwise, however, the presence of the U_x term may in some way affect the transition probability at the point T. (In such cases the state x is sometimes called a *virtual state*.) Halpern and Orgel[509] defined two mechanisms: *single exchange* in which the transition probability is a function of overlap of orbitals of atoms A and B only, and *double exchange* in which the overlap of orbitals of X is significant. Together these two mechanisms are sometimes called 'resonance transfer', but this is a misnomer, perhaps arising from the fact that Halpern and Orgel's original treatment was founded on the model of gas-phase atom—atom reactions.

In case (b), (*Figure 7.1 (b)*) the term U_x has the same energy as the intersection point of U_p and U_s. Thus, the activation energy of the reaction p → s is the same as before, but the transition probability at the crossing may be increased by mixing in the wavefunction of state x. The effectiveness of this mixing will depend on the degree of overlap of the relevant orbitals, but for a given overlap, it will be greatest when the energies are equal, as here. This is the *superexchange mechanism* as defined by Halpern and Orgel[509]. In case (c), (*Figure 7.1 (c)*) the higher of the two intersections of U_x and U_p, or U_x and U_s,

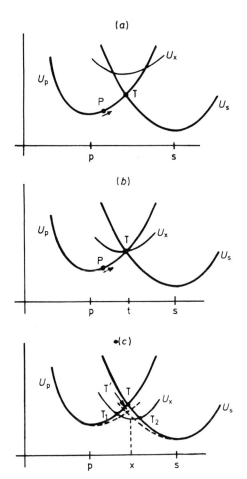

Figure 7.1 Energy terms for electron transfer in a three-state system showing three possible mechanisms according to the energy of the intermediate zero-order state. See text, p. 224

is lower than the lowest intersection of U_p with U_s. This opens a pathway with two successive transitions, T_1, T_2 and if the U_x surface has a minimum between these, the state x is a true intermediate. This defines the *chemical* mechanism[455, 509]. It should be noted that in case (c), the alternative superexchange and resonance mechanisms will, in general, also occur, though with lower probability. The reaction coordinates most favourable to each or the three paths, in a given system, need not necessarily coincide. That is to say, suppose that one set of atomic motions favours the reaction p → x, and another set favours the reaction x → s; then a continuation of these two sets of motion might lead to a crossing point T (*Figure 7.1 (c)*) but a different set of motions might lead to a similar crossing at lower energy, point T'. It has been argued, however, that a single reaction coordinate can be used as a first approximation so that the alternative reaction paths can be represented using only three parabolas. Dogonadze and co-workers have developed this idea and have considered the effects of various relative energies[1061].

As in the two-state model we distinguish 'adiabatic' reactions with transition probabilities close to unity, from 'non-adiabatic' reactions with low transition probability. It seems reasonable to suppose that adiabatic processes will be more prevalent among bridged than among non-bridged electron transfer reactions, and we have already seen that for the latter, adiabatic or at least weak-overlap conditions are consistent with most experimental results. Nevertheless, most authors discussing bridged electron transfer so far have assumed non-adiabatic transitions.

In cases (a) and (b), if the overalp of A and B orbitals is sufficient to give $W_{ps} \simeq 1$, then the further interaction of the bridge can have little effect. In case (c), if both probabilities are high, $W_{px} \simeq W_{xs} \simeq 1$, and the effect of varying the U_x term is to change the chemical stability of the radical-intermediate and hence the free energy of activation. Under these conditions, the energy profile can be replaced by a free energy profile, and analysed as described below, p. 231.

7.1.3 Applicability of the models

When the bridging group is coordinated to both centres and consists of one atom (Cl^-) or one atom with a side chain (OH^-, or NCS^- with adjacent attack), it seems likely that direct or resonance, rather than chemical, transfer will be preferred. As will be seen later, the proven cases of chemical transfer all involve polyatomic bridging groups and remote attack. It is also intuitively clear that some reactions involve only direct (two-state) transfer while others involve participation of a third state. For example, in the exchange process $Cr^{3+}.Cl^-.Cr^{2+} \rightarrow Cr^{2+}.Cl^-.Cr^{3+}$, the state $Cr^{2+}.Cl.Cr^{2+}$ is probably too high in energy to be considered even as a contributory structure in the transition state. The free energy difference for the reaction between separate species. $Cr^{3+} + Cl^- \rightleftharpoons Cr^{2+} + Cl$ is about 66 kcal mol^{-1}. However, for the corresponding reaction with cobalt in place of chromium, the corresponding free energy difference is only $\simeq 12$ kcal mol^{-1}, and the virtual state must therefore contribute significantly[a].

Similar considerations apply to the other variant mechanisms in which the bridging group is coordinated to only one reactant centre, or to neither, and again it is true that chemical transfer has been demonstrated only for molecular, not atomic, bridges. The cationic bridges postulated in the reactions $MnO_4^- + MnO_4^{2-}$ and $Fe(CN)_6^{3-} + Fe(CN)_6^{4-}$ probably do not play any direct role as electron acceptors. The fact that the order of effectiveness of alkali cations (see p. 166) agrees with the order of increasing polarisability has been pointed out, but against this is the fact that $Co(NH_3)_6^{3+}$ is a still more effective catalyst. It seems likely that the primary role of the cations in these reactions is to lower the electrostatic repulsion between the reactant ions.

(a) Taube and Gould estimate activation energies for the reaction $CrF^{2+} + Cr^{2+}$, assuming either $Cr^{3+}.F^{2-}\ Cr^{3+}$ or $Cr^{2+}.F^0.Cr^{2+}$ as an intermediate. Even using quite gross approximations they obtain $\Delta G^{\ddagger} \simeq 70$ and 100 kcal mol^{-1} respectively, compared with experimental values $\Delta G^{\ddagger} = 19.6$ kcal mol^{-1} [1015]

7.2 Review of theories

7.2.1 Single exchange: the two-state model

If the term U_x (*Figure 7.1*) is at sufficiently high energy, the bridging group does not influence the electronic structure of the transition state and the problem of calculating the activation energy reduces to that of the two-state description reviewed in the previous chapter.

Dogonadze, Ulstrup and Kharkats[333] have given calculations for a simple model of a trinuclear linear system, with two electronic configurations $(A^+ - X \ldots B)$ and $(A \ldots X-B^+)$, which in our notation may be written p, s. The A–X or X–B bond vibrations are considered to be simple harmonic with different frequencies, before and after electron transfer, and they are coupled in different ways depending on the relative masses of the atoms A, X, B. The non-bridging inner-sphere ligands are considered as independent harmonic oscillators, and solvent polarisation is introduced in terms of the phonon model, though later re-written in terms of the continuum model.

Since there are many independent oscillating systems, the energy terms U_p and U_s are multi-dimensional and this leads to the problem of finding the minimum energy of interaction (*see* section 6.4, p. 205). The authors give approximate solutions appropriate to several limiting cases. However, they use an expression for the outer-sphere solvation energy which is tantamount to saying that solvent coordinates remain fixed throughout the reaction event; hence, they consider only the interaction between movement of the bridge and movement of non-bridging inner-sphere ligands. The general result is that, even for a symmetrical reaction, with $A \equiv B$, $\Delta G^{\ddagger} = 0$, the transfer coefficient α is not necessarily equal to 0.5. The detailed calculation is complicated, but several limiting cases have been worked out.

As already mentioned, the theory takes different forms according to the relative masses of the atoms A, X and B. In the case where the bridging ligand X is taken to be of relatively small mass, the atoms A^+ and B can be considered to remain fixed during the reaction event and the motion of X can be represented by single coordinate. This model seems appropriate for most typical systems, such as $(Fe^{III}ClFe^{II})^{4+}$, provided that the masses of the coordinated water molecules are included. In other cases, the bridging ligand is of large mass, and is considered to remain fixed while atoms A^+ and B vibrate, and again only one coordinate is needed.

Here we shall quote only one set of results. In the case of a symmetrical reaction $(A^+-X \ldots A) \to (A \ldots X-A^+)$, with the A^+ and A atoms fixed, the activation energy $\Delta U^{\ddagger}$ is found to be

$$\Delta U^{\ddagger} = (E_S + P_L + P_X)^2 / 4(E_S + P_L) \tag{7.1}$$

where E_S and P_L are the reorganisation energies of solvent and of non-bridging inner-sphere ligands respectively. P_X is the reorganisation energy associated with movement of the bridging atom, i.e. it is the energy change for the process

$$(A^+-X \ldots A) \rightarrow (A^+ \ldots X-A) \quad (7.2)$$
$$ p p^*$$

in which the bridging atom moves from initial to final location, while the electronic configuration remains unchanged. In the harmonic approximation, this is

$$P_X = \tfrac{1}{2} L m_X \omega_{AX}^2 (\Delta r)^2 \quad (7.3)$$

where m_X is the mass of atom X, ω_{AX} the bond stretching frequency and Δr the distance traversed in the process of equation (7.2). For practical calculations, frequencies are taken from the infra-red spectra of mononuclear complexes A^+-X, and the distance Δr is assumed to be the difference between A^+-X and $A-X$ bond distances, again in mononuclear complexes.

Taking the complexes $Co(NH_3)_5 X^{2+}$ where X is F, Cl, Br, the frequencies ω_{AX} decrease in the order $F > Cl > Br$[783] but this is more than offset by the increase in mass $F < Cl < Br$ giving the trend in P_X as F, 1.9; Cl, 3.0; Br, 3.5 kcal mol^{-1}. Hence, if vibration of the bridge were the only factor to consider the normal order of rate would be $F > Cl > Br$. On the other hand, outer-sphere solvation effects run in the opposite direction owing to the increasing molecular volume. From detailed calculations on the system $Fe^{III}X^{2+} + Fe^{2+}$, Dogonadze et al.[333] deduced the order of activation energies as $F > Cl > Br$.

An alternative to this treatment is to view the bond-stretching process as analogous to that which precedes a substitution reaction. This presumably involves both inner- and outer-sphere reorganisation, and leads to the intuitive prediction that rates of reaction will correlate with lability of the A^+-X bond, as measured, for example, by the aquation rate constant or (in a series of complexes with the same metal but different ligands X) with the thermodynamic dissociation constant. We return to this suggestion in section 7.5 below.

7.2.2 Resonance transfer

Halpern and Orgel[509] considered only the transition probability at the crossing point. For the probability P of a transition within time t they wrote

$$P(t) = \sin^2 \pi v t = \sin^2 (\pi \Delta E_\pm t / h) \quad (7.4)$$

where $E_\pm$ is the separation of the energy surface at the crossing (see Chapter 1, equation 1.16). For $\Delta E_\pm$, a first-order treatment gives

$$\Delta E_\pm = 2(H_{AB} - S_{AB} H_{AA})/(1 - S_{AB})^2 \quad (7.5)$$

where

$$H_{AB} = \int \phi_A H \phi_B^* d\tau, \quad S_{AB} = \int \phi_A \phi_B^* d\tau,$$

and ϕ_A, ϕ_B are one-electron wavefunctions; or, when $S_{AB} \simeq 0$,

$$\Delta E_\pm = 2H_{AB} = 2\beta \quad (7.6)$$

For transfer mediated by the bridging group, Halpern and Orgel expressed the wavefunctions ψ_p and ψ_s, for states p and s, as determinants formed from the one-electron wavefunctions ϕ_A, ϕ_X, ϕ_B:

$$\psi_p = [6(1 - S_{BX}^2)]^{-1/2} |\phi_B \phi_X \overline{\phi}_X| \qquad (7.7)$$

$$\psi_s = [6(1 - S_{AX}^2)]^{-1/2} |\phi_A \phi_X \overline{\phi}_X| \qquad (7.8)$$

where ϕ_X and $\overline{\phi}_X$ refer to electrons with opposite spins, and then the probability is $P(t) = \sin^2 \pi v_1 t$, with

$$hv_1 = 2(H_{ps} - S_{ps} H_{pp})/(1 - S_{ps})^2$$
$$\cong 2(H_{ps} - S_{ps} H_{pp}) \qquad (7.9)$$

where

$$S_{ps} \cong (S_{AB} - S_{AX} S_{XB})$$

and

$$H_{pp} = \int \phi_B(1)\phi_X(2)\phi_X(3) H \phi_B(1)\phi_X(2)\phi_X(3) d\tau -$$
$$- \int \phi_B(1)\phi_X(2)\phi_X(3) H \phi_X(1)\phi_B(2)\phi_X(3) d\tau \qquad (7.10)$$

$$H_{ps} = \int \phi_B(1)\phi_X(2)\phi_X(3) H \phi_A(1)\phi_X(2)\phi_X(3) d\tau -$$
$$- \int \phi_B(1)\phi_X(2)\phi_X(3) H \phi_X(1)\phi_A(2)\phi_X(3) d\tau \qquad (7.11)$$

The two terms in H_{ps} correspond to direct exchange of the electron between atoms A and B, and double exchange, i.e. simultaneous exchanges B → X and X → A, respectively. By a calculation using hydrogen 1s orbitals for all three one-electron wavefunctions, Halpern and Orgel showed that both mediated and unmediated exchange frequencies diminished with increasing A–B distance, but the unmediated frequency fell off more rapidly than the mediated frequency. They considered also the effect of bridging groups with more than one closed shell of electrons and found that for conjugated systems the exchange frequency increased with increasing mobile bond order. However, it must be stressed that these are quantum-mechanical exchange processes taking place at a fixed nuclear configuration in the transition state.

7.2.3 Superexchange

Still using the resonance model, Halpern and Orgel[509] considered the effects of mixing intermediate states such as $A^+.X^-.B^+$ and $A.X^+.B$. They considered the effect of various orbitals, ϕ_{xj}, of the bridge but noted that the most relevant would be the lowest unoccupied and highest occupied orbitals, respectively. In general, the transition probability would increase with increased mixing of these states with the p and s states: that is, with increased overlap of orbitals ϕ_A, ϕ_B and ϕ_{xj}, and lower energy of ϕ_{xj}. They noted that where the bridging group had many orbitals not too distant in energy (as in organic conjugated systems), the symmetry relations might be such that some would contribute positively, and some negatively, to the transition probability.

7.2.4 The chemical mechanism

(a) Definition

The chemical mechanism may be written symbolically as

$$p \underset{-1}{\overset{1}{\rightleftharpoons}} x \underset{-2}{\overset{2}{\rightleftharpoons}} s \tag{7.12}$$

giving the specific rate

$$k = k_1 k_2 / (k_{-1} + k_2) \tag{7.13}$$

with limiting forms $k = k_1$ or $k = k_1 k_2 / k_{-1} = k_2^*$ according to whether step 1 or step 2 is rate determining. Since the average lifetime of the intermediate is given by $\tau = 1/(k_{-1} + k_2)^{-1}$ the criterion for chemical transfer is that τ must exceed the lifetime of the fastest possible unimolecular reaction event, or approximately

$$k_{-1} + k_2 < RT/Lh \tag{7.14}$$

The intermediate may be either of two forms depending on whether electron donation by B precedes electron acceptance by A^+

$$A^+ . X . B \underset{-1}{\overset{1}{\rightleftharpoons}} A^+ . X^- . B^+ \underset{-2}{\overset{2}{\rightleftharpoons}} A . X . B^+ \tag{7.15}$$

or vice versa

$$A^+ . X . B \underset{-1}{\overset{1}{\rightleftharpoons}} A . X^+ . B \underset{-2}{\overset{2}{\rightleftharpoons}} A . X . B^+ \tag{7.16}$$

The two mechanisms may be termed 'electron transfer' and 'hole transfer'. Alternative names[1049] are 'push-pull' and 'pull-push'–a nomenclature which focusses attention on the motion of the electron. The two mechanisms are kinetically indistinguishable but in particular cases it is possible to guess which mechanism operates from the redox characteristics of the reacting species.

Since chemical transfer consists simply of two successive electron transfers, each of these could be discussed separately on the general principles which govern all other electron transfer reactions. It is possible, however, that the formation of the intermediate may be especially favoured by the presence of both groups A and B (or A^+ and B^+ as the case may be) so that the electron, or hole, is simultaneously repelled from one reactant centre and attracted to the other. A possible example is the reaction $Ru(NH_3)_5 L^{3+} + Cr^{2+}$, where L is iso-nicotinamide, discussed below (p. 241). In the extreme limit of this argument, the intermediate with reduced or oxidised ligand might become more stable than either the precursor or the successor complex. So far, this has not been observed.

(b) Theory – non-adiabatic case

Vol'kenshtein and Dogonadze have considered reaction profiles of the type (c) (*Figure 7.1*) using non-adiabatic expressions for all three curve crossings and have obtained the corresponding rate expressions. In their first paper they represented the energy profiles as V-shaped intersecting straight lines, but in subsequent work parabolic profiles were used. As in the adiabatic case, considered in more detail in the next subsection, the overall activation energy is determined by the higher of the two curve intersections, but the pre-exponential factor contains the product of two Landau–Zener factors, one for each intersection, reflecting the fact that a complete reaction requires two consecutive transitions[618]. The theory has been elaborated to cover multi-state systems with closely-spaced energy levels on the bridge[335] or on the donor and acceptor ions A, B[336], and has been extended to electrode reactions[334].

Schmidt[923] has considered the case of an adiabatic reaction p → x, followed by a non-adiabatic step x → s, arguing an analogy with the theory of predissociation in molecular spectroscopy.

(c) Theory – adiabatic case

When both transition probabilities are close to unity, the energy profiles may be replaced by conventional free energy profiles and these may be used to predict in detail the variations of rate with the redox properties of donor, acceptor, and bridging group.

Referring to equation (7.12) above, we define free energy differences ΔG_1°, ΔG_2°, and activation free energies $\Delta G_2^\ddagger$ and $\Delta G_2^\ddagger$, for the successive steps. The overall free energy change is

$$\Delta G^\circ = \Delta G_1^\circ + \Delta G_2^\circ \quad (7.17)$$

and the net activation free energy of step 2 is

$$\Delta G_2^{\ddagger *} = \Delta G_1^\circ + \Delta G_2^\ddagger \quad (7.18)$$

The activation free energy actually measured, $\Delta G^\ddagger$, is then $\Delta G_1^\ddagger$ or $\Delta G_2^{\ddagger *}$, whichever is the greater, or in other words, according to whether step 1 or step 2 is rate-determining. If the Marcus theory is assumed to apply to both steps, then changes in the standard free energy differences will effect changes in the activation free energies. Hence, if the chemical mechanism occurs, with electron transfer (equation 7.15), and if step 1 is rate-determining, the rate should be affected by change in the reductant B, but not by change in the oxidant A^+. For hole transfer (equation 7.16) the reverse is true. This prediction has been examined, to a limited extent, as a criterion of the chemical mechanism (section 7.3.2 below). Here we examine in more detail the predicted dependence of rate upon free energy.

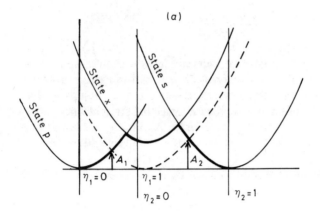

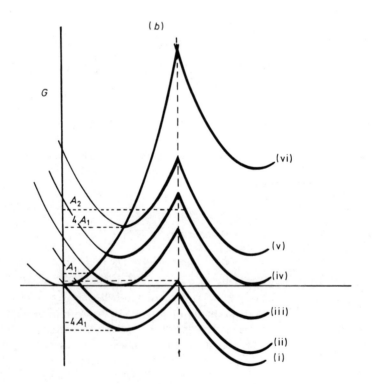

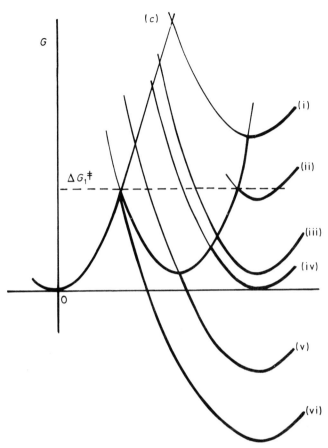

Figure 7.2 (a) *Potential energy profile for a 'chemical' electron transfer reaction* (see text, equations 7.12). *The three full-line parabolas, identical in shape, give the energies of wavefunctions of the pure configurations* p, x, s. *In this case, the ground states of* p *and* s *have the same energies, but the intrinsic barriers to the two reaction steps are unequal* ($A_2 > A_1$) *and this is shown by making the horizontal scales of reaction coordinates* η_1 *and* η_2 *unequal.*
(b) *Profiles for chemical electron transfer reactions with intrinsic barriers* A_1 *and* A_2 *as in* (a), *but with variation of* $\Delta G°$ *due to changes in* $\Delta G_1°$: (i) $\Delta G_1 = -4A_1$, *below this profile, step 1 lies in the 'anomalous region'*; (ii) $\Delta G_1^{\ddagger} = \Delta G_2^{\ddagger *}$, *below this profile, step 1 is rate-determining; above, step 2 is rate-determining;* (iii) $\Delta G_1 = 0$, *in this example, owing to choice of parameters, the profile lies between* (ii) *and* (iv)—*it could alternatively lie between* (i) *and* (ii); (iv) $\Delta G° = 0$; (v) $\Delta G_1 = +4A_1$, *above this profile, step 1 is again in the anomalous region;* (vi) $\Delta G_1^{\ddagger} = \Delta G_2^{\ddagger}$.
(c) *Profiles for chemical electron transfer reactions, showing the effects of variation of* $\Delta G^{\ddagger}$ *due to changes in* $\Delta G_2°$. *For ease of comparison with* (b), *the intrinsic barriers of* A_1 *and* A_2 *have the same relative magnitudes but in reverse, i.e. here* $A_1 > A_2$, *and the profiles are numbered to correspond with those in* (b). (i) $\Delta G_2 = 4A_2$, *above this profile, step 2 lies in the anomalous region;* (ii) $\Delta G_2^{\ddagger *} = \Delta G_1^{\ddagger}$, *above this profile, step 2 is rate determining; below, step 1 is rate determining;* (iii) $\Delta G_2° = 0$, *in this example, owing to the choice of numerical values of parameters, the profile lies between* (ii) *and* (iv), *it could alternatively lie between* (i) *and* (ii); (iv) $\Delta G° = 0$ (cf. (a) *reversed.*); (v) $\Delta G_2 = -4A_2$, *below this profile, step 2 lies in the anomalous region;* (vi) $\Delta G_2^{\ddagger *} = \Delta G_1^{\ddagger}$

The Marcus equations for the two steps are

$$\Delta G_1^\ddagger = A_1(1 + \Delta G_1^\ominus/4A_1)^2 \tag{7.19}$$

$$\Delta G_2^\ddagger = A_2(1 + \Delta G_2^\ominus/4A_2)^2 \tag{7.20}$$

where A_1, A_2 are the intrinsic free energy barriers. *Figure 7.2(a)* shows the profile for a possible two-step reaction, taking $\Delta G^\ominus = 0$ and $A_2 > A_1$. In order to apply the Marcus model it is necessary to have all three parabolas identical in shape, so the difference in intrinsic barrier is obtained by adjusting the relative horizontal placings. This is legitimate as long as the reaction coordinates for the two steps may be considered independent.

If there are variations in $\Delta G^\ominus$ due to changes in $\Delta G_1^\ominus$ only (i.e. changes in redox potential of the reductant if the reaction involves reduction of the bridge, or of the oxidant if the reaction involves oxidation of the bridge) these are represented by the middle and right-hand parabolas in *Figure 7.2(a)* moving together vertically, as shown in *Figure 7.2(b)*. The resulting variations in $\Delta G_1^\ddagger$ and $\Delta G_2^{\ddagger*}$ are shown in *Figure 7.3(a)*. Values of $\Delta G_1^\ddagger$ lie on the parabola defined by equation (7.19) above; values of $\Delta G_2^{\ddagger*}$ lie on the straight line of gradient 1, equation (7.18). The profiles numbered (i) to (vi) in *Figure 7.2(b)* illustrate various 'critical' values of $\Delta G^\ominus$, and the points in *Figure 7.3(a)* are correspondingly numbered. For step 1, the range of 'normal' crossings is from points (i) to (v), $\Delta G^\ominus = \Delta G_2^\ominus \pm 4A_1$. At $\Delta G^\ominus$ values below this range step 1 would presumably be diffusion controlled; above this range the reverse of step 1 would be diffusion controlled. The range of $\Delta G^\ominus$ values for which step 2 is rate-determining is from points (ii) to (vi) but the latter always falls in the anomalous region; hence step 1 is rate-determining only below point (ii), and step 2 is rate-determining for all $\Delta G^\ominus$ above point (ii). The position of this point depends on the intrinsic barriers. As A_1 tends to zero, the point (ii) moves to the point (v) and the region of $\Delta G_2^\ddagger$ rate-determining vanishes; but if $A_1 \leq \Delta G_1^\ominus/4$, point (ii) moves to point (i) or beyond and the parabolic portion of the plot of $\Delta G^\ddagger$ against $\Delta G^\ominus$ vanishes.

Variations in $\Delta G^\ominus$ due to changes in $\Delta G_2^\ominus$ only (i.e. changes in the oxidant for electron transfer, or in the reductant for hole transfer) are expressed by moving the right-hand parabola only, as shown in *Figure 7.2(c)*. In this case $\Delta G_1^\ddagger$ remains constant and $\Delta G_2^\ddagger$ varies quadratically. The predicted plot of $\Delta G^\ddagger$ against $\Delta G^\ominus$ is shown in *Figure 7.3(b)*. The 'critical' points are numbered to correspond with those in *Figures 7.2(b)* and *7.3(a)*. For step 2 the range of normal crossing is from points (i) to (v), $\Delta G^\ominus = \Delta G_1^\ominus \pm 4A_2$. At $\Delta G^\ominus$ values more positive than this range, the reverse of step 2 is presumably diffusion controlled, and thenceforth the plot of $\Delta G^\ddagger$ against $\Delta G^\ominus$ becomes a straight line of slope 1. Below this range, step 2 is diffusion controlled. The range of values of $\Delta G^\ominus$ for which step 1 is rate-determining is from points (ii) to (vi), but the latter always falls in the anomalous region, hence step 1 is rate-determining below point (ii) and step 2 is rate-determining above point (ii). The position of point (ii) depends as before on the values of the intrinsic barriers.

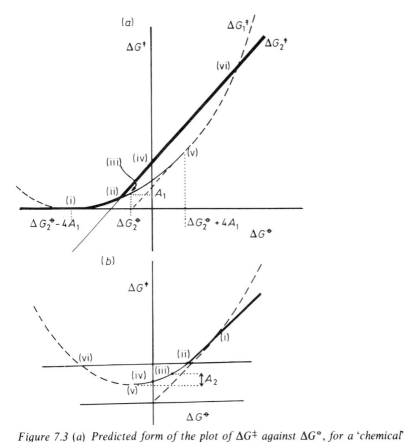

Figure 7.3 (a) Predicted form of the plot of $\Delta G^{\ddagger}$ against $\Delta G^{\ominus}$, for a 'chemical' electron transfer reaction, with variation in $\Delta G^{\ominus}$ due to changes in $\Delta G_1^{\ominus}$, and with $A_2 > A_1$. The parabola shows the variation of $\Delta G_1^{\ddagger}$ with $\Delta G^{\ominus}$ according to equation (7.19). The full-line portions correspond to normal curve crossing in step 1 as shown in Figure 7.2; the broken portions correspond to the anomalous regions. The diagonal line shows the variation of $\Delta G_2^{\ddagger*}$ with $\Delta G^{\ominus}$ according to equation of the text. The thickened line shows the predicted overall dependence of $\Delta G^{\ddagger}$ on $\Delta G^{\ominus}$. Values of $\Delta G^{\ominus}$ at the numbered points are as follows: (i) $\Delta G_2^{\ominus} - 4A_1$; (ii) $\Delta G_2^{\ominus} + 4A_1 - 4(A_1 \Delta G_2^{\ddagger})^{1/2}$; (iii) $\Delta G_2^{\ominus}$; (iv) 0; (v) $\Delta G_2^{\ominus} + 4A_1$; (vi) $\Delta G_2 + 4A_1 + 4(A_1 \Delta G_2^{\ddagger})^{1/2}$. Thus, the horizontal coordinates of (i) and (v) are equidistant from that of (iii), and those of (ii) and (vi) are equidistant from that of (v). The tangent to the parabola at (v) cuts the horizontal axis at the coordinate of point (iii).
(b) Predicted form of the plot of $\Delta G^{\ddagger}$ against $\Delta G^{\ominus}$, for a chemical electron transfer reaction, with variation in $\Delta G^{\ominus}$ due to changes in $\Delta G_2^{\ominus}$, and with $A_2 < A_1$. The parabola shows the variation of $\Delta G_2^{\ddagger}$ with $\Delta G^{\ominus}$, according to equations (7.18) and (7.20). The full-line portions correspond to normal curve crossing in step 2, as shown in Figure 7.2; the broken curves correspond to the anomalous regions. The horizontal full line shows the value of $\Delta G_1^{\ddagger}$, a constant for the reaction series. The thickened lines show the predicted overall dependence of $\Delta G^{\ddagger}$ on $\Delta G^{\ominus}$. Values of $\Delta G^{\ominus}$ at the numbered points are as follows: (i) $\Delta G_1^{\ominus} + 4A_2$; (ii) $\Delta G_1^{\ominus} - 4A_2 + (\Delta G_1^{\ddagger}/A_2)^{1/2}$; (iii) $\Delta G_1^{\ominus}$; (iv) 0; (v) $\Delta G_1^{\ominus} - 4A_2$; (vi) $\Delta G_1^{\ominus} - 4A_2 - (\Delta G_1^{\ddagger}/A_2)^{1/2}$. Thus the horizontal coordinates of (i) and (v) are equidistant from that of (iii), and those of (ii) and (vi) are equidistant from that of (v). The tangents to the parabola at points (i) and (v) intersect at the point where $\Delta G^{\ominus} = \Delta G_1^{\ominus}$.

These diagrams summarise the predicted changes in rate as functions of thermodynamic driving force. For given values of $\Delta G_1^\ominus$, A_1 and A_2, the two diagrams could be combined as one, showing two functional dependences of $\Delta G^\ddagger$, as either the oxidant or reductant is varied. This produces a rather complicated diagram, however. A simpler result is obtained by considering a reaction which is symmetrical in the sense that $A_1 = A_2$. (The oxidising and reducing centres need not be identical). The result is shown in *Figure 7.4*. Starting with the case $\Delta G^\ominus = 0$, this gives $\Delta G^\ddagger = \Delta G_1^\ddagger = \Delta G_2^\ddagger{}^* = \Delta G_1^\ominus = A(1 + \Delta G_1^\ominus/4A)^2$, marked as point X on the diagram. The effects of changes in $\Delta G^\ominus$ due to changes in $\Delta G_1^\ominus$ are shown by the two full-line segments marked 1; one of these segments is a parabola, the other a straight line of slope 1.0. The effects of changes in $\Delta G^\ominus$ due to changes in $\Delta G_2^\ominus$ are shown by the two full-line segments marked 2; one of these is parabolic the other horizontal indicating no change in $\Delta G^\ddagger$. The 'critical' points numbered (ii), (iii) and (iv) in *Figures 7.2* and *7.3* all coincide at X. The other points are numbered to correspond with *Figures 7.2* and *7.3*.

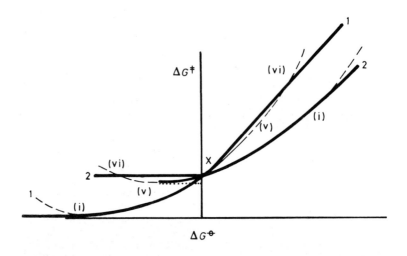

Figure 7.4 Predicted forms of the plots of $\Delta G^\ddagger$ against $\Delta G^\ominus$ for chemical electron transfer reactions (equation 7.12) with the same intrinsic barrier A for each of the two steps. The curves marked 1 and 2 show the effects of variation of $\Delta G^\ominus$ due to changes in $\Delta G_1^\ominus$ (see Figure 7.2(b)) and $\Delta G_2^\ominus$ (see Figure 7.2(c)). The parabola (1) has equation $y = A[1 + (B + x)/4A]^2$; the straight line (1), $y = A(1 + B/4A)^2 + x$; the parabola (2), $y = B + A[1 + (x - B)/4A]^2$; the straight line (2), $y = A(1 + B/4A)^2$; where $x = \Delta G^\ominus$, $y = \Delta G^\ddagger$, B is the value of $\Delta G_1^\ominus$ when $\Delta G^\ominus = 0$. All four intersect at the point $x = 0$, $y = A(1 + B/4A)^2$. The numbering of the points, and the broken and full curves correspond to Figures 7.2 and 7.3.

The effects of changes in the oxidising or reducing power of the bridging group (other things being equal) are expressed by varying the 'height' $G_1^\ominus$ of the central parabola. The resulting family of curves, for the general case with $\Delta G^\ominus < 0$ and $A_2 > A_1$, is shown in *Figure 7.5* and the predicted dependence of

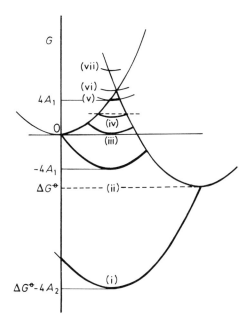

Figure 7.5 Free energy profiles for chemical electron transfer reaction (equation 7.12), with $\Delta G^\circ < 0$, $A_2 > A_1$ showing the effect of varying the electronic energy of intermediate. (i) $\Delta G_1^\circ = \Delta G^\circ - 4A_2$, below this profile, step 2 lies in the anomalous region; (ii) $\Delta G_1^\circ = -4A_1$, below this profile, step 1 lies in the anomalous region; (iii) $\Delta G_1^\circ = 0$, (iv) $\Delta G_1^\ddagger = \Delta G_2^{\ddagger *}$; (v) $\Delta G_1^\circ = 4A_1$, above this profile, step 1 lies in the anomalous region; (vi) $\Delta G_1^\ddagger = \Delta G_2^{\ddagger *} = \Delta G_1^\circ$, (vii) $\Delta G_1^\circ = \Delta G^\circ + 4A_2$, above this profile, step 2 lies in the anomalous region

$\Delta G^\ddagger$ on ΔG° is shown in *Figure 7.6*. The crossing energies $\Delta G_1^\ddagger$ and $\Delta G_2^{\ddagger *}$ both vary quadratically with ΔG_1°. For the sake of geometrical completeness, negative values of ΔG_1° have been included in the diagrams, corresponding to the situation of a thermodynamically stable intermediate. The range of normal curve crossings for step 1 lies between the points marked (ii) and (v). The range of normal crossings for step 2 is between points (i) and (vii). The case of $\Delta G_1^\circ = 0$ gives $\Delta G_1^\ddagger = A_1$, $\Delta G_2^{\ddagger *} = A_2(1 + \Delta G^\circ / 4A_2)^2$, either of which may be rate-determining depending on the parameters (point iii). The two parabolas giving $\Delta G_1^\ddagger$ and $\Delta G_2^{\ddagger *}$ as functions of ΔG_1° intersect in two points (iv) and (vi); of these points, (vi) corresponds to the intersection of all three energy curves at the same point (*Figure 7.5*), and must always lie in the anomalous region for one step or the other. The point (iv) is at lower energy and may occur in various positions relative to the other points. Assuming diffusion-controlled steps as appropriate in the anomalous regions, the complete free energy dependence consists of two straight portions joined by two parabolic portions (*Figure 7.6*).

The picture is somewhat simplified in the case where $\Delta G^\circ = -4(A_2 - A_1)$. Geometrically this means that in *Figure 7.5* the axis of the central parabola

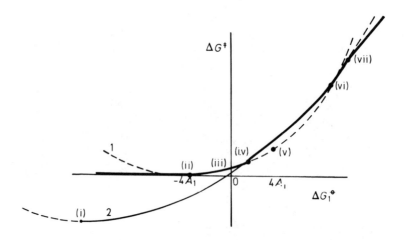

Figure 7.6 Predicted variation of activation free energy, according to Figure 7.5. The parabolas 1 and 2 show the variation of $\Delta G_1^\ddagger$ and $\Delta G_2^\ddagger{}^*$, respectively. As ΔG_1° is made more positive the rate controlling step passes from the diffusion-controlled step 1 to the electron-transfer-controlled step 1, at point (ii); then to step 2 at the intersection point (iv). It does not revert to step 1 at the intersection point (vi), since step 1 is now in the anomalous region. It changes to the linear dependence with slope $+1$ at point (vii) when step 2 passes into the anomalous region

passes through the intersection point of the other two parabolas, hence points (v), (vi) and (vii) coincide. This gives $\Delta G_1^\ddagger \geq \Delta G_2^\ddagger$, so that step 1 is rate-determining throughout.

7.3 Evidence for chemical mechanisms

7.3.1 Detection of intermediate

(a) *Inner-sphere reactions*

In favourable cases, the transient intermediate can be directly detected either visually, or better, by its e.s.r. spectrum. In the reaction

$$\text{pyrazine-COOCo(NH}_3)_5 + \text{Cr}^{2+} \longrightarrow \text{intermediate} \longrightarrow \text{pyrazine-COOH-NCr(OH}_2)_5 + \text{Co}^{2+} \quad (7.21)$$

Spiecker and Wieghardt[962a] have observed an e.s.r. spectrum which they attribute to the free-radical intermediate containing both cobalt and chromium in the trivalent state, and the odd electron delocalised in the π-electron system of the bridging ligand.

An intensely coloured green product is also formed, which may also be a radical complex, although its e.s.r. spectrum could not be observed. This reaction was previously studied by Gould[478] who first proposed the mechanism, but on the supposition that the green substance was the intermediate. The free ligand is also reduced by chromium(II) to give the chromium(III) complex of the pyrazine-2-carboxylate radical ion[962a]. The same ligand is involved in the reactions

$$(Co^{III})_2L + Cr^{2+} \rightarrow (Co^{III})_2(\dot{L})Cr^{III} \rightarrow (Co^{III})LCr^{III} + Co^{2+} \quad (7.22)$$

$$(Co^{III})LCr^{III} + Cr^{2+} \rightarrow (Co^{III})Cr^{III}(\dot{L})Cr^{III} \rightarrow Cr^{III}LCr^{III} + Co^{2+} \quad (7.23)$$

where $(Co^{III})_2$ denotes the binuclear unit $[(H_3N)_3Co(OH)_2Co(NH_3)_3]$, (Co^{III}) denotes $Co(NH_3)_3(OH_2)_2$, L denotes pyrazine-2-carboxylate and $\dot{L}$ denotes the radical ion. Both radical intermediates were detected as strongly-absorbing species which decayed at measurable rates, but the e.s.r. spectra could not be detected[962a].

In some cases the ligand suffers a net two-electron reduction. Thus the chromium(II)-reduction of complex I produces an intensely coloured species formulated as complex II, which is reoxidised to complex III on exposure to air.

I   II   III

Presumably the formation of complex II occurs in successive one-electron steps via a radical intermediate[475]. In other cases reduction of the ligand is inferred from the fact that more than one mole of reductant is consumed per mole of oxidant[476].

(b) *Outer-sphere reactions*

A clear example of stepwise transfer is shown in the equation[550]

$$Co(NH_3)_5OOC\text{-}\langle\ \rangle\text{-}NO_2^{2+} + e_{aq}^-$$

$$\xrightarrow{k_1} Co(NH_3)_5OOC\text{-}\langle\ \rangle\text{-}\dot{N}O_2^+$$

$$\xrightarrow{k_2} Co^{II} + HOOC\text{-}\langle\ \rangle\text{-}NO_2 \quad (7.24)$$

The reductant may be the solvated electron, as shown, or various one-electron radical reducing agents such as CO_2^-, generated by pulse-radiolysis. The initial redox reaction, which may be classed as outer-sphere, is followed by intramolecular electron transfer. Recently, Cohen and Meyerstein[240] have measured the rate of the analogous reaction with the free radical ion as reductant:

$$Co(NH_3)_5OOC\text{-}\langle\rangle\text{-}NO_2 + HOOC\text{-}\langle\rangle\text{-}\dot{N}O_2^-$$
$$\rightarrow Co^{II} + 2HOOC\text{-}\langle\rangle\text{-}NO_2 \quad (7.25)$$

This is second-order, and when allowance is made for the estimated stability constant of the precursor complex the specific rate of the elementary step is found to be similar to, if not greater than, k_1 of equation (7.24).

Presumably the odd electron in the intermediate is mainly localised in the nitro group, perhaps with some alteration of the O–N–O bond angle, and this rate comparison confirms that the carboxyl group and benzene ring are not an effective electron transfer bridge. The intramolecular process can in fact be considered as a resonance transfer of the electron between localised centres.

In a similar experiment with the benzoate complex (equation 7.26), no intermediate was detected[550].

$$Co(NH_3)_5OOCC_6H_5^{2+} + e_{aq}^- \rightarrow Co^{II} + HOOC.C_6H_5 \quad (7.26)$$

Since the reduced ion $C_6H_5CO_2^{2-}$ is a known species with a known spectrum, an upper limit for the concentration of any intermediate $Co(NH_3)_5OOCC_6H_5^+$ could be calculated, and a lower limit for the rate of decomposition of such an intermediate, $k_2 > 10^7 \text{ s}^{-1}$.

Symons, West and Wilkinson[1001] have reported an analogous series of reactions (equation 7.27) of the nitroprusside ion $[Fe(CN)_5(NO)]^{2-}$ (complex IV).

$$(NC^-)_5Fe^{2+}(NO^+) \xrightarrow[77K]{e^- \text{ in MeOH glass}} (NC^-)_5Fe^{2+}(NO)$$
$$\text{IV} \qquad\qquad\qquad\qquad\qquad\qquad\qquad \text{V}$$

with arrows showing e^- in solution from IV, and anneal in MeOH glass from V, both leading to:

$$(NC^-)[(NC^-)_4Fe^+](NO^+)$$
$$\text{VI} \quad (7.27)$$

In the ground state this can be classed[707] as an iron(II) complex (low-spin d^6) with an NO^+ group coordinated in a straight line to the ion atom, like the iso-electronic CN^- ions. When the ion is trapped in a methanol glass at low temperature and irradiated with γ rays an unstable, reduced species (complex V) is formed. From esr evidence the single unpaired electron is located on the nitrosyl group, now in a bent conformation. On warming this changes to the more stable complex (complex VI) containing iron(I) (low-spin d^7) with strong tetragonal distortion. The same sequence occurs on irradiating crystalline sodium nitroprusside $Na_2Fe(CN)_5NO.2H_2O$ [129]. The product VI is also obtained under equilibrium conditions in solution, either by reduction or by electrolysis of $Fe(CN)_5NO^{2-}$ [1001]. The reaction sequence of complexes IV → V → VI constitutes a chemical electron transfer, remarkable for the fact that in the second step the electron is effectively transferred between neighbouring atoms. The activation energy arises from the changes in the Fe–N–O bond angle and rearrangement of the *trans* cyano group.

An example of reversible metal-to-ligand electron transfer has recently been found in a nickel(III)–posphysin complex[338].

7.3.2 Free-energy relationships

As shown in section 7.2.4, if the rate of electron transfer remains unaffected by a change in either oxidant or reductant, a chemical mechanism is indicated; if the rate is affected by both, resonance transfer is indicated. Examples of reactions characterised in this way are shown in *Table 7.1*, classed according to the bridging ligand.

(a) Iso-nicotinamide complexes

Changing the oxidising centre from $Co(NH_3)_5$ to $Cr(H_2O)_5$ reduces the rate of reduction by Cr^{2+} by factors of only 10 and 40, whereas with other bridging groups (F^-, OH^-, NCS^-, Cl^-) the corresponding ratios are in the range 3×10^4 to 3×10^7. The fact that the ratio differs from unity emphasises two important qualifications in this type of argument:

(a) it is not clear that either of the two steps (equation 7.12) is strictly rate-determining; the general rate equation (7.13) admits a dependence of k_{obs} on both K_1 and K_2.
(b) even so, the rate constant k_1 can be significantly affected by the nature of the oxidant, if a sufficiently strong interaction exists between the orbitals of the oxidant and the bridge.

Such interactions can be described in various ways, differing only in degree. With a weak interaction, the reducibility of the ligand may be affected either by changes in solvation or by changes in the energy of the lower orbital which accepts the electron. The reducibility of molecules such as pyrazine and the fumarate ion is greatly enhanced by protonation of the basic sites. With a strong interaction, the orbital system might be so affected that an electron donated into the ligand orbitals would be delocalised on to the oxidising centre without further activation, so that the reorganizational barrier is also lowered. This seems to occur in the ruthenium(III)–*iso*-nicotinamide system. Although the oxidation potentials of the $Ru^{III}(NH_3)_5$ and $Co^{III}(NH_3)_5$ centres are similar, the ruthenium complex is reduced some 10^4 times faster than the cobalt complex[454]. The acceptor orbital of the metal ion is a t_{2g} orbital, with

Table 7.1 Specific rates of some inner-sphere reactions $A^+ \cdot X + B \to A + B^+ \cdot X$, showing effects of variation of oxidant and reductant. (Table entries are $\log(k/M^{-1} s^{-1})$, 25 °C, and literature reference.) Figures in bold type indicate probable 'chemical' mechanisms.

Precursor complex $A^+ \cdot X \cdot B$	$\dfrac{A^+}{B}$	$Ru(NH_3)_5^{3+}$	$Co(NH_3)_5^{3+}$	$Cr(OH_2)_5^{3+}$	$Cr(NH_3)_5^{3+}$
$A^+ \!-\! N\!\!\bigcirc\!\!-\!\!\underset{NH_2}{\overset{O-B}{C}}$	Cr^{2+}	5.59 (454)	1.24 (811)	**0.25** (811)	
$A^+ \!-\! O \cdots HO \!-\! B \cdots COOH$ (fumarate-type)	Cr^{2+}		1.78 (311)	**0.5** (311)	
$A^+ \!-\! O \!-\!\!\bigcirc\!\!-\! O \!-\! B$ (with OH)	Cr^{2+}		2.20 (821)	**0.60** (821)	−1.75 (290)
$A^+ \!-\! O \!-\! \underset{B}{C} \!-\! COCH_3$	Cr^{2+}		4.04 (860)		
	V^{2+}		1.01 (860)		
(a)	Cr^{2+}		5.30 (574)	−0.89 (963)	−2.19 (290)
	Fe^{2+}		−3.39 (574)		
$A^+ \!-\! Cl^- \!-\! B$	Cr^{2+}		6.4 (190)	0.79[c] (1015a)	−1.29 (817)
	Fe^{2+}		1.52 (990a)	−11.7[b] (348)	

(a) Oxidants are $Co(NH_3)_4^{3+}$, $Cr(OH_2)_4^{3+}$, $Cr(NH_3)_4^{3+}$
(b) Calculated from the reaction $FeCl^{2+} + Cr^{2+}$, using equilibrium data from reference 950
(c) 0 °C

π symmetry, overlapping the empty antibonding π-orbital of the ligand. From studies of ruthenium(II) complexes with such ligands, there is evidence of extensive back-donation. Hence, it is suggested that as soon as step 1 is complete 'the electron is accepted into an orbital largely centred on the metal ion but spreading on to the ligand. The reaction is no longer stepwise and the rate-determining act ... involves electron transfer from the chromous ion directly to the final acceptor orbital'[454]. The rate-determining act has been called resonance transfer[454, 1013], or tunnelling[1015].

(b) Maleate complexes

In an extensive study of the reaction

$$(H_3N)_5CoOOC{-}COOH + Cr^{2+}$$

$$\downarrow (a) \qquad \searrow (b)$$

$$Co^{II} + (H_2O)_5Cr^{III}OOC{-}COOH \underset{(c)\ Cr^{2+}}{\rightleftarrows} Co^{II}$$

$$+ \left[\begin{array}{c} COO \\ COO \end{array} Cr^{III}(H_2O)_4 \right]^{+} + H^{+} \qquad (7.28)$$

Olson and Taube[821] detected two parallel inner-sphere pathways, leading to monodentate and bidentate chromium(III) products in the ratio 20% to 80% (steps (a), (b)), and a subsequent chromium(II)-catalysed isomerisation (step (c)). Equation (7.28) expresses the fact that under conditions where the predominant cobalt complex is the protonated form $Co(NH_3)_5LHCr^{5+}$ (where L^{2-} is the maleate ion), the two reactions with chromium(II) are zero-order in hydrogen ion; similarly the monodentate chromium(III) complex is presumed to be protonated, and the principal isomerization pathway, step (c) is zero-order in the hydrogen ion in the forward direction, first-order in the reverse direction. From the structures of the products, steps (b) and (c) both have cyclic transition states, which can plausibly be formulated as complex VII (where M is $Co(NH_3)_5$ or $Cr(H_2O)_5$).

MO—⟨ring⟩—OH
       Q O
       Cr^{II}

VII

Davies and Jordan[290] studied the analogous system with $Cr(NH_3)_5$ as oxidising centre.

$$(H_3N)_5CrOOC\overset{\frown}{}COOH + Cr^{2+} \rightleftharpoons (H_3N)_5CrOOC\overset{\frown}{}COO + H^+ + Cr^{2+}$$

$$\searrow \text{(d)} \qquad \swarrow \text{(f)}$$

$$Co^{II} + (H_2O)_5Cr^{III}OOC\overset{\frown}{}COOH \underset{\text{(e)}}{\rightleftharpoons} Co^{II} + \left(\begin{array}{c}COO\\COO\end{array}\right)Cr^{III}(OH_2)_4^+ \qquad (7.29)$$

They obtained mixtures of monodentate and bidentate chromium(III) products, but owing to slowness of the preceding reactions, the equilibrium step (e) was fully established within the time-scale of the experiments. They also found a pathway inversely first order in the hydrogen ion, corresponding to reduction of the deprotonated complex $Cr(NH_3)_5L^+$ (step (f)). Steps (d) and (f) presumably also involve cyclic transition states, although this is not proved by the data.

Comparison of steps (b) and (c) (equation 7.28) shows that again the rate ratio is small compared with the 'normal' ratios for $Co^{III}(NH_3)_5X$ and $Cr^{III}(H_2O)_5X$. On the other hand, comparison of reactions (c) and (d), shows a ratio comparable with other pairs of reactions where $Cr^{III}(H_2O)_5$ is replaced by $Cr^{III}(NH_3)_5$. Davies and Jordan argue that there is a difference in mechanism: the cobalt(III) complex possibly reacting by the chemical mechanism but the two chromium(III) complexes reacting by resonance transfer. The activation parameters are consistent with this: $\Delta H^\ddagger = 2.9, \simeq 9,$ and 10.4 kcal mol^{-1} for $Co^{III}(NH_3)_5$, $Cr(H_2O)_6^{3+}$, $Cr(NH_3)_5^{3+}$ respectively. However, the same data are consistent with the chemical mechanism if it is assumed that the rate-determining step changes from step 1 in the cobalt(III) case to step 2 in the two chromium(III) cases. This explanation seems preferable, and is consistent with energy profile diagrams constructed from intersecting parabolae, as shown in *Figure 7.7*. These show that the rate data are at any rate not inconsistent with the 'chemical' mechanism. It is clear that the energy minimum for the presumed intermediate cannot be far below the energy maxima for the reaction $Cr^{III}.LH.Cr^{II} \to Cr^{II}.LH.Cr^{III}$, hence, whether or not the rate-determining step changes on replacing the $Cr^{III}(H_2O)_5$ centre by $Co^{III}(NH_3)_5$, the rate of reaction cannot change greatly.

From the potential data and the assumptions used in constructing the diagram, the equilibrium concentration and the lifetime of the radical intermediate can be estimated. For the chromium(III)–chromium(II) reaction

$$Cr^{III}(LH^-)Cr^{II} \underset{-1}{\overset{1}{\rightleftharpoons}} Cr^{III}(LH^{2-})Cr^{III} \qquad (7.30)$$

we thus obtain $K_1 \simeq 10^{-9}, k_{-1} \simeq 10^{10}$ s^{-1}. It is interesting to note that if the reactions were carried out in aqueous ammonia, with $Cr(NH_3)_5^{2+}$ as reductant, the corresponding equilibrium constant would be as high as 10^{-5}. In the case of the $Co(NH_3)_5$ complex, attempts to detect the radical intermediate by

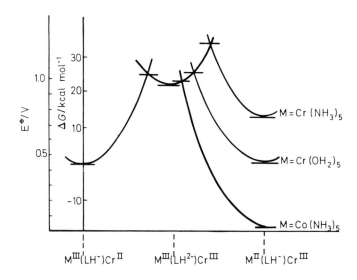

Figure 7.7 Free energy profiles of the reactions $M^{III}(LH^-)Cr^{II} \rightarrow M^{II}(LH^-)Cr^{III}$ via assumed intermediates $M^{III}(LH^{2-})Cr^{III}$, where M is $Cr(NH_3)_5$, $Cr(OH_2)_5$, $Co(NH_3)_5$, and L^{2-} is the maleate ion. The potentials E° of the reactants, intermediate and products are shown relative to the state $M^{III}(LH^-)Cr^{III} + e^-$, assuming that the potentials for reduction at the three centres are the same as for the isolated species M^{III}, LH^-, Cr^{III}. For the inorganic systems thermodynamic data are taken from reference 950, assuming in addition $\log K(Cr^{3+} + 5NH_3) = 25$, $\log K(Cr^{2+} + 5NH_3) = 5$. For HL^- the potential was taken as the mean of the values for L^{2-} and $(C_2H_5)_2L^{371}$. The four transition states are shown with activation free energies determined by experiment (References cited in the text), and the three parabolae have been drawn with approximately the same shape, consistent with the required intersection points

e.s.r. were inconclusive[565], but no other experiments along these lines seem to have been attempted yet.

The reduction chemistry of maleic acid and its ions is of particular interest in comparison with these reactions. Direct one-electron reduction is well-established: the radical ion H_2L^- has been generated in glass matrices by the action of electrons from high energy irradiation[593, 1044], and in solution by reactions such as $H_2L + CO_2^-$ [32] and $H_2L + e^-(aq)$[529]. It has been characterised by its visible and e.s.r. spectra[529]. Electrochemical reduction in protonic solvents is however strongly acid-catalysed[18, 19, 371, 513, 637, 714–716, 855] and the protonated radical ion H_3L has been characterised in solutions of pH ⩽ 4 [32, 855]. Maleic acid is less stable thermodynamically than fumaric acid[289]. The isomerisation reaction is very slow[289, 668]. It is accelerated slightly by radiation[467, 898a (b)]; more efficiently by catalysis with electron donors such as V^{2+} ion[1066, 1068], and by photoelectron transfer from Cu^+

(b) For a detailed study of the isomerisation of the diethylester, see reference 510b ion[402]. The radical ion H_2L isomerises more easily than the unreduced molecule[1044]. Further reduction yields succinic acid[245], and the kinetics of the reactions of maleic acid with chromium(II)[718, 939], vanadium(II)[1067, 1068] and titanium(III)[426] have been studied in some detail. The maleate-bridged electron transfer reactions show hydrogen-ion catalysis, but no isomerisation to fumaric acid nor any reduction to succinic acid[821]. Evidently the time-scale of proton attachment is shorter than or comparable with the lifetime of the reduced-ligand intermediate, so the reduction potential of the conjugated bond system is effectively lowered, but the time-scales of bond rotation and further reduction are longer. Possibly the mode of attachment of the protons in the bridged-electron transfer transition state is different from that in the intermediate observed in the electrochemical processes.

The direct reactions, maleic acid with chromium(II) and europium(II), are much slower than the maleate-bridged metal-to-metal reactions, and this too suggests that the bridging ligand loses its electron very rapidly. In the case of the more easily reducible ligand, *iso*-nicotinamide, the reactions $Cr^{III}L + Eu^{2+}$, $Co^{III}(NH_3)_5L + Eu^{2+}$, and $HL^+ + Eu^{2+}$ are all of similar rate.

All these considerations place severe restrictions on the possible lifetime of the maleate-radical ion intermediate: an upper limit of 10^{-7} s has been inferred[402], but the true value may be much less. It seems clear that the maleate-bridged reactions are near the borderline between chemical and resonance mechanisms.

(c) α-carbonylcarboxylate complexes

Price and Taube[860] compared rates of reaction of a series of complexes $Co(NH_3)_5OCOCOR$ with Cr^{2+} and V^{2+}. Reactions with chromium(II) covered a wide range of rates, increasing in the order R = OH, NH_2, H, $C(CH_3)_3$, CH_3, O^-. Allowing for the influence of chelation in some cases (important when R is O^-) the order corresponds with increasing reducibility as judged by various criteria. Reactions with vanadium(II) covered a narrow range of rates, consistent with substitution control. It follows that if the electron transfer (whether resonant or chemical) had been rate-determining, the rates with vanadium(II) would have been higher than those observed. Price and Taube then argue in favour of resonance transfer mechanisms, as follows: if the chromium(II) reaction proceeded by chemical transfer, and step 1 were rate-determining, the rates with chromium(II) would be expected to be greatly in excess of those with vanadium(II), since it is observed that Cr^{2+} reduces the ligands themselves much more rapidly than does V^{2+}. The actual rate ratios vary, however, from $\simeq 10^3$ down to $\simeq 10^1$, and for the reason just given, these are actually upper limits on the relevant ratio. Alternatively, if step 2 were rate-determining, the rate-ratio, $k(Cr^{2+})/k(V^{2+})$ would be expected to be approximately constant, $k(Cr^{2+})/k(V^{2+}) \simeq 10^3$, determined by the relative reducing strengths of the two ions. Again, this is inconsistent with the observations. Thus, resonance transfer seems to apply at least in some cases.

(d) Bidentate oxalate and chloride complexes

In these cases there are data on the variation of both oxidant and reductant. Both changes produce changes in rate, in directions which correlate with standard free energies. Hence, by the arguments of section 7.2.4 above, a resonance transfer is indicated. The same is true with chloride ion as bridge, for which only a resonance transfer seems plausible in any case (p. 226).

7.3.3 Kinetic isotope effect

Itzkowitz and Nordmeyer[591] studied the reaction

$$Co^{III}(NH_3)_5L + Cr^{2+} \rightarrow Co^{2+} + LCr^{III} \qquad (7.31)$$

with a series of bridging ligands L. They found that, in general, replacement of the NH_3 molecules by ND_3 led to a lowering of the rate. The absence of this effect in certain other cases was then taken to imply that the rate-determining process does not involve effective transfer of the electron to cobalt. Thus, rate ratios k_H/k_D are close to unity for the reaction $Co(NH_3)_5N_3^{2+} + V^{2+}$ which is substitution-limited. Ratios $k_H/k_D \simeq 1$ are also found for the reactions $Co(NH_3)_5L + Cr^{2+}$ where L is *iso*-nicotinamide, 4-carboxypyridine and fumarate, which are now believed to have chemical mechanisms, in contrast to the case where L is acetate, for which a resonance mechanism seems assured. A high ratio is also observed for the nicotinamide complex, and on this basis a resonance transfer mechanism is suggested[591]. This method may be regarded as a variant of the preceding one, with $Co^{III}(ND_3)_5$ and $Co^{III}(NH_3)_5$ as different oxidants.

7.4 Bridging by organic ligands: reducibility of bridging group[1015]

Since 1964, Gould, Taube and co-workers have published many measurements of rates of electron transfer reactions with organic molecules as bridging groups, with a view to identifying the factors which make for effective electron transfer. The most comprehensive set of data is for the series of inner-sphere bridged reactions of the type

$$Co(NH_3)_5OOCR^{2+} + 5H^+ + Cr^{2+} = Co^{2+} + 5NH_4^+ + CrOOCR^{2+} \qquad (7.32)$$

with different 'pendent groups' R. The data vary in quality from comprehensive kinetic studies establishing the order of reaction with respect to oxidant, reductant and hydrogen ion, together with characterisation of the chromium(III) product, to more cursory examinations based on two or three kinetic runs only. Among the factors which have been identified as influencing rates are chelation in the transition state, steric factors (for which see also reference 932) and, less certainly, coulombic forces. Besides these, however, there are some striking variations in rate, almost certainly related to redox properties of the ligand, and it is this aspect which will be reviewed in the following sections.

7.4.1 Qualitative comparisons

Figure 7.8 contains a summary of rate constants classified according to the structural characteristics of the pendent group, to show the range of variation observed with each structure-type. It is clear that there are characteristic ranges of rates, according to the nature of the group adjacent to the carboxyl, and significant variations between classes. Rapid rates are associated with heterocyclic ring systems, with *ortho* and *para* formyl groups; and with α-carbonyl systems (RoOCOCOX), and to a lesser extent with conjugated double bond systems; they are not noticeable with benzenoid aromatic systems, or small heterocyclic rings. Some smaller effects seem well established but are harder to explain, for example the slight but significant trend to increased rates along the series of structure-types

$$-C_6H_4X < -CH_2X < \underset{Y}{\bigcirc} X \text{ (with various hetero atoms Y)}$$
$$< -CH{:}CH{-}X$$

At first this was believed to be steric in origin, but a later study[670] has revealed acid catalysis in some reactions of the –CH:CHX type, which (if real and not due to medium effects[1023]) may be a sign of ligand reduction. With suitably chosen pairs of systems, the differences are indeed striking. Five examples are listed in *Table 7.2*. In examples 1, 2, the right-hand complex possesses a reducible functional group which is absent in the left; in examples 3, 4, both complexes possess reducible conjugated or aromatic systems, but electrochemical and other data show that the right-hand ligand is more reducible than the left. It is clear from these examples, and more especially from *Figure 7.8*, that a double-bond system conjugated to the carboxyl group (RoOOC.CH:CHX) is not itself a sufficient condition for an effective bridge. What does seem necessary is that the bridge orbital which carries the electron (by either the chemical or resonance mechanism) should have an appreciable density close to the oxidising centre, and for this effective conjugation is required. In example 5, the left-hand ligand is more reducible than the right, but the reaction with Cr^{2+} is slower. This is attributed to the triple bond being less able to delocalise π-electrons than the double bond[540].

7.4.2 Quantitative correlations

From the preliminary discussion above (p. 223) one rough general prediction can be made: as the electron affinity of the bridging group is increased, the mechanism will change from resonance transfer to superexchange, and thence to the chemical mechanism. The first of these transitions should be marked by a 'threshold' where the plot of $\log k$ against ΔG_1° changes from a

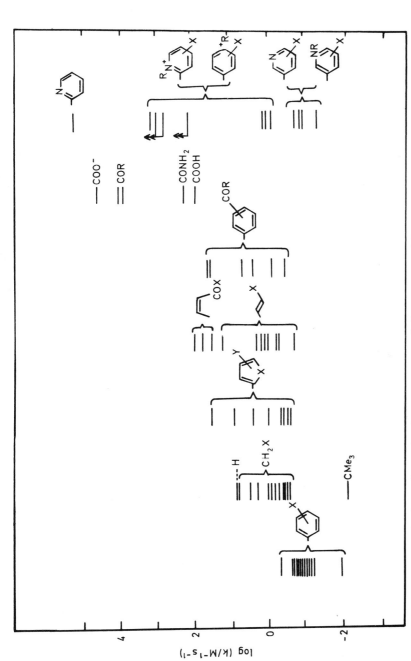

Figure 7:8 Distribution of values of second-order rate constants k for reactions of the type of equation (7.32) (25 °C, ionic strength 1.0 M). Data from various publications of Gould, Taube and co-workers

Table 7.2 Specific rates (k/M^{-1} s^{-1}, 25 °C) for reactions Co^{III}(NH$_3$)$_5$OOCR + Cr^{2+} → Co^{2+} + Cr^{III} OOCR

	R	k	Reference	R	k	Reference	Ratio of rates
1	–CH$_2$OH	3.06	180	–C(H)=O	1.0×10^2	860	33
2	–CH$_3$	0.35	76a	CH=CH–COOH	1.32	935	3.8
3	(tolyl)	0.15	76a	N$^+$–CH$_3$ (pyridinium)	1.4	479a	9.3
4	(o-COOH phenyl)	0.075	935	CH=CH–COOH	1.6×10^2	821	2.1×10^3
5	–C⋮C–COOH	0.67$^{(a)}$	540	CH=CH–COOH	2.64$^{(a)}$	540	3.9

(a) Oxidising centre is [(H$_3$N)$_3$Co(OH)$_2$Co(NH$_3$)$_3$]. Reductant attacks the remote carboxyl group

horizontal line to a rising curve. Whether or not the second transition should be marked by any kind of discontinuity is unknown: predictions of the dependence of $\log k$ upon ΔG_1° have been worked out for the chemical mechanism (p. 231) but not for the superexchange mechanism. However, as will be seen from *Figure 7.9*, the data available so far are too scattered to show any such effect. Undoubtedly the major reason for this is the difficulty of finding a single quantitative measure of ligand reducibility, applicable to the whole range of molecules. Ideally, we require the energies of the lowest antibonding orbital of each ligand. This might be obtained from gas phase electron affinities or from charge-transfer energies of complexes of the relevant ligands with some standard electron donor. In practice a variety of other data must be used. Two of the most comprehensive sets are reviewed in the following paragraphs.

(a) *Electrochemical data*

The most directly applicable parameter is the one-electron reduction potential

$$L(aq) + e^-(electrode) \rightleftharpoons L^-(aq) \tag{7.33}$$

where L denotes the bridging ligand, but this is not always observed. More commonly, the dominant reduction process is a two-electron transfer, such as equation (7.34):

$$HOOC.CH=CH_2(aq) + 2e^-(electrode) + 2H^+(aq) \rightleftharpoons HOOCCH_2CH_3(aq) \tag{7.34}$$

These reactions can sometimes be avoided by using aprotic solvents such as acetonitrile, and by comparison with suitable systems, the corresponding electrode potential in aqueous medium can be estimated[814]. Other systems show complicated electrokinetic behaviour. It is essential to know the hydrogen-ion dependence of the potential and the degree of reversibility. Irreversible behaviour means that some chemical reaction involving reactants or products proceeds at a rate less than or comparable with the rate of the electrode process itself. Far from merely invalidating the results (as some of the earlier discussions in the literature might suggest) this is actually a source of additional information about the mechanism. There are good reviews and text books in this field[27, 62, 355, 749, 772, 1023, 1064, 1065, 1125–1127]. From a study of this literature, data have been obtained for 15 bridging molecules. Some of the data are redox potentials for reactions of the same type as equation (7.33) and (7.34); others are for corresponding reactions of related compounds such as the esters $RCOOCH_3$. Altogether seven series of data were used, and pooled to arrange the ligands in rank order of increasing reducibility, as shown. The plot of $\log k$ against rank, shows some correlation, rates increasing with increasing reducibility (*Figure 7.9 (a)*).

(b) *Spectroscopic data*

Transitions from some lower-lying orbital into the lowest antibonding orbital

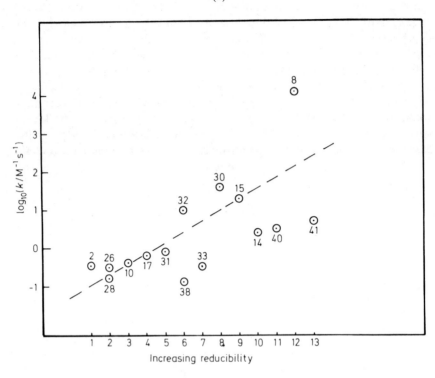

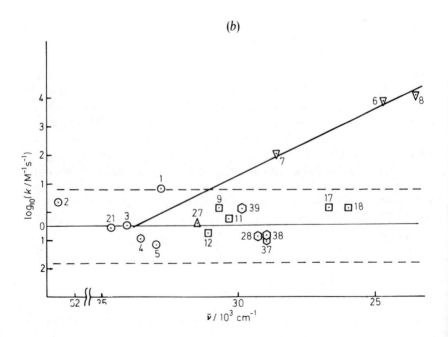

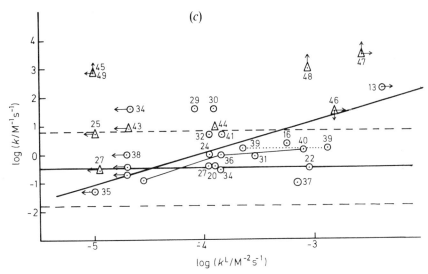

Figure 7.9 Correlation of second-order rate constants for reactions of the type of equation (7.32) (25 °C, ionic strength 1.0 M), with reducibility of the pendant group R of the ligand, as judged by various criteria.
(a) Correlation with electrochemical reducibility, ligands arranged in order of increasing ease of reduction. The order is obtained by collating E° values for the following reactions: $RCOO^- + e^-$, $RCOOR' + e^-$, $RCOOR' + e^- + H^+$, $RCOOR + 2e^- + 2H^+$, $RR' + e^-$, $RR' + e^- + H^+$, $RR' + 2e^- + 2H^+$ (R' = H, CH_3 or C_2H_5). Data from references 27, 62, 355, 749, 772, 1064, 1065, 1124–1127.
(b) Correlation with wavenumbers $\tilde{v}_{max}$ of $n \to \pi^*$ transitions in the aldehydes RCHO. Data from reference 822b. The symbols refer to the classification of pendant groups R by structure: ⊖ R is H, NH_3, CH_2X; △ R is ⟨○⟩ ; ⊡ R is $-CH:CHX$ or $-C:CX$; ⊙ R is $-C_6H_4X$ or $-C_5H_4NX$; ∇ R is $-COX$. The horizontal lines represent the mean of log k, and limits of ± 2 standard deviations, for all points except numbers 6, 7 and 8.
(c) Correlation with rate constants k^L for reduction of the free ligand RCOOH by chromium(II) in acid solutions. The data in the literature (references 475, 476, 480) consist of the extent of reaction for specified concentrations of oxidant, reductant and acid, in a given time. To obtain specific rates the rate equation
$-d \ln[Co^{III}]/dt = k^L[Cr^{2+}][H^+]$ has been assumed in all cases. Points marked with arrows represent only upper or lower limits. Points △ indicate ligands with oxygen, nitrogen or sulphur atoms in positions to form five-membered rings with chromium(II). The three points joined by full lines refer to the isometric $CH_3(C_5H_3N)COOCo^{III}(NH_3)_5$ complexes mentioned in the text. The two points joined by dotted lines indicate different published rates for the same ligand. The horizontal lines are set at the same levels as in Figure 7.9(b). The sloping line has gradient $d \log k/d \log k_L = 1$. In (a), (b) and (c), pedant groups R are identified by numbers, as follows:

1, H; 2, NH_2; 3, CH_3; 4, CH_2Cl; 5, $CHCl_2$; 6, CHO; 7, COOH; 8, $-COCH_3$; 9. $C \equiv CH$; 10. $C \equiv CCH_3$; 11. $CH = CH_2$; 12. $CH:CHCH_3$-*trans*; 13, $CH:CHCOOH$-*cis*; 14, $CH:CHCOOCH_3$-*cis*; 15, $CH:CHCOOCH_3$-*trans*;

16, $CH:CHCH:CHCOOH$; 17, $CH:CHCH:CHCH_3$; 18, ⟨furan-vinyl⟩;

(*continued*)

of the ligand are a convenient measure of the availability of the antibonding orbital, provided that a common donor orbital can be found for a sufficiently wide range of ligands, and that the relevant bands can be reliably identified in the spectrum. One suggestion[1013] is that the metal-to-ligand $(t_{2g} \to \pi^*)$ transitions[422] in the corresponding $Ru^{II}(NH_3)_5X$ complexes might be used. Alternatively we may choose transitions from some suitable non-bonding orbital of the ligand itself, for example the $n \to \pi^*$ transition from the oxygen of the free carbonyl group[595, 782]. The most satisfactory set of data is from spectra of the aldehydes RCHO. Care must be taken to distinguish the $n \to \pi^*$ transition from the more intense $\pi \to \pi^*$ bands. Often the desired band appears as a shoulder with extinction coefficient $\varepsilon \simeq 10^{1 \pm 0.5}$ M^{-1}cm^{-1}. The range of energies is very wide, wavenumbers v covering about 30×10^3 cm^{-1}, equivalent to 85 kcal mol^{-1}. Clearly the differences in Lhv between pairs of ligands cannot be equated with the absolute magnitude of difference in redox potentials but presumably there is a correlation. The available data are plotted in *Figure 7.9(b)*. It is clear that over a very wide range of energies there is no

Figure 7.9 caption (continued)

19, [structure]—OH; 20, HO—[structure]; 21, CH_2CH_2COOH;

22, $CH_2CH_2COCH_3$; 23, $CH_2CH_2CH_2COOH$; 24, $(CHOH)_4COOH$; 25, $CH_2SCH_2C_6H_5$; 26, C_3H_4N; 27, C_3H_3O; 28, C_6H_5; 29, C_6H_4CHO-o; 30, C_6H_4CHO-p; 31, $C_6H_4COCH_3$-o; 32, $C_6H_4COC_6H_5$-o; 33, $C_6H_4COC_6H_5$-p; 34, $C_6H_4[COC_6H_4COOH$-$p]$-o; 35, $C_6H_4NH_3^+o$;

36, [structure]; 37, [structure]; 38, [structure];

39, [structure]; 40, [structure]; 41, [structure];

42, [structure]; 43, [structure]; 44, [structure];

45, [structure]; 46, [structure]; 47, [structure];

48, [structure]; 49, [structure];

significant correlation between reaction rate and reducibility. The majority of rates lie in the range $\log k = -0.5 \pm 1.3$ and within this range steric factors seem more important in determining rates, rates decreasing with increasing bulk of the pendent group R in the order H > alkyl > aryl. The only significant correlation is found among the α-carbonyl acids. The line drawn through those points in *Figure 7.9(b)* has slope $\partial \log k/\partial \bar{v} = 0.46 \times 10^{-3}$ cm. The theoretical slope, if an increase in the n→π* energy gap produced an equal increase in activation free energy, would be $\partial \log k/\partial \bar{v} = Lhc/2.3 RT = 2.09 \times 10^{-3}$ cm (c is the velocity of light).

(c) Rates of ligand reduction

Perhaps the most direct measure of ligand reducibility is the rate of actual reduction of the free ligand by the reducing agent in question. Chromium(II) reduces many organic molecules. Its applications to synthetic organic chemistry have been reviewed[512], and in a few cases the mechanisms of reaction have been studied[939]. Most of the data amassed by Gould, Taube and co-workers, however, consist simply of measurements of the extent of reaction when ligand and chromium(II) are mixed under controlled conditions for a fixed time. When the comparison is made between compounds of closely related structure, the correlation is quite clear, as in the sequence of isomeric N-methylated pyridine derivatives $CH_3(C_5H_4N)COOCo(NH_3)_5^{3+}$. The rates of both reactions, cobalt(III)–chromium(II) and ligand–chromium(II) increases in the order 3-carboxy < 2-carboxy < 4-carboxy[480]. On a wider view, however, the picture is not so clear, though there does appear to be some correlation (*Figure 7.9(c)*).

7.4.3 Acid catalysis

Several reactions of the type of equation (7.32) show a positive hydrogen-ion dependence which at first sight seems unexpected (*Table 7.3*). The site of attachment of the proton has not been proved, but the most likely position is one of the oxygen atoms of the coordinated carboxyl group. This would be analogous to the acid-catalysed aquation of $Co(NH_3)_5OOCR^{2+}$ complexes[761], and is supported by the comparison of $Co(NH_3)_5OOC.C_6H_4CHO^{2+}$, which shows acid-catalysis, with $Co(NH_3)_5NCC_6H_4CHO^{2+}$, which does not[68]. (Both reactions involve attack of Cr^{2+} at the remote formyl group.) On the other hand, the reaction of $Co(NH_3)_5OOCCH_3^{2+}$ with Cr^{2+} is not acid-catalysed, and it seems clear that where the effect is found, the additional proton has some specific effect on the electron transfer process. Most of the bridging ligands concerned have been considered reducible, by one or more criteria, and all have the feature of a conjugated bond system. Presumably the proton lowers the energy of the relevant acceptor orbital, which would facilitate transfer by either the superexchange or the chemical mechanism. (Compare the effect of coordinated H^+ ion in lowering the thermodynamic reduction potentials of pyrazine and

Table 7.3 Examples of proton-assisted electron transfer

Reactants[a] A^+	B	Rate expression $R/[A^+][B]$	References
RoOOC–CH=CH–COOH	Cr^{2+}	$k_1 + k_2[H^+] + k_3[H^+]^{-1}$	935
RoOOC–CH=CH–CH=CH–COOH	Cr^{2+}	$k_1 + k_2[H^+]$	480
RoOOC–CH=CH–COOCH$_3$	Cr^{2+}	$k_1 + k_2[H^+]$	567, 935
RoOOC–C$_6$H$_4$–CHO	Cr^{2+}, Eu^{2+}, V^{2+}	$k_1 + k_2[H^+]$	400, 1118
RoOOC–C$_6$H$_4$–COC$_6$H$_5$	Cr^{2+}, Eu^{2+}	$k_1 + k_2[H^+]$	480
RoOOC–CH=CH–C$_6$H$_4$–CHO	Cr^{2+}, Eu^{2+}	$k_1 + k_2[H^+]$	400, 479
RoOOC–C(Cl)=C(Cl)–COOH	Eu^{2+}	$k_1 + k_2[H^+]$	400

(a) Ro is Co(NH$_3$)$_5$

the maleate ion.) Other descriptions of the effect have also been given, such as[935] that the electrons in the bridging ion are redistributed so as to improve the conjugation—compare structures VIII and IX.

$$\text{Co(NH}_3)_5\text{O}-\underset{\text{VIII}}{\overset{\text{O}}{\text{C}}}-\text{CH=CH}-\text{COOH} \qquad \text{Co(NH}_3)_5\text{O}-\underset{\text{IX}}{\overset{\text{OH}}{\text{C}}}=\text{CH}-\text{CH=C}-\text{COOH}$$

No doubt the attachment of H^+ also weakens the cobalt–oxygen bond. This is supported by comparison of the fumarato and maleato complexes. The hydrogen-ion term does not occur in the latter case[474] and a similar difference is found between the two ester hydrolysis reactions,

$$CH_3OOCCH:CHCOOH + H^+ \rightarrow CH_3OH + HOOCCH:CHCOOH \tag{7.35}$$

Hydrolysis of methyl hydrogen fumarate is mainly first order in acid, while hydrolysis of methyl hydrogen maleate is independent of acid[835]. Both maleic acid[273, 274] and the monohydrogen maleate ion[835] have a chelate hydrogen-bonded structure. On a final note of caution, we observe that in some cases, especially the Eu^{2+} reactions, the specific rates for the H^+-catalysed pathway, in relation to the uncatalysed path, are small and hardly distinguishable from so-called 'medium-effects'.

7.5 Bridging by halide ions

Evidence that halide ions can function as bridging groups has been cited in Chapter 5. From the earliest observations it was clear that the rate of reaction varied widely according to the nature of the halide ion. The data now available are summarised in *Table 7.4*. For the symmetrical exchange processes

$$Cr^{III}X^{2+} + {}^*Cr^{2+} = Cr^{2+} + {}^*Cr^{III}X^{2+} \tag{7.36}$$

Ball and King[69] found the order $F < Cl < Br < I$ and Ogard and Taube[817] found the same sequence in the unsymmetrical reactions

$$Cr^{III}(NH_3)_5X^{2+} + Cr^{2+} = Cr^{2+} + 5NH_3 + Cr^{III}X^{2+} \tag{7.37}$$

At first[69, 817] it was thought that this order was general for electron transfer processes, and reflected some kind of interaction between the transferring electron and the bridging group, such that the conductivity or electron-permeability might be said to increase in the order $F < Cl < Br < I$. Such an interaction certainly cannot be ruled out, but at the present time it seems equally clear that it is not strictly required by the experimental data, and that other explanations are still sufficient.

The first evidence of this was the discovery that in some systems the order of reactivity is inverted: $I < Br < Cl < F$. In the reactions $Rh(NH_3)_5X^{2+} + Cr^{2+}$ [1002], this order could be related to the order of

Table 7.4 Equilibrium and rate parameters for halide-bridged electron transfer reactions

Reactant central ions		$\log K^{(a)}$ $A^+ + B \rightleftharpoons A + B^+$	$\log$ (equilibrium constant)$^{(a)}$ $A^+.X^- + B^+ \rightleftharpoons A^+ + B^+.X^-$				$\log$ (rate constant) $A^+.X^- + B \rightarrow A + B^+.X^-$				References$^{(c)}$
A^+	B		F	Cl	Br	I	F	Cl	Br	I	
$Co(NH_3)_5^{3+}$	U^{3+}	15	7.4	0.7	0.7	0.5	5.7	4.5	4.1		1073
	Cr^{2+}	13	3.0	−1.1	−2.3	−4.3	5.4	6.8	6.1	6.5	190
	Eu^{2+}	13	2.0	−0.2	−0.2		4.4	2.6	2.4	2.1	193
	$Co(CN)_5^{3-}$	?	?	−0.7	0.3	2.3	3.3	7.7			192
	Ti^{3+}	7	5.3$^{(b)}$	0.6$^{(b)}$			2.3	1.1	0.2	0.6	251
	Cu^+	3					0.0	4.7	5.7		829
	Fe^{2+}	−7	3.8	0.5	0.1		−2.1	−2.8	−4.0		315
Fe^{3+}	Cr^{2+}	20	−0.8	−1.6	−2.4		5.9	7.3	>7.3		208–9, 348
	Eu^{2+}	20	−1.8	−0.7	0.1		7.3	6.3	6.1		207–8, 348
	Fe^{2+}	0	0	0	0		1.8	1.0			885

Oxidant	Reductant											Ref.
Cr^{3+}	U^{3+}	2	4.4	1.8	3.0	4.8	1.2	<0				393
	Cr^{2+}	0	0	0	0	0	−2.6	1.0	>1.8			69, 596
	Eu^{2+}	0	−1.0	−1.1	2.5		−3.2	−2.9	−2.5	−1.4		16
Cr(NH$_3$)$_5^{3+}$	Yb^{2+}	?					1.6	1.0	1.1			226
	Cr^{2+}	−20					−3.6	−1.3	−0.5	0.7		817
trans-Cr(NH$_3$)$_4^{3+}$	Cr^{2+}	—						0.1		2.2		299, 556
Rh(NH$_3$)$_5^{3+}$	Cr^{2+}	<0		−3.3	−5.0	−7.7		0.0	−1.9	−2.2		1002
Ir(NH$_3$)$_5^{3+}$	Cr^{2+}	—						−0.9	−1.8	−1.7		1002
cis-Co(en)$_2^{3+}$	Fe^{2+}	−3						−1.5	−3.5			677, 669
Ru(NH$_3$)$_5^{3+}$	Cr^{2+}	10		−2.9	−4.1			4.5	3.3	2.4		974
	Eu^{2+}	10		−2.0	−1.6			4.4	4.1			974

(a) Calculated from stability constants of A$^+$.X$^-$ and B$^+$.X$^-$ complexes mainly from reference 950
(b) Ti^{IV} species is TiO^{2+}
(c) Where one reference is shown this applies to all the rate data; where more than one is shown, they apply respectively

strengths of Rh X bond, assuming that the $Rh(NH_3)_5^{3+}$ structural unit is a 'soft' Lewis acid or 'class (b)' acceptor, while $Cr(H_2O)_6^{3+}$ and $Co(NH_3)_6^{3+}$ are of class (a). But the reactions $Co(NH_3)_5X^{2+} + Fe^{2+}$ [315, 377] also show the inverted order. In this case (assuming the bridged mechanism for the whole series) one may point to the iron(III) product: Fe^{3+} is a class 'a' metal, hence the stabilities of the products increase in the order $I < Br < Cl < F$, and evidently this is the predominating factor. One suggested reason why this might be so is that the Fe–X bonds in the transition states of these reactions may be more fully formed than the Cr–X bonds in the transition states of, say, equations (7.36). This is probably true, in the sense that the oxidation states of Fe are closer to those of the product Fe^{III} than to the reactant Fe^{II}. However, before seeking an explanation along these lines, we must consider the standard free energy changes of the reactions. Assuming that the electron transfer step $p \rightarrow s$ is rate-determining in every case, the relevant parameter is $\Delta G_{is}^\ominus$ and it is reasonable to suppose that for a given oxidant–reductant pair the trend in $\Delta G_{is}^\ominus$ will be dominated by the trend in $\Delta G_{ps}^\ominus$ (equation 7.38).

$$A^+.X^-.B \rightleftharpoons A.X^-.B^+, \qquad \Delta G_{ps}^\ominus \qquad (7.38)$$

These data are not available, but it is probable that they in turn follow the trend of values of $\Delta G'$

$$\underset{i}{A^+.X^-} + B = A + \underset{f}{B^+.X^-} \qquad \Delta G' \qquad (7.39)$$

(When the reaction involves group transfer, $\Delta G'$ is identical with $\Delta G_{if}^\ominus$. It has, however, been argued that in some cases, e.g. with Eu^{2+} as reductant, the group X is not transferred[209]).

Introducing the formation equilibria

$$A^+ + X^- \rightleftharpoons A^+.X^-, \qquad \Delta G^{AX} \qquad (7.40)$$

$$B^+ + X^- \rightleftharpoons B^+.X^-, \qquad \Delta G^{BX} \qquad (7.41)$$

and the relative oxidation strengths as defined by

$$A^+ + B \rightleftharpoons A + B^+, \qquad \Delta G^E \qquad (7.42)$$

we have

$$\Delta G' = \Delta G^E + (\Delta G^{BX} - \Delta G^{AX}) \qquad (7.43)$$

Figure 7.10 shows a plot of $\Delta G^{\ddagger}$ against $\Delta G'$, for all known halide-bridged reactions. Although the scatter is wide, there is an overall positive correlation, consistent with the Marcus equation. The curve shown in the diagram is a parabola with slope 0.5 at $\Delta G' = 0$, drawn through the points with its maximum set at an assumed diffusion-controlled limit, $\log k \simeq 7$. With this curve as guideline it can now be seen that in the majority of cases the rates correlate with the free energies, regardless of the actual order in relation to the halogen sequence. The only examples of negative correlation are in the sequences $RoX^{2+} + Cr^{2+}$ and $FeX^{2+} + Cr^{2+}$, and these are among the fastest

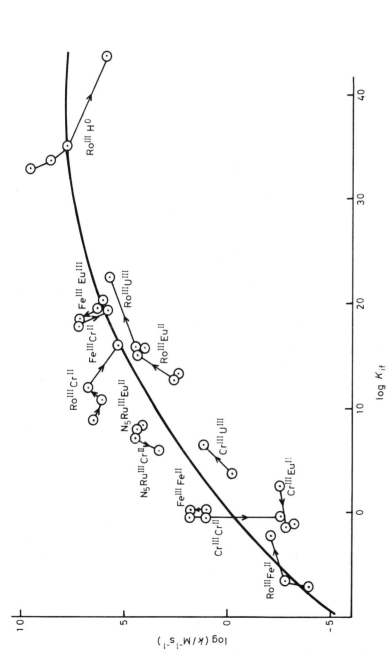

Figure 7.10 Correlation of $\log k$ with $\log K$, for bridged electron transfer reactions with transfer of halide ion, $A^+X + B \to A + XB^+$. The arrows indicate the trends from $X = I$ to $X = F$. Data from Table 7.4. The broken curve is a parabola with slopes $\frac{1}{2}$ at $\log K = 0$, and zero at $\log k = 8$. $Ro = Co(NH_3)_5 \, N_5Ru = Ru(NH_3)_5$

rates, where the assumed Marcus curve is approaching the horizontal limit.

A different approach to these data has been adopted by Haim[497] who proposed combining equations (7.39) and (7.40) to obtain the net activation process

$$A^+ + X^- + B \rightleftharpoons (A.X.B)^{\ddagger} \tag{7.44}$$

This formulation is independent of which of the two metal ions originally carried the bridging group, and it also makes possible direct comparison with other reactions such as

$$Co(NH_3)_6^{3+} + Cr^{2+} + Cl^- \rightarrow Co^{2+} + 6NH_3 + CrCl^{2+} \tag{7.45}$$

$$Eu^{3+} + {}^*Eu^{2+} + F^- \rightarrow Eu^{2+} + {}^*Eu^{3+} + F^- \tag{7.46}$$

in which the halide ion appears separately in the third-order rate law

$$\text{Rate} = k[A^+][B][X^-] \tag{7.47}$$

Reaction (7.45) is necessarily of the outer-sphere type, but reaction (7.46) is generally assumed to be inner-sphere. When applied to substitution-inert systems such as equations (7.36), (7.37), Haim's procedure gives the third-order rate constant which would have been measured if the halide complexes had been labile and the rate law had been as equation (7.47). The ratio of two such rate constants, k_{Cl} and k_F for X=Cl and X=F, measures the relative stability of two transition states

$$(A^+.F.B)^{\ddagger} + Cl \xrightleftharpoons{k_{Cl}/k_F} (A^+.Cl.B)^{\ddagger} + F^- \tag{7.48}$$

and the sequence of rate constants k_x may be used to classify transition states as (a)-type or (b)-type according to whether $k_F > k_I$ or vice versa.

Figure 7.11 shows the variation of ratios k_X/k_F with halogen X, for all known systems. The horizontal coordinate is arbitrary, consisting of equal spacings with increasing atomic number of X; and in the case where all four data are available the resulting plots are zig-zag in shape, like the corresponding plots of metal–ion complex stabilities. It is noteworthy that most systems fall well into the class (a) category. The sequence reading down the diagram represents a qualitative scale of (a)/(b) character. To interpret this sequence we have to bear in mind that each transition state contains two metal ions in fractional oxidation states, the actual effective valency depending on the relative electron affinities. If a quantitative measure of (a)/(b) character were applied to each to the two metal ions, the (a)/(b) character of the binuclear complex could be predicted to be the mean of the two measures, and any further effects due to the intrinsic 'conductivity' of the halide would be superimposed on this mean. Clearly it is not possible with the present data to detect such 'conductivity' unambiguously. It is of interest nonetheless to note the sequences of (a)/(b) character which can be deduced from the data. The diagram contains five systems involving the $Cr(H_2O)_5^{3+/2+}$ couple (variously shown with chromium(III) or oxidant or chromium(II) as reductant).

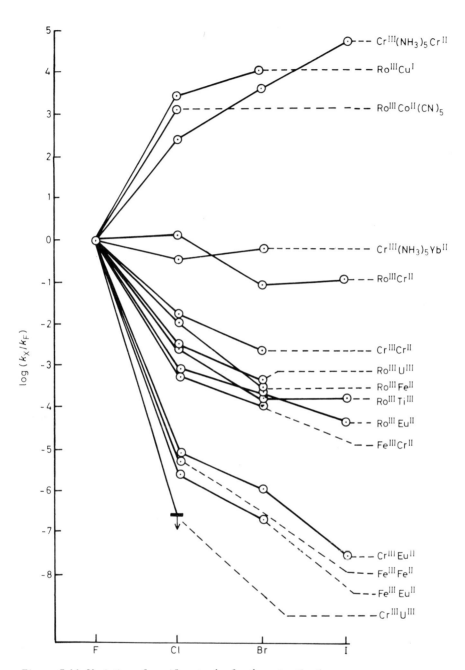

Figure 7.11 Variation of specific rates k_X for the net activation process $A^+ + X^- + B \rightleftharpoons [A^+.X^-.B]^{\ddagger}$, with different redox couples $A^+ \ldots B$, and different bridging ligands $X^- = F^-$, Cl^-, Br^-, I^-, relative to $X^- = F^-$. (Data from Table 7.4) The symbols at the right denote oxidation states of the metals in the precursor complex, $A^+.X.B$. The non-bridging ligands are H_2O except where shown otherwise. Ro is $Co(NH_3)_5$

From these, the other couples involved may be placed in the sequence of decreasing (b) and increasing (a) character, as $Cr(NH_3)_5^{2+/3+}$ $> Co(NH_3)_5^{2+/3+} > Cr(H_2O)_5^{2+/3+} > Eu(aq)^{2+/3+} > U^{III/IV}(aq)^{3+/4+}$. These seem reasonable in view of the known chemistry of the metals concerned. It should be noted that the relevant cobalt(II) state is the low-spin form, which may well be a b-type acceptor, like the other known low-spin cobalt(II) complexes. Similarly from the systems with iron(III) as oxidant or iron(II) as reductant we obtain the sequence $Co(NH_3)_5^{2+/3+} > Cr(H_2O)_5^{2+/3+} > Fe(H_2O)_5^{2+/3+} > Eu(aq)^{2+/3+}$. The two systems with $Cr(NH_3)_5^{3+}$ as oxidant give $Cr(H_2O)_5^{2+/3+} > Yb(aq)^{2+/3+}$. Finally the reactions involving $Co(NH_3)_5^{3+}$ as oxidant give $Cu(aq)^{+/2+} > Co(CN)_5^{3-/2-} > Cr(H_2O)_5^{2+/3+} \geqslant U(aq)^{3+/4+} > Fe(H_2O)_5^{2+/3+} > Ti(aq)^{3+/4+} > Eu(aq)^{2+/3+}$. In this last sequence the uranium couple seems out of place, especially in comparison with the iron couple, but this can be attributed to the relative overall free energy changes. The valencies in the transition states are probably close to $Co^{III}U^{III}$ in the one case but $Co^{II}Fe^{III}$ in the other.

It might still be argued that there is evidence of an intrinsic difference in conductivity, in the sequence of symmetrical exchange reactions such as equation (7.36) but even this is not clear since the same trend can be predicted by a different argument, as follows.

Consider the symmetrical intramolecular reaction

$$A^+.X^- \ldots A \to A \ldots X^-.A^+ \qquad (7.49)$$
$$\quad\; p \qquad\qquad\qquad\qquad s$$

where the single and triple dots indicate short and long bond distances, respectively. Assuming that the A–A atom distance remains constant throughout, the reaction coordinate can be defined by the position of the X^- ion on the A–A axis. In the precursor complex p, the bond energy can be regarded as the sum of energies due to the A^+–X^- and X^-–A interaction, and the dependence of these on x, the reaction coordinate, can be represented by Morse curves as shown in *Figure 7.12(a)* (curves 1, 2). Similarly, the bond energies in the successor complex are represented by curves 3 and 4. We assume that the A^+–X bond is stronger than the A–X bond at all distances so that curve 1 lies entirely below curve 3, and curve 4 below curve 2.

The sum of the energies of curves 1 and 2 gives the energies of reacting pairs in the precursor electronic configuration (*Figure 7.12(b)*, curve P). Depending on the details of the system concerned this may have a minimum A at the right-hand side as well as at the left, B, and therefore a maximum as shown at point D. The minimum would represent a complex $A^+ \ldots$ XA, and point D would represent the transition state for the reaction of atom transfer without electron transfer.

$$A^+.X \ldots A \to A^+ + X^- + A \qquad (7.50)$$

Curves 3 and 4 together give the energy of the reacting pairs in the successor configuration (*Figure 7.12(b)*, curve S). The crossing point C denotes the

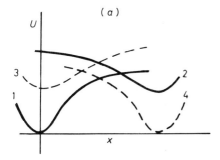

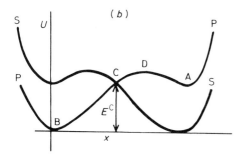

Figure 7.12 Energy profiles for a symmetrical bridged electron transfer reaction $A^+.X^-.A \rightarrow A.X^-.A^+$. The distance between A atom centres is presumed to remain fixed and the reaction coordinate, x, is the position of the X^- ion centre on the A–A axis. In (a) curves 1 and 2 show the energies of A^+-X and X^--A and together they give the energy curve P of the reacting pair in the precursor electronic configuration, (b). Curves 3 and 4 show the energies of $A-X^-$ and X^--A^+ and together they give the energy curves for the successor electronic configuration. Points A, B, C, D, are defined in the text

transition state for equation (7.49) and the activation energy $E^\ddagger$ is given by

$$E^\ddagger = E^C - E^{Res} \qquad (7.51)$$

where E^{Res} is the resonance energy. The effect of varying 'conductivity' of the bridging ligand may now be defined as the effect of varying resonance energy; or of varying transmission coefficient if the resonance energy is small. It seems likely, however, that different bridging ligands will lead to different crossing energies E^C, and that the variation in E^C may dominate the observed variations in $E^\ddagger$.

When the ligand X^- is changed, say from Cl^- to F^-, the binding strength in both the A^+X^- and AX^- complexes increases (assuming both A and A^+ to be of the (a)-type), but more so in the former. Thus the minima at B and B' are lowered, but the crossing point C is lowered to a lesser extent, hence E^C

increases. Thus in general, rates k_x with various bridging groups should correlate inversely with stabilities of the corresponding A^+X complexes, and where both oxidised and reduced forms of the metal ion are of (a)-type, we expect a trend to slower rates in the direction $I > Br > Cl > F$, as observed.

A more detailed discussion of this problem would require more data than are available at present. Diagrams of the type of *Figure 7.12* could be made more quantitative with the aid of the relevant force constants. As yet there is no fully characterised example of a single-atom-bridged, mixed valence complex. Studies of anharmonicity of the vibrations of the bridging ion, in a mixed valence complex, and in the corresponding symmetrical complex $A^+.X^-.A^+$ and $A.X^-.A$, could in principle give evidence of the resonance effect; so also would comparison of the intervalence charge transfer energy with the thermal activation energy (*see* chapter 8)[c].

(c) An attempt [491] to correlate electron transfer rates with metal–halogen bond polarities omits consideration of the 'inverted order $I < Br < Cl < F$

Chapter 8
Optical electron transfer

8.1 Introduction

Electron transfer reactions have their counterpart in electronic spectroscopy. The analogue of the general electron transfer reaction

$$(A^+ \ldots B)(\text{env}) \to (A \ldots B^+)(\text{env}) \qquad (8.1)$$
$$\quad\;\; p \qquad\qquad\qquad\; s$$

is the photo-excited process

$$(A^+ \ldots B)(\text{env}) \xrightarrow{h\nu} (A \ldots B^+)(\text{env}^*) \qquad (8.2)$$
$$\quad\;\; p \qquad\qquad\qquad\; s^*$$

In these equations, A and B denote single atoms and the positive sign denotes the higher of two available oxidation states; the state symbol (env) denotes the *total* environment of the central atoms comprising inner-sphere ligands, bridging molecules if present, outer-sphere ligands and solvent molecules. The species in equation (8.1) are precursor and successor complexes with the same inner-sphere ligands but with appropriate differences in bond lengths and bond angles. Equation (8.2) is governed by the Franck–Condon principle so that states p and s* have identical molecular dimensions and differ only in the configuration of the transferring electron.

Strictly speaking, any observable electronic transition must involve charge transfer, since the intensity is related to the transition dipole moment (*see* section 8.3.4); but conventionally, terms such as 'charge transfer absorption' are reserved for cases in which a charge comparable to that of one electron is transferred in one direction over one or more interatomic distances. The recognition of such transitions may be said to date from the work of Hilsch and Pohl on the spectra of metal halides[546]. They demonstrated the general occurrence of transitions of the type $A^+.X^- \to A.X$. The energies of such transitions correlated with the ionisation energies of the halide ions $(F^- > Cl^- > Br^- > I^-)$ and of the metal atoms $(Na > Fe^{2+} > Cu^+$, for example). Such correlations were, and still are, the most general criteria for identification of a charge transfer transition. This and other early work was reviewed by Rabinowitch in 1942[872]. A further review by Orgel in 1954[823] stressed the chemical implications of optical charge transfer spectroscopy, explaining such familiar observations as the yellow colour of ferric chloride, and the non-existence of cupric iodide, in terms of ligand-to-metal charge transfer. Orgel also pointed out, though he did not develop, the close relationship between optical charge transfer and thermal electron transfer. When the two central atoms are of the same element (usually a metal) the above complexes come into the category of 'mixed valence compounds', which has attracted a good deal of attention since the publication of reviews by Robin and Day[896], and by Allen and Hush[26,571] in 1967. Prior to that time several hundred examples of such compounds were known, but they were scarcely recognised as a class and there had been relatively little systematic study. Most examples had been reported and discussed (if at all) merely as peculiarities within the descriptive chemistry of the elements concerned. Notable exceptions to this generalisation were the studies of electrical conduction in solids by Williams and others (see Chapter 9), and the work of Davidson, McConnell et al.[280,681,697–701,1093] who studied both charge-transfer spectroscopy and electron transfer kinetics in some mixed valence systems in solution. The first detailed interpretation of the spectrum of a mixed valence compound was Robin's study of Prussian Blue, published in 1962[895]. The term 'intervalence transfer'[571] for charge transfer between metal ions is now widely used.

8.2 Classification of optical electron transfer systems

8.2.1 Structural classification

Since the ground state p in the optical transfer system is also the precursor complex of an electron transfer reaction, possible structures may be classified as in Chapter 5 (*Table 8.1*). The major difference is in the amount of information recorded for each class. Directly bonded systems are by far the most extensively studied. Inner-sphere bridged systems are fairly numerous and include the earliest metal–metal electron transfer systems to be recognised, such as the iron(II)–iron(III) oxide systems studied by Weyl. Outer-sphere systems are the least common. The first to be studied was the complex $Co(NH_3)_6^{3+}.I^-$ [681]. Since most studies have been made in the solid state and structures have been established by x-ray diffraction, distinguishing inner- and outer-sphere systems is straightforward. So, too, is the recognition of electron transfer over various distances, ranging from single atom bridges, through two-atom bridges as in Prussian Blue, to long chains as in the carotene molecules and numerous synthetic dyestuffs. An interesting possibility, which as yet has no counterpart in solution chemistry, is electron transfer over different distances in the same molecule. In the complex $SiW^V W^{VI}_{11}O_{40}^{5-}$, absorption bands at 7.8 and 13.7×10^3 cm^{-1} are attributed to electron transfer from W^V to nearest and next-nearest neighbouring W^{VI} ions[446].

Table 8.1 Classification of mixed-valence and related complexes[a]

	Symmetric		Asymmetric			
			Homonuclear ΔZ = 1		Homonuclear ΔZ > 1		Heteronuclear
FINITE STRUCTURES							
Directly bonded	$Rh_2(OAc)_4^+$ (206, 1099a)	IIIA			Hg_2^{2+}; H_2 etc.	IIIA	HCl etc
Inner-sphere (monatomic bridge)	$Co^{III}Co_2^{II}(OAc)_2(Aa)_4$ (117); $Q_2(V^VO)(V^{IV}O)Q_2^-$ (879); $Fe_4S_4(SR)_4^{2-}$ (552)	I; II; III			$(CF_3)_2,P^{III}SP^V(S)F_2$ 340; $O_3Cr^{VI}OCr^{III}aq_5$	I; II	
Inner-sphere (polyatomic bridge)	$A_5Ru^{III}(pypy)Ru^{II}A_5^{5+}$; $A_5RuNC.CNRuA_5^{5+}$	II; IIIA					$[(NC)_5Co^{III}NCFe^{II}(CN)_5]^{6-}$
Outer-sphere							$[Co(en)_3^{3+}][Fe(CN)_6^{4-}]$; $[Ru(NH_3)_6^{3+}][I^-]$
INFINITE STRUCTURES							
Directly bonded	Ag_2F (611)	IIIB					
Inner-sphere (monatomic bridge)	Cr_2F_5; FeO_2H_{1+x} (155)	I; II	$La_{1-x}Ca_xMnO_3$		$K_xPt(C_2O_4)_2 2\tfrac{1}{2}H_2O$; Pb_3O_4; $Pt(NH_3)_2Cl_3$	IIIB; I; II	$[MnO_4^-][Ag^+]$
Inner-sphere (polyatomic bridge)			$Fe_4[Fe(CN)_6]_3 14H_2O$ [b]	II			$Fe(CN)_6Tl_3$
Outer-sphere			$[FeF_5aq^{2-}][Feaq_6^{2+}]$ (1070)	I	Cs_2SbCl_6 177	II	

(a) Examples are taken from references 26, 571, 896 except where indicated otherwise. For homonuclear systems, the Robin–Day electronic classification is shown in bold type. A is NH_3, Aa is acetylaceton, aq is H_2O, py is 4,4'-bipyridyl, Q is 8-quinolinol

(b) The formula of 'insoluble Prussian blue', formerly written $KFe_2(CN)_6.xH_2O$, is now[178, 179, 179a] established as shown. A different compound, ferrous ferricyanide, has also been characterised[254]

8.2.2 Classification by symmetry

Hush[571] proposed a classification of intervalence transitions according to the symmetry of ground and excited states. The following nomenclature is a modification of Hush's scheme. *Symmetrical* transitions are transitions between atoms of the same element which differ in valency by one unit, and which have the same inner-sphere ligand attachments, so that in equations (8.1), (8.2), A ≡ B, and ΔG° is identically zero. The complex itself is not symmetrical owing to slight differences in bond lengths around the A^+ and A atoms. *Asymmetrical* transitions include all others; but the sources of the asymmetry may be subdivided into three categories, as follows. (i) Homonuclear transitions where the valencies differ by one unit, but the ligand environments of the A^+ and A atoms differ. For example, Prussian Blue contains $Fe^{II}C_6$ and $Fe^{III}N_6$ units. (ii) Homonuclear transitions where the ground state valencies differ by more than one unit, so that a one-electron transfer generates both atoms in high-energy states thus, $Sb^{III} \ldots Sb^V \to Sb^{IV} \ldots Sb^{IV}$. (It is noteworthy that wherever the comparison has been made, the reaction in solution is two-electron transfer). (iii) Heteronuclear transitions. These are still somewhat less common than the others—probably for no other reason than that they have not been systematically searched for. Most of the precursor and successor complexes listed in *Table 5.2 and 5.3* would be expected to show such transitions, but not all have been studied in the relevant wavelength range. In some cases the intervalence band is probably overlaid by other intense absorptions.

8.2.3 The Robin and Day classification

From an extensive review of the descriptive chemistry of mixed-valence systems, Robin and Day[896] proposed a classification based on readily observable physical properties, notably the colours of the compounds—often the only property recorded in the literature. This classification was provided with an outline theoretical basis in terms of the mixing of ground state wavefunctions as discussed below.

Class I compounds behave in effect as if the whole is the sum of the parts. For example the solid $CrF_{2.5}$ can be written $Cr^{II}Cr^{III}F_5$. Its green colour is intermediate between those of 'blue green' CrF_2 and 'yellow-green' CrF_3. The structure, from x-ray crystallography, shows the two forms of chromium, symmetrical octahedrally coordinated Cr^{III} and tetragonally distorted Cr^{II}. In Class II, the two metals preserve their identities (as defined by appropriate physical and chemical tests), but some additional properties appear which cannot be attributed to either metal alone. The charge-transfer transition is allowed and occurs at fairly low energy. Hence, the compounds show strong colours not characteristic of either oxidation state separately and may be completely black. In Class III, the metals, though formally of different valencies according to the empirical formula, are indistinguishable by any physical or chemical test. For these compounds, the term 'average-valence' rather than 'mixed valence' is preferable. They generally show chemical and physical properties peculiar to the average valence state and different from either of the supposed component valencies. The delocalisation of an electron over two or more centres can be described in the usual way by 'resonance'. The additional resonance energy could be calculated by valence-bond or molecular orbital methods, and could be assessed experimentally by making suitable comparisons with 'normal' compounds, though actually this has rarely been done. The delocalisation of the electron leads to a set of energy levels different from those of the isolated metal ion in either of its valencies, and hence to a characteristic absorption spectrum. Class III compounds are often intensely coloured, though not always so.

In terms of the two-centre, two-state description of section 8.3, mixed-valence compounds of Class I correspond to the conditions of low overlap ($\alpha \simeq 0$ or 1) and/or wide energy gap ($E_A \gg E_B$); Class III corresponds to the opposite extreme, $\alpha = \frac{1}{2}$, $E_A = E_B$; and Class II to a range of intermediate cases.

Solid materials of Classes II and III can usefully be further subdivided into two groups on the basis of structure. In group A the oxidising and reducing centres are contained in a finite molecule: normally binuclear, though some more complicated examples are known; in group B the structure is repeated infinitely in one, two or three dimensions. (Robin and Day actually made this subdivision only for Class III compounds.) Infinite structures give rise to a number of collective physical properties not characteristic of the individual complex ion. Class II B are usually semi-conductors, as discussed in the next chapter. Class III B compounds are typically 'metallic' in character. They are generally black in colour or else highly reflecting 'silver'- or 'bronze'-like. They are relatively good electrical conductors, the conductivity increasing with decreasing temperature, as with true metals, and passing into superconductivity below a certain temperature. Some are ferromagnetic. Theoretical and experimental work on these interesting materials is at present rapidly advancing[585], but is outside the scope of this book.

8.3 Theory

8.3.1 Charge localisation: the two-state description

The electronic structure of a Class II complex can be described most simply by taking the precursor and successor states as zero-order wave-functions ψ_p and ψ_s and constructing one-electron functions for the delocalised electron by the LCAO approximation. With the further approximation of zero overlap, the wavefunctions of ground and excited states are obtained as:

$$\psi_+ = \frac{\alpha}{(1+\alpha^2)^{1/2}} \psi_p + \frac{1}{(1+\alpha^2)^{1/2}} \psi_s \quad (8.3a)$$

$$\psi_- = \frac{-1}{(1+\alpha^2)^{1/2}} \psi_p + \frac{\alpha}{(1+\alpha^2)^{1/2}} \psi_s \quad (8.3b)$$

and the energies E_+, E_- are given by the two solutions of

$$(E_p - E)(E_s - E) = \beta^2 \tag{8.4}$$

where β is the resonance integral, $\int \psi_p H \psi_s d\tau$. Thus

$$E = \tfrac{1}{2}(E_p + E_s) \pm \tfrac{1}{2}[(\Delta E)^2 + 4\beta^2]^{1/2} \tag{8.5}$$

where $\Delta E = (E_s - E_p)$; and when β is small compared with ΔE, the two solutions are

$$E_+ \cong E_p - \beta^2/\Delta E \tag{8.6}$$

$$E_- \cong E_s + \beta^2/\Delta E \tag{8.7}$$

where the term $\beta^2/\Delta E$ is the resonance energy. The mixing coefficient α is given by

$$\alpha = \beta/\Delta E \tag{8.8}$$

In the case $E_p = E_s$, we have $\alpha = 1$; the wavefunctions of the two states are

$$\psi_+ = 2^{-1/2}(\psi_p + \psi_s) \tag{8.9}$$

$$\psi_- = 2^{-1/2}(-\psi_p + \psi_s) \tag{8.10}$$

and the resonance energy is β.

8.3.2 Connection with electron transfer kinetics

In the general electron transfer reaction (equation 4.2), the initial and final states, i and f, are formally of the Robin and Day Class I, with zero interaction. In some reactions the precursor and successor states p and s may also be of this type. This might be expected for example in the case of the $CrF^{2+} + Cr^{2+}$ bridged reaction, by analogy with the solid compound Cr_2F_5. However, most precursor complexes are presumably of the Class II type, and the identification of such species is an important objective of current research. The transition state t‡ on the other hand may be regarded as a Class III complex, with indistinguishable central ions of averaged valency.

The energies E_p and E_s depend partly on the electron affinities of the ions A^+ and B^+, and partly on the inner-sphere ligand environments. Mayoh and Day have discussed the case of a symmetrical system (A = B) using the diagram shown in *Figure 8.1*[746a]. The coordinate x is the reaction coordinate for the thermal electron transfer reaction p → s. Using the various theories outlined in Chapter 6, the zero-order curves are represented as parabolae, with equations

$$E_p = A'x^2 \tag{8.11}$$

$$E_s = A'(1-x)^2 + \Delta U_0 \tag{8.12}$$

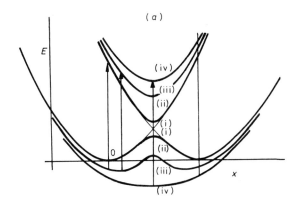

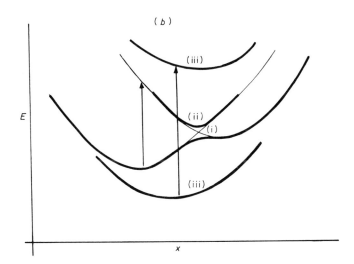

Figure 8.1 Electronic energies E, for electron-transfer complexes $A^+ \ldots B$ as functions of the electron-transfer reaction coordinate x, for different degrees of interaction: (a) Symmetrical homonuclear case ($A \equiv B$), curves (i)–(iv) showing zero, weak, moderate and large interaction; (b) Unsymmetrical case, curves (i), (ii), (iii) showing zero, weak and large interaction (Reference 746a)

In the limit of no interaction, $\beta = 0$ for all x and the energy curves are represented by the zero-order curves, (i). The case of a weak interaction is shown by the curves (ii). The lower curve has a double minimum corresponding to the alternative equivalent structures p and s. The resonance energy is negligible in these structures but it varies strongly with x, reaching its maximum at the intersection point. Curves (iii) and (iv) represent the cases of moderate and strong electronic interactions, respectively.

As the interaction is increased in some systematic way (for example, by varying the internuclear distance, or using different bridging groups), the minima in the ground state curve will come closer together and the height of the intervening barrier will become less. Since this barrier is the activation energy $E^{\ddagger}$ for the thermal electron transfer reaction, this means that the extent of atomic rearrangement needed to effect electron transfer becomes less, while at a certain critical point the activation energy becomes zero and the two minima coalesce into one. Taking the resonance integral β as constant, the equations of the first-order curves are given by substituting equations (8.11) and (8.12) into equation (8.5):

$$E = \tfrac{1}{4}A'[(2x-1)^2 + 1] \pm \tfrac{1}{2}[A'^2(2x-1)^2 + 4\beta^2]^{1/2} \qquad (8.13)$$

where the plus and minus refer to the excited state and the ground state respectively. The two minima in the ground state energy curve occur with coordinates $x = \tfrac{1}{2}[1 \pm (1 - 4\beta^2/A'^2)^{1/2}]$. The criterion for the existence of the two minima, and for the trapping of the valence states is therefore[a]

$$\beta < A'/2 \qquad (8.14)$$

The activation energy when it occurs is given by

$$E_{act} = \tfrac{1}{4}A' - \beta + \beta^2/A' \qquad (8.15)$$

which, when β is small compared with A', reduces to

$$E_{act} = \tfrac{1}{4}A' - \beta \qquad (8.16)$$

Analogous curves for an unsymmetrical case, $\Delta U_0 < 0$, are shown in *Figure 8.1(b)*.

8.3.3 The optical transition

The optical charge transfer process (equation 8.2) is represented by one of the vertical transitions shown in *Figure 8.1*. There is an important qualitative difference between the symmetrical and unsymmetrical cases. In a symmetrical complex the transition loses its charge transfer character as the resonance interaction increases along the sequence (i), (ii), (iii), (iv). In the final case, with full delocalisation of valencies, the electronic density distribution is symmetrical about the mid-point of the nuclei, in both ground and excited states. In the unsymmetrical case ($\Delta U \neq 0$) on the other hand, the ground and excited states always have different equilibrium reaction coordinates, and the vertical transition retains some charge-transfer character.

General expressions for the energy of the optical transition ($hv = E_- - E_+$) can be deduced from the equations above. In the symmetrical case, when the valencies are localised, equation (8.13) leads to $hv = A'$, for all resonance

---

[a] Mayoh and Day[746a] put forward the criterion $\beta < A'/4$, suggested by the approximate equation (8.16)

energies $\beta < A'/2$, and when the valencies are delocalised, to $hv = 2\beta$. In the unsymmetrical case, more complicated expressions apply. When β is negligibly small, the energy ΔU_{CT} for the transition p → s* is given by

$$\Delta U_{CT} = \Delta U_0 + A' \tag{8.17}$$

while the activation energy for the corresponding thermal reaction p → s (*see* Chapter 6, equation (6.10)) is

$$\Delta U^{\ddagger} = \tfrac{1}{4}A'(1 + \Delta U_0/A')^2 \tag{8.18}$$

Eliminating A' between these equations gives

$$\Delta U^{\ddagger} = (\Delta U_{CT})^2/4(\Delta U_{CT} - \Delta U_0) \tag{8.19}$$

which in the symmetrical case ($\Delta U_0 = 0$) reduces to

$$\Delta U^{\ddagger} = \tfrac{1}{4}\Delta U_{CT} \tag{8.20}$$

These equations, of which equation (8.20) is especially noteworthy, were derived by Hush[571]. The corresponding energies for the reverse processes, s → p* and s → p, are

$$\Delta U'_{CT} = -\Delta U_0 + A' \tag{8.21}$$

$$\Delta U'^{\ddagger} = \tfrac{1}{4}A'(1 - \Delta U_0/A')^2 \tag{8.22}$$

whence

$$\Delta U'^{\ddagger} = (\Delta U_{CT})^2/4(\Delta U_{CT} + \Delta U_0) \tag{8.23}$$

8.3.4 The charge transfer absorption band

For the weak interaction case, Hush[571,572] derived expressions for the width and intensity of the absorption band. At temperatures close to room temperature, the shape of the band is approximately Gaussian, of the form

$$I/v = (I_{max}/v_{max})\exp[-(hv_{max} - hv)^2/4A'k_B T] \tag{8.24}$$

where I is the absorption intensity at frequency v, and A' is the reorganisation energy already defined. Defining the half-width of the band as the energy $\Delta E_{1/2} = |hv_{max} - hv_{1/2}|$, when the frequency $v_{1/2}$ is such that $I/v_{1/2} = \tfrac{1}{2}(I_{max}/v_{max})$ yields

$$\Delta E_{1/2} = 4(\ln 2)^{1/2}(A'k_B T)^{1/2} \tag{8.25}$$

and when $\Delta U_0 = 0$,

$$\Delta E_{1/2} = 4(\ln 2)^{1/2}(\Delta U_{CT} k_B T)^{1/2} \tag{8.26}$$

$$\Delta v_{1/2} = a(v_{max})^{1/2} \tag{8.27}$$

where $\Delta v_{1/2}$ and v_{max} are wavenumbers, and the constant $a = 4(\ln 2)^{1/2}(k_B T/hc)^{1/2}$ has the value 47.9 cm$^{-1/2}$ at 25°C[(b)].

To calculate the intensity of absorption, the *oscillator strength* f is defined[773] by

$$f = \frac{4\varepsilon_0 \ln 10 \, mc^2}{e^2 L} \int \varepsilon \, dv \qquad (8.28)$$

where $\int \varepsilon \, dv$ is the area under the absorption band, using the decadic extinction coefficient ε, and the constant preceding the integral has the value 4.32×10^{-9} mol dm^{-3} cm^2. This in turn is related to the transition dipole moment M by

$$f = (8\pi^2 mc/3h)\bar{v}_{max}|M|^2 \qquad (8.29)$$

where the constant in parentheses has the value 1.085×10^{11} cm^{-1}.

For a single electron, considered to move independently of all others in the molecule

$$M = e \int \psi_+ r \psi_- \, d\tau \qquad (8.30)$$

where r is the radial vector coordinate of the electron and e is the electronic charge. Using the wavefunctions given by equations (8.9) and (8.10) this gives

$$M = e\alpha R_{AB} \qquad (8.31)$$

where R_{AB} is the internuclear distance. Some authors have introduced the term *dipole strength*, but with conflicting definitions: Hush[571] defines dipole strength as $D = M/R_{AB} = \alpha e$; Mulliken[773] used the same name for the square of this quantity.

For interactions in the solid state, these equations have to be modified to allow for electron transfer from a single site on to a symmetrical group of nearest-neighbour sites. For example, in the compound $(NH_4)_4[Sb^{III}Br_6][Sb^V Br_6]$, the Sb^{III} and Sb^V ions form a super-lattice in

(b) The low temperature forms of these equations are important for solid-state systems. A full treatment involves consideration of the overlap between different vibrational states of the p and s electronic configurations. When the p and s states are assumed to be harmonic oscillators, with the same vibrational frequency ω, and the p state is populated according to the Boltzmann distribution, the band-width is given by

$$(\Delta E_{1/2})^2 = 8(\ln 2)\hbar\omega A' \coth(\hbar\omega/2k_B T)$$

where $\hbar = h/2\pi$. This reduces to equation (8.25) in the high-temperature limit. Atkinson and Day measured the charge transfer spectra of the salts $(CH_3NH_3)_2Sb_xSn_{1-x}Cl_6$ and $(CH_3NH_3)_2Bi_xSn_{1-x}Cl_6$ and obtained good agreement with this equation, thus estimating $v = \omega/c = 210$ and 200 cm^{-1} respectively[46]. See also reference 404a.

which each Sb^V complex is surrounded by eight Sb^{III} and four Sb^V, and each Sb^{III} is surrounded by eight Sb^V and four Sb^{III}. The lowest excited states can be described in terms of molecular orbitals constructed from the atomic orbitals of the antimony ions, combined subject to the overall symmetry of the thirteen ion group[c]. However, binuclear species also occur in solid state systems and can be detected by methods analogous to those of solution chemistry. When two ions, A^+ and B are doped into a host lattice, the concentration of $A^+ \ldots B$ pairs is proportional to the product $[A^+][B]$ and a spectral band showing this concentration dependence is diagnosed as an interaction band. Examples studied in this way include $Sb^{III} \ldots Sb^V$, in Cs_2SnCl_6, and $Ti^{IV} \ldots Fe^{II}$ in $Al_2O_3{}^{360}$. If the concentration-dependence cannot be studied, as with naturally occurring minerals, but the interaction bands can be identified from other evidence, the measured concentrations of the relevant ions may be such as to ensure that nearest-neighbour $A^+ \ldots B$ pairs occur largely in isolation from other A^+ or B ions.

Using equations (8.28)–(8.31), the degree of delocalisation of the transferring electron can be estimated from the intensity of the spectrum (*Table 8.2*). Mayoh and Day[747] have calculated values of α theoretically using perturbation theory. In some cases, estimates of electron localisation have been checked against other experimental evidence, from e.s.r. (for the heteropolyions $PVW_{11}O_{40}^{5-}$ and $PVMo_{11}O_{40}^{5-}$) and from the solvent-dependence of ΔU_{CT} (for the pyrazine-bridged ruthenium complexes).

8.3.5 Thermodynamic considerations

The energy difference A′ between states s and s*, or p and p*, may be regarded as the Franck–Condon barrier to electron transfer. It is made up of vibrational and solvent reorganisational (rotational and 'librational') contributions and, as will be shown below, it can be estimated from experimental data and calculated theoretically. A problem which arises here is that some of the experimental data are available as energy or enthalpy differences, while some, notably the activation energies of thermal reactions, are expressed in terms of free energy. Moreover, some theories of the activation process, such as the bond-stretching model (Chapter 6, p. 186) are based on energy, while others, such as the Marcus and Hush continuum treatments, are expressed in terms of free energy. The root of the problem (which we shall not attempt to solve) is that the excited 'states' p* and s*, and the transition state t‡, are not thermodynamic states and cannot strictly be assigned thermodynamic properties. In the solution at equilibrium, the precursor complexes p actually comprise a population of species all having the electronic configuration $(A^+ \ldots B)$ but with ligand and solvent molecules in various positions defining the vibrational etc. energy levels populated according to the Boltzmann distribution. When the solution is under irradiation, the electronically excited

---

(c) In an earlier discussion by Hush a different superlattice was assumed[571]

Table 8.2 Calculations of electron localisation in trapped valence states[a]

System	$\bar{v}_{max}/10^3$ cm^{-1}	$D/10^{-10}$ m	$R/10^{-10}$ m	α^2	References
$N_5Fe^{III}NCFe^{II}C_5$ (in $Fe_4^{III}[Fe^{II}(CN)_6]_3 \cdot 14H_2O$)	14.1	1.23	5.1	1.6×10^{-3}	514, 571, 895
$Ti^{III}ClTi^{IV}$ (in HCl.aq)	20.41	0.11	5.0	4.4×10^{-4}	571
$Sb^VCl_6^- \ldots Sb^{III}Cl_6^{3-}$ (in (MeNH$_3$)$_2$SnCl$_6$)	17.7	—	7.1	4×10^{-4}	46
$(NC)_5Fe^{III}NCFe^{II}(CN)_5^{6-}$	7.7	0.80			692
$O_4Fe^{II}(OH)_2Fe^{III}O_4$ (in biotite mica)	13.9	0.134			894
$O_4Fe^{II}(O,OH)Fe^{III}O_4$ (in biotite mica)	13.9	0.080			894
$P^VV^{IV}W_{11}^{VI}O_{40}^{5-}$	20.0	—	3.3	0.015	28
$P^VV^{IV}Mo_{11}^{VI}O_{40}^{5-}$	15.2	—	3.3	0.034	28
$Fe^{3+} \cdot Fe^{II}(CN)_5dmso^{3-}$	16.1	0.86	5.1	6×10^{-3}	1039
$(NC)_5Fe^{III}(pyz)Fe^{II}(CN)_5^{5-}$	8.3	—	—	0.01	404a
$Tl_3Fe^{III}(CN)_6$	21.0	—	—	1.6×10^{-3} [c]	537
$(bipy)_2ClRu^{III}(pyz)Ru^{II}Cl(bipy)_2^{3+}$	7.9	—	7.0[b]	2×10^{-3}	186
$(NH_3)_5Ru^{III}(pyz)Ru^{II}Cl(bipy)_2^{4+}$ (in CH$_3$CN)	10.4	—	7.0	$\simeq 3 \times 10^{-3}$	185

(a) Temperature 300 K (some authors have also reported temperature-dependences). Aqueous medium unless otherwise stated
(b) Estimated (reference 263, quoted in reference 186)
(c) From n.m.r. measurements

state is generated and this consists of a population of species having the electronic configuration (A ... B$^+$), and with ligand and solvent molecules distributed as before, defining another set of vibrational energy levels. These levels, however, are not populated in accordance with the Boltzmann distribution: their populations depend on the respective transition probabilities, or Franck–Condon factors (*see* Chapter 2). Thus, although the s* states together have a defined total energy, they cannot be said to have a free energy or entropy. This problem has been discussed by Adamson in connection with the photoexcited ligand field states of complexes[3]. Adamson distinguished between the 'excited state' which is formed immediately on excitation, and the thermally equilibrated excited state,—'thexi-state' formed by non-radiative relaxation. In his nomenclature the state s* is the charge-transfer excited state of p, and s is the thexi-state.

Following previous authors, we shall ignore this problem and assume that the excited states do have thermodynamic properties. We may then write the equations for the two cycles corresponding to forward and reverse electron transfer

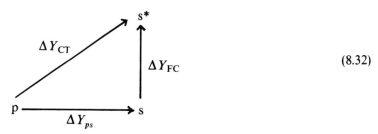 (8.32)

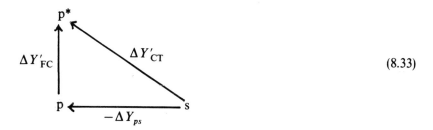 (8.33)

where the symbol Y denotes any thermodynamic property U, H, G, S, etc. and deduce the following general relationships

$$\Delta Y_{CT} = \Delta Y_{ps} + \Delta Y_{FC} \tag{8.34}$$

$$\Delta Y'_{CT} = -\Delta Y_{ps} + \Delta Y'_{FC} \tag{8.35}$$

whence

$$\Delta Y_{CT} - \Delta Y'_{CT} = 2\Delta Y_{ps} + (\Delta Y_{FC} - \Delta Y'_{FC}) \tag{8.36}$$

$$\Delta Y_{CT} + \Delta Y'_{CT} = \Delta Y_{FC} + \Delta Y'_{FC} \tag{8.37}$$

A further assumption commonly made, again based on the assumption that s* can be treated as a thermodynamic state, is that the entropy changes ΔS_{CT} and $\Delta S'_{CT}$ associated with the spectroscopic transitions are zero (apart from changes in spin multiplicity, which we shall ignore). This assumption does not seem to have been extensively discussed, though it has been held to be intuitively obvious as a consequence of the Franck–Condon principle[967]. It leads immediately to

$$Lh\nu_{CT} = \Delta U_{CT} = \Delta H_{CT} = \Delta G_{CT} \tag{8.38}$$

and also, from the above cycles,

$$\Delta S_{FC} = -\Delta S_{ps} = -\Delta S'_{FC} \tag{8.39}$$

Moreover, any reaction model involving energy curves with the same radix in both precursor and successor states (i.e. the same constant A in equations (8.11) and (8.12)) leads to

$$\Delta Y_{FC} = \Delta Y'_{FC} \tag{8.40}$$

for all thermodynamic properties Y. Thus equation (8.36) simplifies to

$$\Delta Y_{CT} - \Delta Y'_{CT} = 2\Delta Y_{ps} \tag{8.41}$$

which in principle could be subjected to direct experimental test. Also, equations (8.39) and (8.40) combined lead to the remarkable conclusion that *all* entropy terms in the above cycles are zero:

$$\Delta S_{FC} = \Delta S'_{FC} = 0 \tag{8.42}$$

$$\Delta S_{ps} = 0 \tag{8.43}$$

Equations (8.42) and (8.43) could be tested by experiment, though this has not yet been done. However, the following arguments provide some support for them, and also indicate their limitations.

(a) The term ΔS_{ps} in equation (8.43) can be identified with the difference in solvation entropies of the complexes p and s. It is well known that for single ions of spherical symmetry, entropies of hydration correlate well with ionic radius and charge[856]; for complex ions, further smaller contributions are to be expected, due to dipoles, quadrupoles, etc. If the charges z on the oxidant and reductant centres differ by one unit in the sense $\Delta z = z(A^+) - z(B) = 1$, then the charge-type remains unchanged on reaction, and the solvation entropy will be little effected. When the charge type changes so as to increase the dipolar contribution to the total field, the entropy change will be negative and vice versa. We may compare the overall reactions $A^+ + B = A + B^+$ as in the following examples[950]:

$$Co^{3+}.aq + Fe^{2+}.aq = Co^{2+}.aq + Fe^{3+}.aq \tag{8.44}$$

$$Fe(CN)_6^{3-}.aq + Fe^{2+}.aq = Fe(CN)_6^{4-}.aq + Fe^{3+}.aq \tag{8.45}$$

with $T\Delta S^\circ = 1.0$ and $-26\,\text{kcal mol}^{-1}$ respectively, in aqueous solution at 25 °C. Presumably, in the second case, equation (8.40) does not apply. In terms of the continuum model, the error in assuming that equation (8.40) holds consists of using the same boundary surface for both initial and final charge distributions, and neglecting the image effects.

(b) Equations (8.42) and (8.43) together require also that $\Delta S^\ddagger = 0$ for the intramolecular process, equation (8.1). This is probably more accurate. It has been noted[506, 545, 808] that for second-order electron transfer reactions in aqueous solution the total activation entropies $(\Delta S^\circ_{ip} + \Delta S^\ddagger_{ps})$ correlate well with the sum of the charges of the reacting ions and this is what would be expected if the total entropy resided mainly in the ΔS°_{ip} term with $\Delta S^\ddagger_{ps}$ close to zero.

In some discussions of this subject, the low entropies of activation characteristic of electron transfer reactions were regarded as evidence in favour of the Franck–Condon mechanism, or more generally as evidence that a highly specific orientation of reactant ions was essential to electron transfer[673]. On the argument put forward here, these explanations are redundant: the low entropies merely reflect the increased orientation of solvent molecules due to the concentration of like charges in the precursor complex.

(c) In cases where solvent reorganisation is a major contribution to ΔG_{FC}, the continuum theory can be applied to give

$$\Delta G_{FC} = \tfrac{1}{2}L\varepsilon_0(\varepsilon_{op}^{-1} - \varepsilon_s^{-1}) \int_V (E_c^p - E_c^s)^2 \, dV \tag{8.46}$$

whence, using $\Delta S_{FC} = -\partial \Delta G_{FC}/\partial T$, we obtain

$$T\Delta S_{FC}/\Delta G_{FC} = -\partial \ln(\varepsilon_{op}^{-1} - \varepsilon_s^{-1})/\partial \ln T \tag{8.47}$$

For water at 25° C, this is calculated to be about -1.5%.

8.3.6 Solvent effects

(a) Inner- and outer-sphere energy terms

As already discussed, the total reorganisation energy can be divided into components of inner- and outer-sphere reorganisations. Methods of calculating these terms have been reviewed in Chapter 6.

$$\Delta G_{FC} = \Delta G_{FC}^{in} + \Delta G_{FC}^{out} \tag{8.48}$$

Several authors[266, 1037] have pointed out that these terms can be distinguished experimentally, by studying the dependence of ΔG_{CT} upon solvent, using symmetrical complexes so that $\Delta G_{CT} = \Delta G_{FC}$. It is assumed that ΔG_{FC}^{in} is independent of the nature of the solvent, while ΔG_{FC}^{out} is proportional to $(\varepsilon_{op}^{-1} - \varepsilon_s^{-1})$ as predicted by continuum theory. Thus when ΔG_{FC} is plotted against $(\varepsilon_{op}^{-1} - \varepsilon_s^{-1})$ a straight line dependence is expected and extrapolation gives the constant term ΔG_{FC}^{in}. The magnitude of ΔG_{FC}^{out} depends on the change in dipole moment associated with the optical transition, and is therefore a measure of the extent of localisation of valence in the p and s* states. When the resonance integral exceeds the limit, $\beta \geqslant A/2$, the charge transfer contribution to ΔG_{FC}^{out}, and therefore ΔG_{CT}, becomes independent of solvent.

(b) Solvent polarity scales

Variations in ΔG_{CT} with solvent have been much studied especially with organic charge transfer systems such as *p*-nitranisole, or complexes such as $[CH_3NC_5H_5^+][I^-]$. In general a transition involving an increase in dipole moment shifts to lower energy with increasing polarity of solvent. In this way, scales of solvent polarity have been set up; the results obtained with different test complexes have been correlated, and connections have been established with other criteria of polarity, such as dipole moments and reaction rates[98, 629, 630, 708, 880]. However, all this work is based on unsymmetrical charge-transfer systems, so that ΔG_{CT} is the sum of the two contributions shown in equation (8.34) above, and it is not in general possible to deduce the magnitude of the Franck–Condon term.

8.4 Examples of optical electron transfer

Elsewhere[197] the present writer has reviewed descriptively the known examples of optical electron transfer, with emphasis on metal–metal systems. Examples have been classified by their structures, and the magnitude of the Franck–Condon energy ΔG_{FC} has been assessed where possible. In some cases it has also been possible to compare the value of ΔG_{FC} with a prediction based on the Marcus–Hush theory.

In this section we discuss some of the examples most directly relevant to electron transfer kinetics.

8.4.1 'Vertical' ionisation and attachment

In discussions of charge transfer complexes, by Mulliken and others, the vertical ionisation energy of an electron donor, and the electron affinity of an acceptor, play an important part. These (in our notation) are the Franck–Condon processes

$$A \rightarrow A^{+*} + e^- \tag{8.49}$$

$$A^+ + e^{-1} \rightarrow A^* \tag{8.50}$$

in which reactants A, A^+ are molecules or molecular ions, and the products are in vibrationally excited states which have the same atomic configurations as the reactants. The ionisation energies are measured spectroscopically, and in many cases are better known than the corresponding equilibrium energy differences. Vertical attachment energies have proved more difficult to obtain, but some have been measured by electron impact studies.

These data relate to the gas phase. For discussion of complexes in condensed phases we require the analogous parameters for the processes

$$A(\text{env}) \xrightarrow{h\nu_\phi} A^+(\text{env}^*) + \{e^-\}_{\text{env}} \qquad (8.51)$$

$$A^+(\text{env}^*) + \{e^-\}_{\text{env}} \xrightarrow{h\nu'_\phi} A(\text{env}^*) \qquad (8.52)$$

where $\{e^-\}_{\text{env}}$ denotes a 'dry' electron delocalised throughout the solvent medium.

Processes of this type were considered by Mulliken[777], and discussed at some length by Platzman and Franck[850, 851] in 1952. More recently, they have become accessible to direct measurement, by the observation of photoemission spectra[d]. These have been reported for a variety of reducing species in solution including the ferrocyanide ion[70, 70a, 74, 137, 307, 790], the solvated electron [49, 51, 75, 76, 324] and a number of other inorganic and organic species[48, 50, 224, 304, 306]. The mechanism of emission has been analysed theoretically by Delahay and others[303, 789]. The theory is based on two consecutive processes: emission of electrons into unbound states within the solvent, and escape from these states into the gas phase. Sometimes the energy ΔU_ϕ is close to that of a peak in the photoemission spectrum but the theory also leads to methods of calculating ΔU_ϕ when this is not the case[303].

For the case of the $Fe(CN)_6^{4-}$ ion, Ballard and Griffiths[70a, 70b] found that below the energy of the band maximum the photoemission current varies exponentially with photon energy. This is taken as evidence that the bandwidth is due to thermal occupation of the vibrational levels in the electronic ground state of $Fe(CN)_6^{4-}$ (env), and it seems likely that the low-energy threshold of the spectrum represents emission from those $Fe(CN)_6^{4-}$ ions which happen to have solvent molecules in configurations similar to those which characterise the equilibrium state of $Fe(CN)_6^{3-}$.

By combining equation (8.51) with expressions for the standard reduction of A^+ and for formation and solvation of H^+, it is possible in principle to

(d) The process defined in equation (8.51) must not be confused with so-called 'charge-transfer-to-solvent'. For example the I^- ion in solution has two absorptions in the near UV spectrum, as assigned to transitions of the type $I^-.\text{aq} \to I.(\text{aq}^-)^*$ where the excited state consists of the electron effectively detached from the I atom, but still localised in a centrosymmetrical orbital due to the field of the orientated solvent molecules. For literature references and discussion of the solvent reorganisation effects, see reference[197]

calculate the reorganisation energy for $\Delta G_{FC}(A^+)$ for the isolated, solvated ion:

$A(env) \xrightarrow{h\nu_\phi} A^+(env^*) + \{e^-\}_{env}$	$\Delta G_\phi(A)$	(8.51)
$A^+(env) + \tfrac{1}{2}H_2(g) \longrightarrow A(env) + H^+(env)$	$\Delta G_E(A^+)$	(8.53)
$H^+(env) + \{e^-\}_{env} \longrightarrow H^+(g) + e^-(g)$		(8.54)
$H^+(g) + e^-(g) \longrightarrow H(g)$	ΔG_H	(8.55)
$H(g) \longrightarrow \tfrac{1}{2}H_2(g)$		(8.56)
$A^+(env) \longrightarrow A^+(env^*)$	$\Delta G_{FC}(A^+)$	(8.57)

The expression for the reverse processes (equation 8.52) could similarly be used to obtain $\Delta G_{FC}(A)$:

$$A(env) \longrightarrow A(env^*) \qquad \Delta G_{FC}(A) \qquad (8.58)$$

The theoretical values, assuming the continuum model, are given by

$$\Delta G_{FC}^{out} = \left(\frac{1}{4\pi\varepsilon_a}\right) Le^2 \left[\frac{1}{\varepsilon_{op}} - \frac{1}{\varepsilon_s}\right]\frac{1}{2r} \qquad (8.59)$$

where r is the radius of A^+ or A.

Comparison of theory with experiment is difficult at present since the available data mostly relate to different solvent media. Experimental values for $Fe(C_5H_5)_2^+$ and $Fe(CN)_6^{3-}$ are 21 and 15 kcal mol^{-1} in water and 'tetraglyme' respectively, to be compared with 18 and 20 kcal mol^{-1} calculated from equation (8.59)[197].

8.4.2 Outer-sphere electron transfer

For electron transfer within an outer-sphere complex, the relevant free energy terms are related by the following cycle.

$$\begin{array}{c}
A^+(env) + B(env) \xrightarrow{\Delta G_W} A^+.B(env) \xrightarrow{\Delta G'_{FC}} A^+.B(env^*) \\
i \uparrow \qquad\qquad p \qquad\qquad p^* \\
\Delta G_E^{pair} \qquad\qquad \Delta G_{PS} \qquad\qquad \Delta G'_{CT} \\
\qquad\qquad\qquad \Delta G_{CT} \\
A(env) + B^+(env) \xrightarrow{\Delta G'_W} A.B^+(env) \xrightarrow{\Delta G_{FC}} A.B^+(env^*) \\
f \qquad\qquad s \qquad\qquad s^*
\end{array} \qquad (8.60)$$

From this we have

$$\Delta G_{CT} = \Delta G_{ps} + \Delta G_{FC} = -\Delta G_E^{pair} - (\Delta G_W - \Delta G'_W) + \Delta G_{FC} \qquad (8.61)$$

$$\Delta G_{FC} = \Delta G_{CT} + (\Delta G_W - \Delta G'_W) + \Delta G_E^{pair} \qquad (8.62)$$

where ΔG_W and $\Delta G'_W$ are the work terms. At the present time there is no system for which ΔG_W and $\Delta G'_W$ are both known, but in practice these can be

Table 8.3 Outer-sphere complexes: energy parameters for the optical charge transfer process, $A \cdot B^+ \rightarrow A^+ \cdot B^{(a)}$

B^+	A	ΔG_{CT}	$-\Delta G_E(B^+)$	$\Delta G_E(A)$	ΔG_W	$-\Delta G'_W$	ΔG_{FC}
$Ru(NH_3)_6^{3+}$	Cl^-	97.3	2.3	-57.4	-1.6	0.0	40.6
$Ru(NH_3)_6^{3+}$	Br^-	88.6	2.3	-45.0	-1.4	0.0	44.5
$Ru(NH_3)_6^{3+}$	I^-	71.2	2.3	-31.5	-1.4	0.0	40.6
$Ru(NH_3)_6^{3+}$	NCS^-	88.2	2.3	-39.6	-1.4	0.0	49.5
$Ru(NH_3)_6^{3+}$	$S_2O_3^{2-}$	71.0					
$Ru(en)_3^{3+}$	I^-	63.6	4.9	-31.5	-0.9	0.0	36.1

(a) See text and equation (8.60). All energies in units of kcal mol^{-1}. For original references, see reference 197

estimated with some confidence for outer-sphere systems. The data shown in *Table 8.3* have been calculated in this way. It will be noticed that the correlation of ΔG_{CT} with the reducing power of the ion is such that ΔG_{FC} is nearly constant for the series of $Rh(NH_3)_6^{3+} \cdot X^-$ complexes. Analogous spectra have been obtained for some outer-sphere cobalt(III) complexes, but the complete calculation is not possible owing to a change in spin state. Cobalt(III) in the complexes shown has the low-spin configuration t_{2g}^6 hence the immediate charge-transfer product must be low-spin also, whereas the thermodynamic data for cobalt(II) all relate to the high spin state $t_{2g}^5 e_g^2$.

If the redox potentials are not too different it should be possible to measure spectra for both precursor and successor complexes for comparison with equations (8.36) etc above. No such complementary measurements have yet been reported, but reversible optical transfer has been observed in the solid state, between the ions Eu^{2+} and Sm^{3+}, doped into crystalline CaF_2[1084].

8.4.3 Bridged complexes

Provided that the valency states are still localised, bridged binuclear complexes may be treated by a similar formalism. If the bridge (X) is associated with the same atom A in both oxidised and reduced forms, we have the cycle

$$A^+.X(env) + B(env) \xrightarrow{\Delta G'_W} A^+.X.B(env) \xrightarrow{\Delta G_{FC}} A^+.X.B(env^*)$$

$$\Big\uparrow \Delta G_E^{pair} \qquad\qquad \Big\uparrow \Delta G_{ps} \qquad\qquad \nearrow \Delta G_{CT}$$

$$A.X(env) + B^+(env) \xrightarrow{\Delta G_W} A.X.B^+(env) \qquad\qquad (8.63)$$

with

$$\Delta G_E^{pair} = \Delta G_E(B^+) - \Delta G_E(A^+.X) \qquad\qquad (8.64)$$

If X is associated with atom B in both forms we have the analogous cycle with

$$\Delta G_E^{pair} = \Delta G_E(B^+.X) - \Delta G_E(A^+) \tag{8.65}$$

Many examples of such complexes are known[26, 197], but not all of them are completely characterised. When both oxidant and reductant centres are substitutionally labile, the complex is formed in an equilibrium mixture with the separate ions and the bridging ligand. The structure remains uncertain and it is not even possible to be sure which ligand forms the bridging group, though in practice this can often be safely guessed. Thus copper(I) and copper(II) ions in methanol show a dark blue interaction colour, attributable to a mixed valence binuclear complex, when acetate ion is present. The colour does not appear when acetate ion is replaced by some non-complexing ligand, nor when a high concentration of the strongly complexing but non bridging CH_3CN is added. This suggests that acetate ion is a bridging group. References to further work of this type will be found in the review cited above.

When one centre is labile and the other inert, the structure is less ambiguous. The complex formed on mixing $Cr(H_2O)_6^{3+}$ and $HCrO_4^-$ ions in acid solution almost certainly contains octahedrally chromium(III) and tetrahedral chromium(VI). We still cannot distinguish between the inner-sphere structure $(H_2O)_5Cr^{III}OCr^{VI}O_3^+$ and the outer-sphere $[Cr(H_2O)_6^{3+}][CrO_4^{2-}]$, and the same is true of complexes such as $Ag^+.MnO_4^-$ and $Fe^{2+}Mo^V(CN)_8^{3-}$ in which the cations are labile while the anions are inert. Measurements of rate of formation could resolve this question, and in the case of the chromium complex, energetic considerations favour the inner-sphere form. As discussed elsewhere[197], it can be concluded from known or estimated redox potential data, that $\Delta G_{FC} + \Delta G'_W = -10 \pm 12$ kcal mol^{-1}, and if ΔG_{FC} is comparable with other known values, $\Delta G'_W$ must be strongly negative.

Only when both centres are inert can the structure be defined with certainty. Examples of such complexes characterised as products of bridged electron transfer reactions are described in Chapter 5. Others have been formed by the reaction of two mononuclear species, without electron transfer. An example is the complex formed by mixing $Fe^{II}(CN)_6^{4-}$ and $Fe^{III}(CN)_5(NH_3)^{2-}$. A strong interaction colour is observed, which is not seen on mixing $Fe(CN)_6^{4-}$ and $Fe(CN)_6^{3-}$, and the rate of formation of the binuclear complex is rapid compared with the rates of replacement of cyanide ion in either of the two monomers. Hence, the most likely structure is the singly bridged $(NC)_5FeCNFe(CN)_5^{6-}$ [466a].

In unsymmetrical systems, the proper assignment of valencies has to be considered. Usually this is obvious on the basis of the redox potentials of comparable mononuclear complexes and of the absorption spectrum, but some cases remain open to doubt. Thus the complex $FeMo(CN)_8^-$ formed both from $Fe^{2+} + Mo(CN)_8^{3-}$, and from $Fe^{3+} + Mo(CN)_8^{4-}$, was originally[717] formulated $Fe^{II} \ldots Mo^V$ but is more probably $Fe^{III} \ldots Mo^{IV}$. (The equilibrium constant $Fe^{2+} + Mo(CN)_8^{3-} \rightleftharpoons Fe^{3+} + Mo(CN)_8^{4-}$ may be calculated[950] as $\simeq 10^3$.) The complex is blue in colour, with absorption at

820 nm[706] presumably due to electron transfer. The complex $(H_3N)_5RuOV^{4+}$ is formed from $Ru(NH_3)_5OH_2^{2+}$ and VO^{2+}. The infra-red spectrum indicates the presence of the V=O group, acting as a ligand analogous to N_2 or CH_3CN. This supports the valency assignment $Ru^{II}\ldots V^{IV}$, but with the possibility of extreme back-donation of electrons from the d_{xy} and d_{yz} orbitals of Ru into empty π orbitals of the VO group. Thus some admixture of the $Ru^{III}\ldots V^{III}$ configuration in the ground state is possible. The most nearly analogous solution equilibrium for which data are available is $Ru(NH_3)_5OH_2^{2+} + VO^{2+} \rightleftharpoons Ru(NH_3)_5OH^{2+} + VOH^{2+}$, with $K \simeq 10^{-3.7}$ [950].

Assignments for solid state systems are less easy to predict. Earlier workers[1080] formulated the chromophores in Prussian Blue as $Fe^{II}N_6$, $Fe^{III}C_6$ on the basis of the solution equilibrium

$$Fe^{2+} + Fe(CN)_6^{3-} \rightleftharpoons Fe^{3+} + Fe(CN)_6^{4-} \qquad K = 10^{-6.5} \qquad (8.66)$$

whereas later work has established the opposite assignment. Crystal lattice forces may be partly responsible for the difference, but it should also be noted that the ion Fe^{2+}.aq is not a good model for the FeN_6 chromophore. More appropriate data, if available, would be the redox potentials of $Fe(NCCH_3)_6^{3+}$ and $Fe(CN)_6^{3-}$ in acetonitrile as solvent. The complex or ion pair existing in Al_2O_3 doubly doped with Fe and Ti has been identified as $Fe^{II}\ldots Ti^{IV}$ in the ground state, mainly on esr evidence[360]. This agrees with the solution equilibrium[950]

$$Fe^{2+} + TiO^{2+} \rightleftharpoons FeOH^{2+} + TiOH^{2+} \qquad K = 10^{-14} \qquad (8.67)$$

8.4.4 Symmetrical inert complexes

A particularly important sub-class of bridged complexes consists of symmetrical complexes with both chromophores inert to substitution. The first example (complex II below) was reported by Creutz and Taube in 1969[265]; several others have since been prepared, and studies are proceeding rapidly at the time of writing. In most cases, analogous compounds with one more and one less electron are known forming the series $A^+.X.A^+$, $A^+.X.A$, $A.X.A$, and the electronic spectrum of the middle member contains a unique band in addition to the bands characteristic of the A^+ and A chromophores. (These latter bands are identified by reference to the terminal members of the series, and also the monomers $A^+.X$ and A.X when these are known.)

When all the three complexes are known, the equilibrium constant K_{com} for the comproportionation reaction gives a direct measure of the resonance stabilisation in the ground state of the middle complex

$$A.X.A + A^+.X.A^+ \xrightleftharpoons{K_{com}} 2A.X.A^+ \qquad (8.68)$$

When complex II was first reported[265], it was thought that the energy of the unique band gave a direct measure of the Franck–Condon energy, but it is now recognised that this is not necessarily the case: it depends also on the degree of valence interaction as discussed above. The three possibilities of weak, moderate and strong interaction are apparently realised in the complexes I, II, III respectively.

$(NH_3)_5RuN\boxed{}\boxed{}NRu(NH_3)_5{}^{5+}$ Complex I

$(NH_3)_5RuN\boxed{}NRu(NH_3)_5{}^{5+}$ Complex II

$(NH_3)_5RuN\equiv C-C\equiv NRu(NH_3)_5{}^{5+}$ Complex III

Complex I [1037] has the characteristic low-energy absorption band with the predicted half-width, and subject to strong solvent shift as expected for long-range electron transfer, but in the ground state there is evidently very little interaction between the ruthenium atoms. K_{com} is close to the statistical value, 4.

Complex II [103, 235, 265, 266, 367, 973] has the characteristic low energy absorption band, but this is relatively narrow and virtually independent of changes in solvent[260], and there is evidence of appreciable delocalisation in the ground state, $(K_{com} \cong 1.3 \times 10^6)$[265]. The exact state of this complex is still uncertain. The photo-electron spectrum was initially interpreted[235] in terms of trapped valencies, but Hush[573] has shown that a delocalised description is equally consistent with the data available so far. Infra-red data are still under discussion but the Raman spectrum has been interpreted[973] in terms of localisation. It has been shown that the rate of valence interchange is rapid compared with the nmr time scale[367]; the possibility that remains is that the rate may be comparable with the frequencies of certain bond vibrations.

Complex III is quite clearly of the average-valence type[1038] (*Figure 8.1a*, curves iv). Again there is a low-energy absorption band, not characteristic of either ruthenium(III) or ruthenium(II), but narrower than expected for an intervalence band, and not subject to solvent shifts. The comproportionation constant is large $(K_{com} > 10^{13})$ and the valencies are localised at least on the infra-red time scale (i.e., the thermal electron exchange rate constant k_{ex} is greater than 10^{13} s^{-1}). In all these complexes the metal–metal interaction is attributed to back donation of electrons from the $d\pi$ orbitals of ruthenium(II) into the anti-bonding π orbitals of the ligand (as also occurs in the mononuclear $Ru(NH_3)_5(pyz)^{2+}$ complex)[264]. The reverse interaction, from ligand π orbitals into empty $d\pi$ orbitals of ruthenium(III) is postulated in

$$\text{IV} \quad \rightleftharpoons_{-H^+} \quad \text{V} \qquad (8.69)$$

complex V, derived from the t-butylmalonitrile complex IV. The strength of the interaction is indicated by the high acidity of the central hydrogen atom. The low energy electronic absorption band (8.55×10^3 cm^{-1}) is narrow, indicating effective delocalisation of valencies. It is subject to solvent shifts but this can be explained by the lack of a centre of symmetry, so that the transition has an effective dipole-moment change[632].

A different criterion of valence localisation has been applied to the complex $(H_3N)_4ClOsN_2OsCl(NH_3)_4{}^{3+}$. The $N \equiv N$ stretching frequency is found in the Raman spectrum, but not in the infra-red. Hence, the complex has a centre of symmetry and the valencies, formally Os^{II} and Os^{III}, must be considered equivalent on the infra-red time-scale[1014].

The effects of varying the non-bridging ligands have also been studied[10, 165, 184–186, 266, 755, 857]. It has been shown that complex VI

$$[Cl(bipy)_2RuN\widehat{}NRu(bipy)_2Cl]^{3+}$$
VI

unlike Complex II, has localised valencies, and this is explained by postulating back donation of the odd electron on to the aromatic ring systems of the bipyridyl ligands.

Studies have also been made of the effect of non-bridging ligands attached unsymmetrically[185, 266, 857], i.e. to only one ruthenium atom. Regardless of the nature of the ligand, the absorption is shifted to higher energy. This is because if the new ligand L stabilises ruthenium(III), the electronic ground state is biased towards the configuration LRu^{III}.X.Ru^{II} while if it stabilises ruthenium(II) the ground state is biased towards LRu^{II}.X.Ru^{III}. In either case the charge transfer energy now includes an addition contribution from the ΔG_{ps} term. Values of ΔG_{ps} for such systems have not been obtained experimentally, but they may be presumed to correlate with reduction free energies for analogous systems such as LRu^{III}X + e$^- \rightleftharpoons$ LRu^{II}X. The plot of ΔG_{CT} against ΔG_E(LRu^{III}X) has been found to be linear with unit slope as expected. (Non-bridging ligand effects have been reported also for some unsymmetrical complexes, e.g. $\{Cu^{II}NC[Fe^{II}(CN)_4]CO^-\}$[1039].) The series of complexes $(NH_3)_5Ru^{III}XRu^{II}Cl(bipy)_2^{4+}$ shows the trend from Class I to Class II behaviour as a function of the bridging ligand. Where X is $NC_5H_4.CH_2CH_2C_5H_4N$, no intervalence charge transfer is observed; where X is $NC_5H_4.CH:CH.C_5H_4N$, $NC_5H_4.C_5H_4N$, and pyrazine, the band appears with decreasing wavenumber and increasing intensity, corresponding to increasing metal–metal interaction[185]. Polynuclear analogues of these complexes include the cluster–cluster dimer $[Ru_3O(OAc)_6py_2]_2$(pyz) and linear polymers such as $(bipy)_2ClRu[(pyz)Ru(bipy)_2]_nRuCl$ (bipy)$_2^{m+11}$,[164, 1100].

Mixed-valence compounds based on ferrocene units have been extensively studied[257, 767]. The complexes (VII–X, with $n = 1$ show an interesting gradation of properties, and detailed comparison[767] also illustrates some of the complexities that can arise.

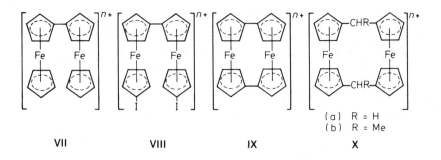

(a) R = H
(b) R = Me

VII   VIII   IX   X

Mössbauer spectra of salts of (Complex VII) in the solid state show clearly the two doublets for Fe^{II} and Fe^{III}, implying that the thermal exchange rate constant k_{ex} is less than 10^7 s^{-1}, even at room temperature. There is, however, some change in the spectrum with temperature and this may be due to the presence at the higher temperature of a second form of the cation, in a different crystal structure, with delocalised valencies. The band in the near infra-red, which is not shown by either the fully reduced $Fe^{II}Fe^{II}$ or fully oxidised $Fe^{III}Fe^{III}$ complex, is attributed to the intervalence transition. The energy of this band varies slightly in CH_3CN solution, and more substantially on going to the solid state at 77 K.

In contrast, the di-iodo-substituted cation (complex VIII) shows the characteristics of valence delocalisation not only on the Mössbauer but also (at 77 K) on the e.s.r. time-scale. Again, it is possible that structural considerations are important: if, for example, the salts of Complex VIII had the *cis*-configuration and complex VII the *trans*, or if complex VIII were *trans* and complex VII had a structure twisted about the central C–C bond by some angle other than 180 °C, the electronic coupling in complex VIII would be stronger than in complex VII, in agreement with the observations. It may be significant that in solution the two ions show rather similar properties: the near IR bands are at practically the same energy, and the successive oxidation potentials (in acetonitrile) are quite similar

$$\text{Complex VII}: (n=0) \xrightarrow{0.42V} (n=1) \xrightarrow{0.70V} (n=2) \qquad (8.70)$$

$$\text{Complex VIII}: (n=0) \xrightarrow{0.31V} (n=1) \xrightarrow{0.72V} (n=2) \qquad (8.71)$$

Presumably in solution there is free rotation about the central C–C bond, and the distribution of conformations is similar for the two complexes.

The bis-ferrocenylium complex IX would be expected to show much stronger metal–metal coupling and certainly the electron transfer is rapid on the Mössbauer time-scale. However, it is apparently not yet possible to place any higher limit on the rate. X-ray photoelectron spectra imply complete localisation ($k_{ex} < 10^{17}$ s^{-1}) but the infra-red spectra show bands in the regions 1200–1300 and 1600–1800 cm^{-1}, reminiscent of the reduced Fe^{II}–Fe^{II}

form of the complex, which could indicate $k_{ex} < 10^{13}\,\text{s}^{-1}$. Again there is a low energy electronic transition ($\simeq 6.5 \times 10^3\,\text{cm}^{-1}$) but it is not necessarily to be interpreted as intervalence charge transfer.

The two [1, 1]-ferrocenophane systems, complex X, show marked differences; again illustrating the importance of seemingly small structural variations in the particular salts used. Complex Xb (I_3^- salt) has localised valences ($k_{ex} < 10^7\,\text{s}^{-1}$) but complex Xa behaves like a mixture of two forms, one localised and one delocalised on the Mössbauer time-scale, the proportions varying with temperature. (X-ray data on the corresponding unoxidised complex have shown two crystallographically different forms of the molecule, with different Fe–Fe distances.) In solution, the two complexes differ again, complex Xa having a band at $13.3 \times 10^3\,\text{cm}^{-1}$ which is not shown by complex Xb; this has been tentatively assigned as intervalence charge transfer.

The relative contributions of direct and ligand-mediated metal–metal interaction have been discussed[768]. Evidence that the latter can operate in isolation is provided by the complex $(C_5H_5)Fe(C_5H_4.C\equiv C.C_5H_4)Fe(C_5H_5)^+$ which has a well-defined intervalence transfer absorption at $6.41 \times 10^3\,\text{cm}^{-1}$ [658]. The comproportionation constant (see equation (8.68)) may be roughly estimated as 3×10^2. Related compounds with bridging groups $-Hg-$, $-C_2H_4-$, $-(CH_3)_2CC(CH_3)_2-$ and $-CH\colon CHC_6H_4CH\colon CH-$ have smaller K_{com} indicating less interaction[768].

Analogous organic species have also been studied. By esr methods[45], the rates of intramolecular electron transfer have been estimated for radical ions of the types $C_6H_5(CH_2)_nC_6H_5$ [1056, 1083], $\alpha-C_{10}H_7(CH_2)_nC_{10}H_7-\alpha$ [947], $p-O_2NC_6H_4XC_6H_4NO_2^--p$ [518a] with various bridging groups X, and $p-C_6H_4[(CH_2)_n][(CH_2)_m]C_6H_4^--p$ [1056, 1083]. Theoretical discussions have been given, based on resonance transfer[695a], and the possibility of electronic interaction being mediated through the chain linkages has been considered[1056], but no intervalence charge transfer spectra have been reported and, in the absence of further information, the odd electron must be assumed to be effectively localised in the ground state.

8.4.5 Directly bonded systems

(a) General

Electron transfer systems in which the donor and acceptor atoms are directly adjacent are extremely common; they were the earliest to be characterised[428, 602, 872] and they cover a wide range of chemical types.

The metal halide complexes are classic examples[609]. When the ground state is predominantly ionic, the excited state is either covalent or only weakly bonded, as presumably in the case $Fe^{3+}.Cl^- \rightarrow Fe^{2+}.Cl$. The excited state may relax to the ground state, or the components may separate[650]. The reactions of radical species produced in this way form an important branch of photochemistry (*see also* the preparation of Tl^{2+} ion, Chapter 3, p. 58). Covalent halides show analogous spectra[872], the transition being from a bonded to a virtually non-bonded state, $CH_3I \to CH_3 \ldots I$. Mulliken[775] calculated the $N \to Q$ transition moments for the spectra of halogen compounds of varying polarities, using ground and excited state wavefunctions of varying degrees of ionicity.

A case of special interest is the diatomic molecule $LiNa^+$ (which may be considered an intermediate in the gas-phase reaction $Li^+ + Na \to Li + Na^+$). The charge transfer spectrum has not been observed, but it has been calculated theoretically[822a]. The transition $X^2\Sigma^+ \to A^2\Sigma^+$ corresponds roughly to the electron transfer $Li \ldots Na^+ \to Li^+ \ldots Na$ and the transition moment is correspondingly large. At the equilibrium bond distance of 3.25 Å, the calculated absorption maximum is at $\bar{v} = 20.3 + 10^3 \text{ cm}^{-1}$, with an absorption cross section $\sigma = 9.2 \times 10^{-17} \text{ cm}^2$, corresponding to an extinction coefficient $\varepsilon = L\sigma/\ln 10 = 2.4 \times 10^4 \text{ M}^{-1}\text{cm}^{-1}$. The transition dipole moment is approximately $1.0 e\text{Å}$. From equation (8.31), the localisation parameter is thus calculated as $\alpha^2 = 0.094$, implying that in the ground state the electron spends about 90% of the time on the lithium atom. Actually, this calculation is not valid since the condition of weak overlap does not hold, but it serves to confirm the greater delocalisation of the electron in this molecule, as compared with the bridged complexes discussed above. As the component atoms become closer in electronegativity, the absorption band loses its charge transfer character. Mulliken calculated the transition moments of the corresponding N–Q transitions of halogen molecules[774].

Transitions to and from extensively delocalised states occur in symmetrical complexes of the type $OsCl_6$, where the ligand-to-metal transition leaves a 'hole' symmetrically distributed over the six chlorine atoms. The energy levels in such systems have been extensively discussed[602].

Both metal-to-ligand and ligand-to-metal charge transfer are well characterised, and correlations have been established between transition energies and thermodynamic redox potentials. Interpretations are complicated in some cases by the fact that the same ligands e.g. NO_2^- [73] can either accept or donate electrons, depending on the metal.

The weak association complexes studied by Mulliken and others and generally known as 'charge transfer complexes' also come into this category[161,425,773,776-8], as for example $C_6H_6.I_2 \to C_6H_6^+.I_2^-$ [439]. The structures of many such complexes have been determined in the solid state, and correlation of absorption spectra in the solid state, in solution and in the gas phase, leave no doubt that the oxidant and reductant components are in direct contact, with no intervening solvent molecules[161,425]. In many such systems, the ground state is virtually non-bonded but the excited state is ionic; however, a range of bonding possibilities has been considered, up to and including strongly dative-bonded species such as Me_3NBF_3[777].

(b) Energy cycles

Mulliken and others have interpreted the spectra of such complexes with the aid of the cycle

$$A(env)^* + B^+(env)^* \xrightarrow{\Delta G'^*_W} A.B^+(env)^*$$
$$\Delta U_\phi \uparrow \qquad\qquad\qquad \uparrow \Delta G_{CT}$$
$$A^+(env) + B(env) \xrightarrow{\Delta G_W} A^+.B(env) \qquad (8.72)$$

giving

$$\Delta G_{CT} = \Delta U_\phi + (\Delta G'^*_W - \Delta G_W)$$
$$= \Delta U_\phi(B) - \Delta U_\phi(A) + (\Delta G'^*_W - \Delta G_W) \qquad (8.73)$$

where $-\Delta U_\phi(A)$ and $+\Delta U_\phi(B)$ are the vertical electron affinity of $A^+(env)$ and the vertical ionisation energy of $B(env)$.

If the forces contributing to the work terms are assumed to be entirely Coulombic, the data can be used to calculate the net quantity of charge transferred in the transition, and thence to infer the amount of electron localisation in the ground and excited states. The assumption is that

$$\Delta G'^*_W - \Delta G_W = \left(\frac{L}{4\pi\varepsilon_0}\right)\frac{x^2 e^2}{R} \qquad (8.74)$$

where xe is the amount of electron charge transferred. The coefficient x may be roughly equated with α of equation (8.8). For the complex of tetracyanethylene (as acceptor) with p-xylan (as donor) in dichloromethane as solvent, Trotter[1044a] calculated $x = 0.90$.

Alternatively, equation (8.74) may be used to calculate the work term $\Delta G'^*_W$ and so to draw conclusions about the nature of the bonding in the excited state. Several authors have presented arguments along these lines. Generally the 'vertical' data used refer to the gas phase: ionisation energies being obtained from absorption spectroscopy and electron affinities from rapid electron attachment experiments. The resulting error is lessened, however, by the fact that the solvents considered are usually non-polar, and by the fact that the typical donors and acceptors are polyatomic molecules. Hence, some of the Franck–Condon energy is due to internal strain, and is not solvent-dependent.

MacConell, Ham and Platt[696] considered a series of complexes of I_2 with various electron donors. Using an equation equivalent to equation (8.74) and taking $\Delta G_W \cong 0$ (the donors being all uncharged), they estimated $\Delta G'^*_{W}$, assuming this to be constant for the series. The value was large and negative ($\Delta G'^*_W \simeq -80$ kcal mol^{-1}) indicating strong binding in the excited states.

The work terms have been treated theoretically, as follows[778]. The states which we have written $A^+.B(env)$ and $A.B(env^*)$ may both be considered as hybrids of the two zero-order states, and may be compared with hypothetical states having the same atomic configurations, but having the pure zero-order electronic configurations. Thus the cycle (8.72) may be expanded as

$$A + B^+ \xrightarrow{\Delta U_C'^*} \{A \ldots B^+\} \xrightarrow{\Delta U_R'^*} \{A^+ \ldots B \longleftrightarrow A \ldots B^+\}$$

$$\downarrow \Delta U_\phi \qquad\qquad \downarrow \Delta U_\pm \qquad\qquad\qquad \uparrow h\nu_{CT}$$

$$A^+ + B \xrightarrow{\Delta U_C} \{A^+ \ldots B\} \xrightarrow{\Delta U_R} \{A^+ \ldots B \longleftrightarrow A \ldots B^+\} \quad (8.75)$$

where the terms ΔU_C, $\Delta U_C'^*$ contain mainly electrostatic and Van der Waals bonding terms, and the terms ΔU_R, $\Delta U_R'^*$ are resonance energies. (The solvent is omitted from equation (8.75) for brevity, and indeed is not explicitly considered in the treatments cited here, apart from the specification that ΔU_ϕ refers to vertical transitions.) Thus we have

$$Lh\nu_{CT} = \Delta U_\pm + (\Delta U_R'^* - \Delta U_R) \quad (8.76)$$

When the energy difference $\Delta U_\pm$ is large, the resonance energies are small and perturbation theory gives:

$$Lh\nu_{CT} = \Delta U_\pm + (\beta_0^2 + \beta_1^2)/\Delta U_\pm \quad (8.77)$$

where

$$\beta_0 = \beta_{01} - W_0 S_{01} \qquad \beta_1 = \beta_{01} - W_1 S_{01}$$

$$\beta = \int \psi_A H \psi_B d\tau \qquad S_{01} = \int \psi_A \psi_B d\tau$$

and W_0, W_1 are the electronic energies of the zero-order states $\{A^+ \ldots B\}$, $\{A \ldots B^+\}$. A better approximation, when $\Delta U_\pm$ is small, is given by

$$(Lh\nu_{CT})^2 = \left(\frac{\Delta U_\pm}{1 - S_{01}^2}\right)\left(1 + \frac{4\beta_1 \beta_0}{(\Delta U_\pm)^2}\right) \quad (8.78)$$

When equation (8.77) is used to correlate variations in $h\nu_{CT}$ for a series of donors with a given acceptor, or vice versa, it takes the form

$$Lk\nu_{CT} = \Delta U_\phi - C_1 + C_2/(\Delta U_\phi - C_1) \quad (8.79)$$

where C_1 and C_2 are constants.

There have been many studies of the variation of ΔG_{CT} with a common oxidant and a series of reductants, or vice versa. The more recent literature has been reviewed by Foster[425]. In most cases, values of ΔG_ϕ are not available, but other parameters, such as gas-phase vertical ionisation or electron attachment energies, gas phase equilibrium energies, redox or polarographic potentials, etc., have been used. In most cases, the plots of ΔG_{CT} against ΔU_ϕ are found to be straight lines of unit slope (and conversely, the existence of such a correlation is often taken as evidence for the assignment of a band and charge transfer), but exceptions are well established. Some authors have gone so far as to say that linearity is surprising—'the almost religious belief ... in this linearity must be cautioned against'[703]. Non-linear correlations have been fitted to equations (8.77)[161,778] and (8.79)[173,778]. Some failures of the correlation have been attributed, by implication, to variations in solvent reorganisation energy: that is to say, in the absence of vertical ionisation data, the adiabatic values were used, and deviations from the expected plot were attributed to this cause. Finally, Mulliken and Person[778] have emphasised the possible influence of the work terms ΔU_C and $\Delta U_C'^*$ which may vary significantly according to the structure of the reactants (when, for example, amines and I_2 are compared as electron donors).

8.5. Reorganisation energies: comparison with theory

Elsewhere the present writer has attempted to compare reorganisation energies calculated from the above thermodynamic cycles directly with predictions from the Marcus–Hush theory[197,198]. The main difficulty in so doing is to take account of the difference between inner- and outer-sphere contributions. It would be possible to measure these two separately for each of the systems described here, by studying the variation of the spectrum with changes in solvent, but so far this has not been done. Although the results would be interesting, the collection of such data would be a lengthy project, since not only the spectra but also the redox potentials and formation constants would have to be obtained for the complex of interest, in every solvent. In an attempt to circumvent this problem the Marcus equation (6.74) is applied in two different integrated forms.

8.5.1 The separate sphere model

The oxidant and reductant ions are considered as spheres of radii a_1 and a_2, respectively, these being the Pauling radii of the central atoms, separated by a distance R estimated from molecular models or otherwise. The continuum model then gives (see equation 6.96)):

$$\Delta G_{FC}(\text{calc}) = \left(\frac{1}{4\pi\varepsilon_0}\right) Le^2 \left(\frac{1}{2a_1} + \frac{1}{2a_2} - \frac{1}{R}\right) \qquad (8.80)$$

Thus the solvent, the inner-sphere ligands and the bridging ligands when present, are all treated together as the dielectric continuum. This is a very gross approximation, since apart from the non-uniformity of the medium, there are various possible electronic interactions. These will be largely of two sorts: the charges on the oxidant and reductant ions will be partially delocalised on to the neighbouring ligands, and also on to each other so that the effective charge transferred amounts to less than one electron. These are aggregated by introducing an empirical factor χ in place of $(\varepsilon_0^{-1} - \varepsilon_s^{-1})$ in equation (8.80), to obtain

$$\Delta G_{FC}(\text{obs}) = \left(\frac{1}{4\pi\varepsilon_0}\right) Le^2 \chi \left(\frac{1}{2a_1} + \frac{1}{2a_2} - \frac{1}{R}\right) \qquad (8.81)$$

For water at 25°C $(\varepsilon_0^{-1} - \varepsilon_s^{-1}) = 0.55$, and the values of χ are all lower than this. The inner-sphere and/or bridging ligands can be arranged in order of decreasing χ in the series $NH_3 \simeq en > O^{2-} > $ pyrazine $\simeq$ bis-cyclopentadienyl $\simeq$ 4-4'-bipyridyl; and this coincides with intuitive ideas of increasing polarisability or covalent bonding. A more refined treatment might attempt to relative values of χ to measurable parameters such as the refractive indices of the ligands.

8.5.2 The ellipsoid model

In this case only the outer-sphere contribution is calculated, taking the boundary of the complex as an ellipsoid of revolution. The electrostatic model gives (see equation (6.99)):

$$\Delta G_{FC}^{out} = \left(\frac{1}{4\pi\varepsilon_0}\right)\left(\frac{Lp^2}{2a^2b}\right)\left(\frac{1}{\varepsilon_{op}} - \frac{1}{\varepsilon_s}\right) S(\lambda_0) \quad (8.82)$$

where the terms have been defined in Chapter 6 (p. 202). The dipole moment change is taken as $p = eR$.

For four outer-sphere complexes, $Ru(NH_3)_6^{3+}.X^-$ (where X is Cl, Br, I) and $Ru(en)_3^{2+}.I^-$, the calculated ΔG_{FC}^{out} are similar to (though in fact slightly larger than) the experimental ΔG_{FC}. This suggests that the outer-sphere part of the reorganisational energy is a substantial fraction of the whole. For the three bridged complexes VI, I, II, the separate terms ΔG_{FC}^{out} and ΔG_{FC}^{in} are known by experiment. The calculated values of ΔG_{FC}^{out} are greater than the experimental values by factors of 2, 8 and ≥ 30 respectively. This trend probably reflects increased delocalisation of the Ru^{III} and Ru^{II} valencies along the series.

8.6 Photo-induced electron transfer reactions

The field of inorganic photochemistry is covered in recent books[4,71]. Here we mention only the relatively recent discovery of direct metal-to-metal electron transfer, induced by the intervalence charge-transfer transition.

The binuclear complex shown in the following reaction

$$(NC)_5Co^{III}NCFe^{II}(CN)_5^{6-} \underset{}{\overset{h\nu}{\rightleftharpoons}} Co^{II}(CN)_5^{3-} + Fe^{III}(CN)_6^{3-} \quad (8.83)$$

has absorption maxima, at 326 nm and 385 nm, assigned to Co^{III} ligand-field and intervalence transfer respectively[1058]. Irradiation into either of these has little effect, presumably because the primary product quickly falls back to the ground state. In the presence of oxygen, however, overall reactions are seen, consistent with the formation of the highly reducing $Co(CN)_5^{3-}$ ion. In alkaline solution, for example, the binuclear superoxo complex is formed via the reactions

$$2Co^{II}(CN)_5^{3-} + O_2 \rightarrow (NC)_5Co^{III}O_2Co^{III}(CN)_5^{6-} \quad (8.84)$$

$$(NC)_5Co^{III}O_2Co^{III}(CN)_5^{6-} + Fe(CN)_6^{3-}$$
$$\rightarrow (NC)_5Co^{III}O_2Co^{III}(CN)_5^{5-} + Fe(CN)_6^{4-} \qquad (8.85)$$

The maximum quantum yields were found on irradiation in the IT band.

Analogous results have been obtained[1057] with the complex $(NH_3)_5Co^{III}NCRu^{II}(CN)_5^-$: irradiation at 375 nm leads to the redox reaction

$$(NH_3)_5Co^{III}NCRu^{II}(CN)_5^- \xrightarrow{h\nu} 5NH_3 + Co^{2+} + Ru(CN)_6^{3-} \qquad (8.86)$$

Photochemically-induced electron transfer has also been reported[402] in a series of Co^{III}–Cu^I complexes, with organic bridging ligands. A typical complex (XI) has an absorption at 318 nm[566] attributable to a $Cu(d) \rightarrow$ ligand (π^*) transition, and irradiation into this band produces Co^{II} and Cu^{II} whereas the corresponding thermal electron transfer is almost immeasurably slow[566]

$$\left[Co(NH_3)_5OOC\underset{CH_2}{\overset{CH}{=\!=}}Cu^I \right]^{3+} \qquad \left[Co(NH_3)_5OOC.CH_2\underset{CH_2}{\overset{CH}{=\!=}}Cu^I \right]^{3+}$$

XI   XII

It is interesting to note that complexes with π-electron systems not conjugated with the carboxyl group react about equally well: in fact the quantum yield is greater for complex XII than for complex XI. Irradiation in the cobalt(III) ligand field region is much less effective. Intermolecular electron transfer is also promoted in this way, as, for example, the reaction $Co(NH_3)_5OAc^{2+} + Cu(CH_2:CHCH_2OH)^+$. It is proposed that the primary excitation product is the singlet state $Cu(d^9).L(\pi^*)$, and that this reacts directly and rapidly in a thermal intramolecular process to give the observed products. This is an example of photon-excited 'chemical' transfer, as distinct from the photo-excited 'resonance' transfer of equation (8.2). An example of photo-induced electron transfer in the solid state is discussed in the next chapter (p. 314).

Chapter 9
Electron transfer in the solid state

In the solid state, electron transfer manifests itself as electrical conductivity. Of the many different conduction mechanisms which have been recognised, the so-called hopping mechanism is particularly closely related to the solution processes discussed previously in this book. An electron localised on a particular atom or molecule is transferred to a neighbouring site by a thermally activated process analogous to the thermally activated intramolecular transfer. In section 9.1 we define some terms and briefly describe the experimental criteria by which a hopping mechanism may be recognised, and in section 9.2 we discuss some examples.

9.1 Outline of theory

9.1.1 Electrical conduction[555]

The resistance R of a sample of material is defined by Ohm's law

$$i = \Delta\phi/R \tag{9.1}$$

where i is the current and $\Delta\phi$ the applied potential difference; and for uniform current, e.g. along a thin wire, the conductivity σ of the material is defined by

$$\sigma = l/AR \tag{9.2}$$

where l is the length of the sample and A is the cross-sectional area. The current density J is then given by

$$J = i/A = \sigma\Delta\phi/l$$
$$= \sigma E \tag{9.3}$$

where E is the field strength, which in this case is uniform. A more general expression applicable to non-uniform fields is

$$J = -\sigma\nabla\phi = \sigma E \tag{9.4}$$

where J and E are the corresponding vectors. For isotropic materials, σ is a scalar quantity; for non-isotropic materials it is a tensor. If the current is thought of as the flow of charged particles, the criterion for the applicability of

Ohm's law is that the average velocity of such particles is proportional to the field strength. In this way we define the mobility u by the equation

$$\langle v \rangle = uE \tag{9.5}$$

whence

$$\sigma = nqu \tag{9.6}$$

where q is the charge per particle and n the particle 'density', i.e. number of charged particles per unit volume. More generally, if different types of charge carrier operate in the same material:

$$\sigma = \sum \sigma_i = \sum (n_i q_i u_i) \tag{9.7}$$

and the *transport number* for each type is defined as the fraction of total current carried, i.e.

$$t_i = n_i q_i u_i / nqu \tag{9.8}$$

In terms of the molar concentration [X] of the charge carrier, equation (9.6) becomes

$$\sigma = L[X]qu \tag{9.9}$$

and similarly for other equations in this and the following sections.

9.1.2 The activation energy

Semiconductors are characterised by values of conductivity in the range $\sigma < 0.1$ ohm^{-1} cm^{-1} and a temperature dependence of the form

$$\sigma = \sigma_0 \exp(-E_a/RT) \tag{9.10}$$

Over a wide temperature range, the temperature dependence may be more complicated, implying perhaps that different mechanisms prevail at different temperatures; even so the plot of $\log \sigma$ against T^{-1} is the most convenient general method of displaying experimental data, and the quantity $-R \, d \ln \sigma / d(T^{-1})$ may be considered as an activation energy analogous to the Arrhenius activation energy of a chemical reaction.

Semiconducting materials have been classed into two types: *intrinsic* semiconductors, in which the charge-carriers originate from the pure material by a thermal excitation process; and *extrinsic* conductors in which the charge carriers are present in constant temperature-independent concentration (either at trace concentrations, 'doped' into the crystal lattice, or in stoicheiometric concentration at high concentration, in a mixed-valence compound). This distinction is not, however, particularly useful from the present point of view. Most of the substances to be discussed in this chapter depend for their conductivity on some foreign chemical component, and could therefore be classed as extrinsic semiconductors, but most if not all also require thermal excitation, the charge carrier being generated by a reaction between the dopant and the bulk constituents of the crystal. In general, it is useful to define two contributions to the overall activation energy E_a, the energy E_f of formation, and the activation energy E_p of propagation of charge carriers so that the concentration and mobility are given by

$$n = n_0 \exp(-E_f/RT) \tag{9.11}$$

$$u = u_0 \exp(-E_p/RT) \tag{9.12}$$

where n_0 and u_0 are constants, and from equation (9.6)

$$E_0 = E_f + E_p \tag{9.13}$$

9.1.3 The hopping mechanism

In a simple treatment, charge carriers are assumed to 'hop' over a fixed distance, i.e. the distance between nearest neighbouring carrier sites and the motion is treated by the theory of diffusion. The result is

$$\sigma = \gamma n e^2 a^2 / k_B T \tau \tag{9.14}$$

where a is the distance between nearest-neighbouring sites and γ is a geometric factor related to the crystal structure. τ is the average time for transfer from an occupied to an unoccupied site, and is in turn assumed to be given by the Eyring theory of reaction rates, as

$$\tau^{-1} = \tau_0^{-1} \exp(-\Delta G^{\ddagger}/RT) \tag{9.15}$$

where τ_0^{-1} is the frequency factor and $\Delta G^{\ddagger}$ is an activation free energy (sometimes identical with E_p). Discussion of these quantities closely parallels the discussions of electron transfer theory reviewed in previous chapters. For adiabatic transfer, τ_0^{-1} is usually identified with the frequency ω_0 of a lattice vibration[531]. Non-adiabatic and low-temperature processes have also been discussed[532].

Equation (9.14) is strictly valid only when the number of charge carriers is small compared with the number of possible sites. A more general equation is

$$\sigma = \gamma n_1 s(1-s) e^2 a^2 / k_B T \tau \tag{9.16}$$

where n_1 is the number density of lattice points available to the charge carriers, and s the probability of such a point being occupied. The expression $\gamma n_1 s(1-s)$ denotes the number density of pairs of neighbouring occupied and unoccupied sites (corresponding to precursor complexes in solution chemistry) and, strictly speaking, it is these pairs which constitute the true charge carriers. The geometric factor γ also depends on the distribution of occupied and unoccupied sites relative to each other (it corresponds to the work term); but usually a random distribution is assumed. Consider, for example, two ions, oxidant A^+ and reductant B, doped into the super-lattice of cations C in a sodium chloride crystal structure to give the formula $A_x^+ B_y C \ldots$ Each A^+ ion has 12 nearest neighbours, of which a fraction y are B ions. The number density of $A^+ \ldots B$ nearest-neighbour pairs is therefore $12n_1 xy = 12xy/a^3$, which, when substituted into equation (9.16) gives

$$\sigma = 12xye^2/ak_B T\tau \tag{9.17}$$

9.1.4 Connection with reaction kinetics

The constant τ^{-1} of equation (9.15) is analogous to the first-order rate constant for electron transfer within a precursor complex. A second-order rate constant for the reaction between donor and acceptor atoms can also be defined. For the case of a symmetrical reaction $A^+ + A = A + A^+$, with A in large excess over A^+, we may write $k = \tau^{-1}[A]^{-1}$. Taking $a = 2 \times 10^{-10}$ m, $T = 300$ K and $\sigma = 0.1$ ohm cm^{-1}, we obtain $\tau^{-1} = 3 \times 10^{10}$ s^{-1}. For a cubic lattice $[A] = L^{-1}a^{-3} = 20$ M, hence $k = 1.5 \times 10^9$ M^{-1}s^{-1}. This is a little below the diffusion-controlled limit for reactions in aqueous solution. In other words, the highest typical conductivity for a solid semiconductor is roughly equivalent to the highest rate constant for a reaction in solution.

When the electron donor and acceptor sites are equivalent and the coupling between them is weak, an expression for $\Delta G^\ddagger$ can be written, by analogy with Marcus–Hush theory, as

$$\Delta G^\ddagger = (\varepsilon_{op}^{-1} - \varepsilon_s^{-1})(\varepsilon_0/2) \int (E_p^c - E_s^c)^2 d\tau \tag{9.18}$$

where the terms are analogous to those in equation (6.74). The variation of conduction activation energy with the optical and static dielectric constants of the medium has indeed been discussed, though not in terms of equation (9.18)[52]. It would be possible to use equation (9.18) to calculate absolute magnitudes of $\Delta G^\ddagger$ for specific crystal structures, and to compare the results with experiment. This has apparently not been done, and the results would in any case have to be received with caution, in view of the simplifying assumptions made. One problem is that not all sites are equivalent even in simple systems—as is mentioned below for the case of the doped nickel oxide system. Another problem is the presence of impurities in the crystal: the charge carriers may preferentially occupy lattice defects, rather than the sites assumed in the theory.

9.1.5 Connection with optical spectroscopy

By the arguments already given in Chapter 8, the optical charge transfer energy, thermal hopping activation energy, and thermal excitation energy are related by

$$E_a = E_{op}^2/4(E_{op} - E_t) \tag{9.19}$$

9.1.6 The sign of the charge carrier

The sign of the predominant mobile charge carrier can be determined experimentally by various methods, of which the two following are the most important.

(a) The Hall effect[576]

When a magnetic field is applied at right angles to the direction of the current, then in general an electrostatic field gradient is set up in the third perpendicular direction; namely

$$E = R_H B J \tag{9.20}$$

where E, B and J are the components of electric field strength, magnetic induction, and current density in the relevant directions, and R_H is the Hall coefficient. An elementary treatment of the motion of charged particles under the electric and magnetic fields, assuming for simplicity that all particles have the same velocity, leads to the result

$$R_H = 1/nq \tag{9.21}$$

The direction of the observed field, expressed by the sign of R, depends on the sign of charge of the carriers. When both positive and negative carriers are present, the sign of the observed effect depends on the balance of contribution from both. The result is then

$$R_H = \frac{1}{e} \frac{(n_p - n_n b^2) + u_n^2 B^2 (n_p - n_n)}{(bn_n + n_p)^2 + u_n^2 B^2 (n_p - n_n)^2} \tag{9.22}$$

where n_p, n_n and u_p, u_n are the number densities and mobilities of positive and negative carriers, and where $b = u_n/u_p$; in the limit of large B this simplifies to

$$R_H = 1/e(n_p - n_n) \tag{9.23}$$

Returning to the case of a single type of carrier, since the charge is known ($q = \pm e$) the number density may be calculated. Likewise from equation (9.6) the mobility is obtained as

$$u = |R_H|\sigma \tag{9.24}$$

the sign of R_H being disregarded since conductivity is by convention always positive.

(b) The Seebeck effect[555]

If a semiconductor is subject to a temperature gradient, an electromotive force also develops, which may be rationalised qualitatively by saying that the charge carriers tend to migrate from the hotter to the colder region. Hence, the sign of the measured e.m.f. indicates the sign of the charge carriers. The Seebeck coefficient is defined by

$$\alpha = \Delta V/\Delta T \tag{9.25}$$

where ΔV is the e.m.f. (as measured, for example, with an infinite impedance voltmeter; or by applying an external e.m.f. so as to just prevent any net flow of current through the sample) and ΔT is the temperature difference. In the steady state, when no current is flowing, the potential V can be regarded as the sum of two contributions: the electrostatic potential ϕ due to the charges of the carriers and the chemical potential μ of the carriers, thus

$$V = \phi + \mu/q \tag{9.26}$$

whence

$$\alpha = \partial V/\partial T = (\partial \mu/\partial T)/q \tag{9.27}$$

Since the chemical potential can be represented as

$$\mu = \bar{U} - T\bar{S} \tag{9.28}$$

where $\bar{U}$ and $\bar{S}$ are the partial molar internal energy and entropy of the charge carriers, then provided that the temperature difference over the length of the sample is small compared with the average absolute temperature, we obtain

$$\alpha = -\bar{S}/q \tag{9.29}$$

When the charge carrier assembly is nearly static, as is the case for the hopping model, this entropy can be represented approximately as the sum of two components: a configurational term $\bar{S}_C$ due to the distribution of charge carriers over the available lattice sites; and a 'lattice relaxation term' S_R related to the distortions produced when the charge carrier is generated in the lattice. In the terms already defined, $\bar{S}_C = k_B \ln[(1-s)/(s)]$; $\bar{S}_R$ varies from system to system, but has been estimated for certain model systems[533]. Thus, we have

$$\alpha = \frac{k_B}{q}\left[\ln\left(\frac{1-s}{s}\right) + \frac{S_R}{k_B}\right] \tag{9.30}$$

Hence the sign of the Seebeck coefficient gives the sign of the charge carrier, and the magnitude can be predicted as a test of the hopping model.

The observed sign can often be rationalised simply from a knowledge of the chemistry of the ions involved[295, 905]. An example of the n-type of semiconductor is titanium dioxide doped with traces of tantalum oxide. The tantalum atoms enter the titanium lattice positions as Ta^{5+} ions, and a corresponding number of Ti^{4+} ions are reduced to Ti^{3+}. The current-carrying process is therefore in effect electron transfer.

$$Ti_A^{4+} + Ti_B^{3+} \rightarrow Ti_A^{3+} + Ti_B^{4+} \tag{9.31}$$

(The subscripts A and B denote different lattice sites.) A p-type semiconductor is exemplified by nickel oxide doped with lithium oxide. The effect of replacing some Ni^{2+} ions by Li^+ is that some others are oxidised to Ni^{3+}. Although the detailed mechanism of conduction is uncertain (see p. 312 below), the fact that the minority species is oxidised relative to the bulk species ensures p-type conductivity.

Materials of this kind have been called 'controlled valency semiconductors'[1054] in that the mean valency of the host metal ion is an experimental variable. In some cases the mean valency can be varied over the whole range without change of lattice structure. An example is the sequence of mixed oxides $MgV_2^{III}O_4$, $Mg_{1+x}V_{2-x}^{III}V_x^{IV}O_4$, $Mg_2V^{IV}O_4$, all having the spinel structure. It will be noted that the sign of the effective charge carrier changes as the composition passes through the equivalence point. When $x < 1$, the charge carriers are V^{4+} ions apparently moving in a 'sea' of V^{3+} ions, and thus in effect are positive holes; when $x > 1$ the carriers are V^{3+} ions in a 'sea' of V^{4+} or in effect, electrons. The Seebeck coefficient is therefore expected to change sign as the composition passes through $x = 1$ and in some cases this phenomenon has been observed[468].

9.2 Examples

9.2.1 Some outer-sphere systems

Day and co-workers have made detailed studies on several mixed valency compounds of the outer-sphere type. One example[47] will be described briefly here: others listed in *Tables 9.1* and *9.2*. The salts formulated $M_2^I Sb_x Sn_{1-x}Cl_6$ (with various alkali metal cations M^+), and $Cs_2In_{(1-2y)/2}Sb_{(1+2y)/2}Cl_6$ have been shown to contain isostructural Sb^{III} and Sb^V complex ions. More explicitly, the formulae are $A_4^I[SnCl_6^{4-}]_{2-2x}[Sb^{III}Cl_6^{3-}]_x[Sb^VCl_6^-]_x$ and $Cs_4[InCl_6^{3-}]_{1-2y}[Sb^{III}Cl_6^{3-}]_{2y}[Sb^VCl_6^-]$. The compositional parameters can be varied widely without change of crystal structure (the $M_2[M'Cl_6]$ lattice is of the anti-fluorite type so that each $M'Cl_6$ complex is surrounded by twelve others at equal distances) and the conductivity is found to vary as x^2 or as y respectively, consistent with the second-order concentration dependence $\sigma \propto [Sb^{III}][Sb^V]$. Charge carrier formation may be written as

$$Sb_A^{III} + Sb_B^V \xrightarrow{\Delta U = E_f} Sb_A^{IV} + Sb_B^{IV} \tag{9.32}$$

followed by one or more of the propagation steps

$$Sb_A^{IV} + Sb_C^{IV} \longrightarrow Sb_A^V + Sb_C^{III} \tag{9.33}$$

$$Sb_A^{IV} + Sb_C^V \longrightarrow Sb_A^V + Sb_C^{IV} \tag{9.34}$$

$$Sb_A^{IV} + Sb_C^{III} \longrightarrow Sb_A^{III} + Sb_C^{IV} \tag{9.35}$$

In these equations the subscripts A, B, C denote different lattice points. Propagation by equation (9.33) requires the coincidence of two Sb^{IV} ions on adjacent sites, and corresponds to equal transport by electrons and holes; equations (9.34) and (9.35) correspond to transport by electrons and holes respectively. Seebeck measurements show that hole transport predominates. From the temperature-dependence of the Seebeck coefficient, and from the observed activation energy of the conductivity, Atkinson and Day[45] deduced the energy ranges $E_p = 0$ to $0.40\,eV$, and $E_f = 0.36$ to $0.76\,eV$. The intervalence charge transfer spectrum[46] gives the energy for the optical analogue of equation (9.32), $E_{opt} = 2.3\,eV$.

9.2.2 Halide-bridged systems

A number of mixed valence compounds of the type Au^I–Au^{III}, Pd^{II}–Pd^{IV}, and Pt^{II}–Pt^{IV}, and some related mixed metal compounds, have chain structures with alternating metal and halide ions. The metal valencies are well defined both by the characteristic coordination number and by the metal–halogen vibrational frequencies. A typical example is the compound of empirical formula $Pt(NH_3)_2Cl_3$, more clearly written as $[Pt^{II}(NH_3)_2Cl_2][Pt^{IV}(NH_3)_2Cl_4]$. Polarised absorption spectra show a strong band consistent with inter-valence transfer along the chains.

$$\ldots Pt^{II} \ldots ClPt^{IV}Cl \ldots \xrightarrow{h\nu} [\ldots Pt^{III} \ldots ClPt^{III}Cl \ldots]^* \quad (9.36)$$

Conductivity measurements on single crystals show a pronounced anisotropy, conduction being mainly along the chains. At first a band conduction mechanism was assumed, but more recent work has been interpreted in terms of a hopping model. The charge carrier formation process may be written as

$$Pt^{II} \ldots ClPt^{IV}Cl \rightarrow Pt^{III}Cl \ldots Pt^{III}Cl \quad (9.37)$$

involving the motion of one halide ion. A sequence of electron transfer between Pt^{II} and Pt^{III} would lead to p-type conduction; while transfer between Pt^{IV} and Pt^{III} would lead to n-type conduction. As yet there appear to be no measurements to distinguish these possibilities experimentally. Interrante et al.[586, 587] find that activation energies for conduction decrease with increasing applied pressure, presumably because the difference between Pt^{II}–X and Pt^{IV}–X bond lengths decreases. In the limit of high pressure it might be expected that the difference in bond lengths would be reduced to zero and the electronic structure would change over from Class II, $Pt^{II} \ldots Pt^{IV}$ to Class III, $Pt^{III} \ldots Pt^{III}$; this would be accompanied by a change to metallic conduction, with negative activation energy. No such change occurs within the pressure range so far studied, however. Further work on these systems would be of great interest in connection with the work on electron transfer in solution, described in previous chapters.

9.2.3 Mixed oxide systems[53, 295]

A representative selection is listed in *Table 9.1*. The nearest-neighbour pairs of oxidising and reducing ions from di-μ-oxo-bridged units $[A^+(O)_2B]$. Hence, they are examples of bridged electron transfer. The correspondence with solution kinetics is not exact, however, since double bridging with a four-membered ring has not yet been demonstrated for any system in solution.

(a) Iron (III)–Iron (II)

The mineral crocidolite (blue asbestos) has the idealised composition $Na_2[Fe_2^{III}Fe_{3-x}^{II}Mg_x]Si_8O_{22}(OH)_2$. The silicon–oxygen framework consists

Table 9.1 Examples of hopping conductivity

Material	Conduction type	Reacting pair	E_a/ kcal mol^{-1}	References	E_{opt}/ kcal mol^{-1}	References
Crocidolite		$Fe^{3+} \ldots Fe^{2+}$	16	685	46.2$^{(a)}$,	23
Fe_2O_3, doped with Ti^{4+}	n	$Fe^{3+} \ldots Fe^{2+}$	27.0	766	50.1$^{(b)}$	
Yttrium–iron garnet, doped with Sn^{4+}	n	$Fe^{3+} \ldots Fe^{2+}$	—		$\geqslant 26$	1105
Yttrium–iron garnet, doped with Ca^{2+}	p	$Fe^{4+} \ldots Fe^{3+}$	—		$\geqslant 26$	1105
$FePO_4 \cdot nH_2O$ etc. (partly oxidised)		$Fe^{3+} \ldots Fe^{2+}$	—$^{(d)}$			515
$Ni^{II}Ni_x^{III}Li_xO$	p	see text	4.6		18.3	25
$Co_x^{II}Co^{III}Li_xO$	p	$Li^+ . Co^{3+} \ldots Co^{2+} . Li^+$	4.6$^{(e)}$	145		
$Co_{3-x}Fe_xO_4$ (spinel) $\begin{cases} x > 2 \\ x < 2: \end{cases}$	n	$\begin{Bmatrix} Fe^{2+} \ldots Fe^{3+} \\ Co^{2+} \ldots Co^{3+} \end{Bmatrix}$	4.3	601		
	p		11.4			
$Mg_{1+x}V_{2-x}^{III}V_x^{IV}O_4$ (spinel) $\begin{cases} x < 1 \\ x > 1 \end{cases}$	p	$V^{3+} \ldots V^{4+}$	$\simeq 3.0^{(f)}$	884		
	n					
$Pr_{1-x}^{III}Pr_x^{IV}O_{1.5+0.5x}$	p	$Pr^{3+} \ldots Pr^{4+}$	10.8$^{(g)}$	375		
$Ce_x^{III}Ce^{IV}_{1-x}O_{2-0.5x}$	n	$Ce^{3+} \ldots Ce^{4+}$	3.7$^{(h)}$	130, 131		
Eu_3S_4	n	$Eu^{3+} \ldots Eu^{2+}$	3.0, 4.8$^{(u)}$	156		
$([Co(NH_3)_5]_2O_2)^{5+}$		$^{(c)}$		412		
$Tl_3Fe^{III}(CN)_6$	p	$Tl^+ \ldots Tl^{2+}$	6$^{(j)}$	157, 537		
$KM^{II}Co^{III}(CN)_6$		$M^{2+} \ldots Co^{3+}$	$^{(k)}$	157		
$[Fe(bipy)_3][IrCl_6]$		$^{(c)}$		685		

307

Compound					
$[Pd^{II}(NH_3)_2Cl_2][Pd^{IV}(NH_3)_2Cl_4]$	(c)		412		
	(r)		1026	1.65$^{(a)}$	1112
		0.33$^{(a)}$	1027		
		0.59$^{(b)}$	1027		
	(r)(t)		587		
$[Pd^{II}(NH_3)_2Br_2][Pd^{IV}(NH_3)_2Br_4]$		0.35$^{(a)}$	586		
$[Pd^{II}(NH_3)_2Cl_2][Pt^{IV}(NH_3)_2Cl_4]$	(r)	0.42$^{(n)(s)}$	587		
	(c)		1027		
$[Pd^{II}(NH_3)_2Br_2][Pt^{IV}(NH_3)_2Br_4]$	(c)	0.22$^{(a)}$	587, 586		
$[Pt^{II}(NH_3)_2Cl_2][Pt^{IV}(NH_3)_2Cl_4]$	(r)(t)		587		
	(c)		586		
$[Pt^{II}(NH_3)_2Br_2][Pt^{IV}(NH_3)_4Br_4]$	(r)(t)		587		
			1027		
			587		
$[Pt^{II}(ea)_2Cl_2][Pt^{IV}(ea)_2Cl_4]\cdot4H_2O^{(q)}$	(c)		587	2.2$^{(a)}$	1113
$[Pt^{II}(ea)_2Br_2][Pt^{IV}(ea)_2Br_4]\cdot4H_2O^{(q)}$	(c)		587	2.4$^{(a)}$	1113
$[Pt^{II}(en)_2][Pt^{IV}(en)_2Cl_2][ClO_4]_4$	(c)		1027	2.4$^{(a)}$	1112
$[Pt^{II}(en)Cl_2][Pt^{IV}(en)Cl_4]$	(c)		586		
$[Pt^{II}(en)Br_2][Pt^{IV}(en)Br_4]$		1.7$^{(n)}$	587		
$[Pt^{II}(en)I_2][Pt^{IV}(en)I_4]$		0.67$^{(a)}$	586		
$Cs_4[Pd^{II}Cl_4][Au^{III}Cl_4]_2^{(m)}$	(r)		469		
$Cs_4[Pd^{II}Br_4][Au^{III}Br_4]_2^{(m)}$	(r)		469		
$Cs_4[Cu^{II}Cl_4][Au^{III}Cl_4]_2^{(m)}$	(r)		469		
$Cs_4[Cu^{II}Br_4][Au^{III}Br_4]_2^{(m)}$	(r)		469		
$Cs_2[Au^II_2][Au^{III}I_4]_2$	(r)		469		
$(NH_4)_6[Au^{III}Cl_4]_3[Ag^I_2Cl_5]$		0.43$^{(a)(l)}$	469	2.4$^{(a)(p)}$	1112
		0.42$^{(b)(l)}$	469		
$(NH_4)_6[Au^{III}Br_4]_3Ag^I_2Br_5]$		0.42$^{(a)}$	469		

Notes to Table 9.1

(a) Parallel to the metal ion chains in the crystal
(b) Perpendicular to the metal ion chains in the crystal
(c) No evidence of semi-conductivity (see original paper for information on the upper limit of σ, permitted by the experiments)
(d) Traces of iron(II) produced in the iron(III) compound by hydrostatic pressure. Temperature dependence of conductivity not observed
(e) Activation energy associated with difference in spin state, Co^{3+} t_{2g}^6, Co^{2+} $t_{2g}^5 e_g^2$
(f) Minimum value, occurring when $x \simeq 0.2$
(g) Value for $x = 0.08$ (E_a decreases with increasing x)
(h) Extrapolated to $x = 0$ (E_a increases with increasing x)
(i) Varies with temperature
(j) See also Table 8.2
(k) E_a increasing with increasing ionisation energy of M^{II}, $Fe < Mn < Co \simeq Ni \simeq Cu < Zn$
(l) The crystal structure[150] features $Au^{III}ClAg^I$ etc. chains; with cross-linking via other gold(III) ions. Although the activation energies are the same, conductivity parallel to the chains is greater than that perpendicular to the chains
(m) Assumed structure
(n) Polycrystalline sample
(p) This compound was formulated $(NH_4)_3AgAuCl_7$ in reference 1112
(q) ea is $(C_2H_5NH_2)$
(r) Semi-conduction detected but temperature dependence not measured
(s) At 10 kbar pressure; E_{act} decreases to minimum of 0.05 eV at 106 kbar
(t) Conductivity measured as a function of pressure
(u) Two values, above and below the transition temperature $T = 175$ K

Table 9.2 Comparison of thermal and optical electron transfer energies for some solid systems

Material	E_a/eV	References	E_{opt}/eV	References	Charge carrier type		Electron transfer process[a]	
Crocidolite	0.69	685	2.0	23	?		$Fe^{III} + Fe^{II} \rightarrow Fe^{II} + Fe^{III}$	
$NiO:Li_2O$	4.6		18.3	25	p	c.f.:	$Ni^{3+}.Li^+ + Ni^{2+} \rightarrow Ni^{2+}.Li^+ + Ni^{3+}$	$E_r = 4.6$
						m:	$Ni^{2+} + Ni^{3+} \rightarrow Ni^{3+} + Ni^{2+}$	$E_m \leqslant 0.01$[b]
$[Co(NH_3)_6]Cu^I_xCu^{II}_yCl_{12+x+2y}$ [c]	0.74	294, 267	2.2	294, 267	?	c.f.:	$Cu^{II}Cl_5^{3-} + Cu^ICl_4^{3-} \rightarrow Cu^ICl_5^{4-} + Cu^{II}Cl_4^{2-}$	
Cs_2SbCl_6	0.76	47	2.3	46	p	c.f.:	$Sb^VCl_6^- + Sb^{III}Cl_6^{3-} \rightarrow Sb^{IV}Cl_6^{2-} + Sb^{IV}Cl_6^{2-}$	
Rb_2SbCl_6	0.87	47	2.43	46	p	m:	$Sb^{III}Cl_6^{3-} + Sb^{IV}Cl_6^{2-} \rightarrow Sb^{IV}Cl_6^{2-} + Sb^{III}Cl_6^{3-}$	
$[Co(NH_3)_6][PbCl_6]$	0.73	296	2.51[d]	24	p	c.f.:	$Pb^{IV}Cl_6^{2-} + Pb^{II}Cl_6^{4-} \rightarrow Pb^{III}Cl_6^{3-} + Pb^{III}Cl_6^{3-}$	
						m:	$Pb^{II}Cl_6^{4-} + Pb^{III}Cl_6^{3-} \rightarrow Pb^{III}Cl_6^{3-} + Pb^{II}Cl_6^{4-}$	

(a) c.f.: carrier formation; m: carrier mobility
(b) This refers to *bound* polaron hopping, $Ni^{3+}.Li^+.Ni^{2+} \rightarrow Ni^{2+}.Li^+.Ni^{3+}$ which contributes only to high frequency a.c. conductivity (see text)
(c) For the structure see also reference 781
(d) Measured for $[Rh(NH_3)_6][PbCl_6]$

of parallel double chains, and the square bracketed cations lie between the chains, ribbon-fashion (*Figure 9.1*). These cations occupy octahedral sites, distinguishable into three types. The M_2 sites are mainly occupied by iron (III), while the iron(II) is distributed over sites M_1 and M_3. However, both the total iron content and the Fe^{III}/Fe^{II} ratio vary markedly from one sample to another, with corresponding isomorphous substitution by Ca^{2+}, Al^{3+} etc. The material varies in colour from blue to brown with increased oxidation[7]. Mixed-valence specimens are dichroic: indigo-blue when viewed along the axis parallel to the chains, and blue-green when viewed across the axis[685]. Polarised absorption spectra show the separate absorption maxima corresponding to intervalence charge transfer in these two directions[23], at 16.13×10^3 cm^{-1}, and 17.5×10^3 cm^{-1} respectively. Littler and Williams[685] measured the conductivities of three natural samples, parallel to the axis. They found that the activation energy remained constant, but the pre-exponential factor decreased as iron(II) was replaced by magnesium(II). Since the iron(II) content of the least conducting sample was too low to provide continuous Fe^{III}–Fe^{II} chains through the crystal, the authors argued that not only nearest-neighbour but also longer electron jumps must be involved in the conduction process.

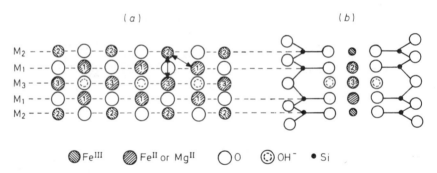

Figure 9.1 Idealised structure of crocidolite showing projections on (a) the 100 and (b) the 001 planes. The metal ions M_1, M_2, M_3 form planar chains sandwiched between the layers of oxygen atoms. The Fe^{3+} ions are located mainly in the M_2 positions, the Fe^{2+} ions mainly in M_1 and M_3. The Fe^{3+}–Fe^{2+} distances are $R_{21} = 3.11$ Å and $R_{23} = 3.25$ Å. (Adapted from references 244, 458, 1094)

The Fe^{III}/Fe^{II} ratio in crocidolite can also be varied by chemical treatment, e.g. atmospheric oxidation at elevated temperature[7, 8]. The main reaction has the stoicheiometry

$$Fe^{2+} + OH^- + \tfrac{1}{4}O_2 = Fe^{3+} + O^{2-} + \tfrac{1}{2}H_2O \tag{9.38}$$

where the OH^- ions of the lattice become converted to O^{2-} at the same positions. The crystal structure remains unchanged apart from some variation of lattice parameters consistent with the conversion of Fe^{2+} to Fe^{3+}. Since the O_2 molecule presumably does not penetrate the structure, the mechanism must involve a surface reaction, together with transfer of the hydrogen ion from the bulk of the crystal to the surface, presumably by successive jumps between OH^- and O ions; and transfer of the electron between Fe^{2+} and Fe^{3+} ions by the same hopping process as is responsible for conduction. The activation energies of the overall reaction are dependent on the surface condition: 33 kcal mol^{-1} when the surface is covered by a film of adsorbed water, 21 kcal mol^{-1} when dry. The activation energy for thermal conduction is 16 kcal mol^{-1} [685]. It would be of great interest to know whether the chemical oxidation of cocidolite is smoothly reversible, and if so, whether the conductivity and optical properties can be reproduced as the oxidation–reduction cycle is repeated.

(b) Nickel(III)–Nickel(II)

Although the two-centre hopping model is attractive in its simplicity and in its close relation to well established mechanisms of electron transfer in solution, the proof that this model is applicable in a particular case is not so simple. This is well illustrated by the case of the nickel oxide system, which at one time was regarded as a classic example of hopping conductivity, but is now described in terms of a more complicated model. Moreover, this is probably the most intensively studied of all mixed-valence semi-conductors, and it raises the suspicion that other systems when studied further will also prove to be more complex.

Pure nickel(II) oxide has the rock salt structure (distorted to rhombohedral below the Néel temperature, 523 K)[900]. It is green in colour and the absorption spectrum is characteristic of isolated Ni^{2+} ions. It has negligible conductivity and the energy gap between the highest filled and lowest unfilled bands is estimated to be $\simeq 4$ eV[1051]. On heating in air, however, the oxygen content increases, the material becomes black, and semi-conduction is observed. The chemical change

$$NiO + \tfrac{1}{2}xO_2 = NiO_{1+x} \tag{9.39}$$

is attributed[155, 758] to the creation of vacancies at metal sites, and a corresponding number of Ni^{3+} ions

$$\tfrac{1}{2}O_2 + 2Ni^{2+} = O^{2-} + \square + 2Ni^{3+} \tag{9.40}$$

(The square sign denotes a cation vacancy). The material can also be doped with lithium oxide, and again a black semi-conducting crystalline solid is obtained. The stoicheiometric formula $Ni_{1-x}Li_xO$ is accounted for by assuming that some Ni^{II} is oxidised to Ni^{III}, as $Ni^{2+}_{1-2x}Ni^{3+}_{x}Li^+_x$ O. In practice the doping is carried out at elevated temperature and equation (9.40) also occurs, which leads to some cation vacancies[a], but we shall ignore this complication.

(a) References cited in reference 25

The absorption spectrum of the doped material has been measured by several groups of workers[a]. With sufficiently low doping levels, the bands corresponding to internal d–d transitions of the Ni^{2+} can be made out, though overlaid by the increased absorption which leads to the black colour. No bands characteristic of Ni^{3+} have been detected, presumably owing to the low concentration. Recent, single crystal measurements[25], however, have established three bands attributable to the intervalence charge transfer, at 6.4, 11.4 and 18.2×10^3 cm^{-1}. The lowest energy band is assumed to correspond to transfer between nickel ions in their electronic ground state:

$$[Ni^{II}(^3A_2) \ldots Ni^{III}] \rightarrow [Ni^{III} \ldots Ni^{II}(^3A_2)]^* \tag{9.41}$$

The higher energies correspond to production of Ni^{II} in the states $^3T_2$ and $^3T_1$ respectively.

At first sight it might appear that the environments of the two nickel atoms are identical, apart from the relaxation effect which determines the Franck–Condon energy, so that, in our notation, $\Delta U_0 = 0$ for equation (9.41). It has been pointed out, however, that the Li^+ and Ni^{3+} ions together form a dipole, since they comprise a negative and a positive charge, relative to the 'sea' of Ni^{2+} ions. Hence, in the ground state the Ni^{3+} ions are to some extent stabilised, and if the charge transfer process takes place between nearest neighbours, the majority of the events will be unsymmetrical, with $\Delta U_0 > 0$:

$$[Ni^{2+} \ldots Ni^{3+} \cdot Li^+] \rightarrow [Ni^{3+} \ldots Ni^{2+} \cdot Li^+]^* \tag{9.42}$$

Against this is the suggestion that, for this very reason, symmetrical transfer will always be preferred, and that transfer distances may therefore exceed the nearest neighbour distance.

Turning to the conductivity experiments, however, we find strong evidence that unsymmetrical events are involved. The possibility was recognised from the start[900, 1051]. The activation energy for semi-conduction is the sum of two components: the energy for charge carrier production, and the activation energy for the mobility of charge carriers (equation 9.11).

Several authors have discussed a model in which charge carrier formation is the thermal analogue of equation (9.41), and this is followed by the movement of positive holes in the partially filled d band of nickel:

$$[Ni^{2+} \ldots Ni^{3+} \cdot Li^+] \xrightarrow{E_f} [Ni^{3+} \ldots Ni^{2+} \cdot Li^+] \tag{9.43}$$

$$Ni^{2+}_A + Ni^{3+}_B \xrightarrow{E^{\ddagger}_p} Ni^{3+}_A + Ni^{2+}_B \tag{9.44}$$

In agreement with this model, the observed activation energy depends strongly on the mole fraction x of Li^+ and Ni^{3+}, falling from 0.46 eV at $x = 3.2 \times 10^{-4}$ to a limiting value, $\simeq 0.2$ eV at $x \geq 0.08$[55, 628, 1051]. It has been pointed out[531] that at still higher ranges of x ($\simeq 0.1$), every Ni^{2+} has on average one nearest-neighbour Li$^+$ ion, so that equation (9.43) may be replaced once again by a symmetrical process

$$[Ni^{2+}.Li^+ \ldots Ni^{3+}.Li^+] \rightarrow [Ni^{3+}.Li^+ \ldots Ni^{2+}.Li^+] \quad (9.45)$$

Initially, Verwey and de Boer[1053] assumed that $E_p^{\ddagger}$ for equation (9.44) must be close to zero, so that this reaction would carry the bulk of the current. Van Houten[1051], however, pointed to the two-centre hopping model outlined above and argued that $E_p^{\ddagger}$ would be finite. Subsequent measurements of the Hall effect[54, 55, 635, 1122] and of the temperature dependence of the Seebeck coefficients[144, 635] have led to modification of this view. These data indicate that the mobility of the charge carriers does not increase with increasing temperature according to Arrhenius' law but rather decreases, and that the observed activation energy must, therefore, be attributed almost entirely to the temperature dependence of charge carrier concentration. Several authors[54, 55, 628] have discussed these data in terms of the same mechanism, except that the conduction process is described in terms of a narrow band model—the Ni^{3+} being polarons of very low stabilisation energy ($\leqslant 0.01$ eV)—and the dependence of E_a on x is attributed at least in part to a variation in the dielectric constant of the material[52]. More recently, a quite different model has been proposed[17, 72]: carrier formation by electron transfer from oxide ions to nickel(III), and conduction by motion of holes in the oxygen $2p$ band:

$$Ni^{3+} + O^{2-} \xrightarrow{E_f} Ni^{2+} + O^- \quad (9.46)$$

$$O_A^{2-} + O_B^- \xrightarrow{E_p^{\ddagger}} O_A^- + O_B^{2-} \quad (9.47)$$

Conduction in undoped NiO has been discussed on a similar basis.

A process nearly equivalent to equation (9.43) can, however, be detected by an analysis of conductivity data at different a.c. frequencies[17]. Band conduction due to a 'large polaron' i.e. a partially delocalised hole or electron, leads to a dependence of conductivity on applied frequency ω of the form

$$\sigma \propto (1 + \omega^2 \tau^2)^{-1} \quad (9.48)$$

where τ is the residence time of the polaron. However, a.c. conduction can also arise by 'bound-polaron hopping' i.e. motion of a positive hole between nickel ions adjacent to one Li$^+$ ion:

$$Ni^{3+} Li^+ Ni^{2+} \rightleftharpoons Ni^{2+} Li^+ Ni^{3+} \quad (9.49)$$

This gives a dependence of the form

$$\sigma \propto \omega\tau/(1 + \omega^2 \tau^2) \quad (9.50)$$

Any effect due to *free* small polarons, as in the hopping model assumed previously, would be a contribution to the d.c. conductivity, independent of frequency. Measurements of frequency dependence at different temperatures indicate that bound hopping predominates at low temperatures, with a transition to band conduction at higher temperatures. The residence time τ for the band mechanism is of the order of 10^{-10} s (though later measurements[449] indicate a dependence on mole fraction x). The activation energy for bound hopping is low, ≤ 0.01 eV, in agreement with the estimates for the process in equation (9.47).

(c) *Niobium (V)–Iron (II)*

An interesting example[234] of solid state photoelectron transfer occurs in the mixed oxide $LiNbO_3$, doped with traces of iron. The redox reaction

$$Fe^{2+} + Nb^{5+} \rightleftharpoons Fe^{3+} + Nb^{4+} \qquad (9.51)$$

is evidently close to equilibrium, since the absorption spectrum shows bands attributable to internal d–d transitions of Nb^{4+} and Fe^{2+} and as oxygen-to-metal charge transfer for Nb^{5+} and Fe^{3+}. An additional broad band at 21.5×10^3 cm^{-1} is assigned to the intervalence transfer, $Fe^{2+} \rightarrow Nb^{5+}$. Irradiation at this wavenumber causes the crystal to bleach, and the colour is restored only gradually after the light is turned off. It is proposed that the electron transferred to the niobium ion finds itself in a delocalised, conduction band. Hence, it migrates out of the light beam and is retrapped by Fe^{3+} elsewhere. By comparing the energies of $O^{2-} \rightarrow Nb^{5+}$ and $O^{2-} \rightarrow Fe^{3+}$ charge transfer, the energy ΔU_0 for equation (9.51) is roughly estimated to be 0.62 eV, while from the intervalence absorption band, $\Delta U_{CT} = 2.66$ eV. Using Hush's equations (8.19) and (8.23), the forward and backward thermal activation energies are thus calculated as $\Delta U^{\ddagger} = 0.87$ eV and $\Delta U^{\ddagger} = 0.25$ eV. The forward energy was independently measured from thermal bleaching experiments, as 1.3 ± 0.2 eV.

References

For some of the less accessible journals an additional reference to *Chemical Abstracts* is given, for example, *CA* **45**, 3735d. Journal titles are abbreviated according to the World List of Scientific Periodicals.

1. ADAMSON, A. W., *J. Am. chem. Soc.*, **78**, 4260 (1956)
2. ADAMSON, A. W., *Discuss. Faraday Soc.*, **29**, 125 (1960)
3. ADAMSON, A. W., *Adv. Chem. Ser.*, **150**, 128 (1976)
4. ADAMSON, A. W. and FLEISCHAUER, P. D. *Concepts of Inorganic Photochemistry* John Wiley: New York (1975)
5. ADAMSON, A. W. and GONICK, E., *Inorg. Chem.*, **2**, 129 (1963)
6. ADAMSON, M. G. and STRANKS, D. R., *Chem. Commun.*, 648 (1967)
7. ADDISON, C. C., ADDISON, W. E., NEAL, G. H. and SHARP, J. H., *J. chem. Soc.*, 1468 (1962)
8. ADDISON, W. E., NEAL, G. H. and SHARP, J. H., *J. chem. Soc.*, 1472 (1962)
9. ADEGITE, A. and KUKU, T. A., *J. C. S. Dalton*, 158 (1976) and references cited therein
10. ADEYEMI, S. A., BRADDOCK, J. N., BROWN, G. M., FERGUSON, J. A., MILLER, F. J. and MEYER, T. J., *J. Am. chem. Soc.*, **94**, 300 (1972)
11. ADEYEMI, S. A., JOHNSON, E. C., MILLER, F. J. and MEYER, T. J., *Inorg. Chem.*, **12**, 2371 (1973)
12. ADIN, A., DOYLE, J. and SYKES, A. G., *J. chem. Soc. (A)*, 1504 (1967)
13. ADIN, A. and ESPENSON, J. H., *Inorg. Chem.*, **11**, 686 (1972)
14. ADIN, A. and SYKES, A. G., *J. chem. Soc., A*, 1518 (1966)
15. ADIN, A. and SYKES, A. G., *J. chem. Soc., A*, 351 (1968)
16. ADIN, A. and SYKES, A. G., *J. chem. Soc., A*, 354 (1968)
17. ADLER, D. and FEINLEIB, J., *Phys. Rev. B*, **2**, 3112 (1970)
18. AFANASEV, B. N., *Colln. Czecho chem. Commun. Engl. Edn*, **33**, 1186 (1968)
19. AFANASEV, B. N., and TIMOFEEVA, T. B., *Elektrokhimiya*, **4**, 1385 (1968); *CA* **70**, 73523c
20. AGRELL, I., *Acta chem. scand.*, **21**, 2647 (1967)
21. AHRLAND, S., CHATT, J., and DAVIS, N., *Quart. Rev.*, **12**, 265 (1958)
22. ALBRITTON, D. L., BUSH, Y. A., FEHSENFELD, F. C., FERGUSON, E. E., GOVERS, T. R., McFARLAND, M. and SCHMELTEKOPF, A. L., *J. chem. Phys.*, **58**, 4036 (1973)
23. ALLEN, G. C., *Transit. Metal Chem.*, *(Weinheim, Ger.)* **1**, 143 (1976)
24. ALLEN, G. C., unpublished measurements quoted in reference 26
25. ALLEN, G. C., and DYKE, J. M., *Chem. Phys. Lett.*, **37**, 391 (1976)
26. ALLEN, G. C., and HUSH, N. S., *Progr. inorg. Chem.*, **8**, 357 (1967)
27. ALLEN, M. J., *Organic Electrode Processes*, Chapman and Hall: London (1958)
28. ALTENAU, J. J., POPE, M. T., PRADOS, R. A., and SO, H., *Inorg. Chem.*, **14**, 417 (1975)
29. ALTMAN, C. and KING, E. L., *J. Am. chem. Soc.*, **83**, 2825 (1961)
30. ANBAR, M. and HART, E. J., *J. phys. Chem.*, **69**, 973 (1965)
31. ANDERSON, A. and BONNER, N. A., *J. Am. chem. Soc.*, **76**, 3826 (1954)
32. ANDERSON, N. H., DOBBS, A. J., EDGE, D. J., NORMAN, R. O. C. and WEST, P. R., *J. chem. Soc., B*, 1004 (1971)

33 APPELMAN, E. H. and SULLIVAN, J. C., *J. phys. Chem.*, **66**, 442 (1962)
34 AQUILANTI, V., *Z. phys. Chem (N.F.)*, **90**, 1 (1974)
35 AQUILANTI, V. and LAGANA, A., *Z. phys. Chem. (N.F.)*, **96**, 229 (1975)
36 ARDON, M., LEVITAN, J. and TAUBE, H., *J. Am. chem. Soc.*, **84**, 872 (1962)
37 ARDON, M. and PLANE, R. A., *J. Am. chem. Soc.*, **81**, 3197 (1959)
38 ARMOR, J. N., SCHEIDEGGER, H. A. and TAUBE, H., *J. Am. chem. Soc.*, **90**, 5928 (1968)
39 ARMSTRONG, A. M. and HALPERN, J., *Can. J. Chem.*, **35**, 1020 (1957)
40 ARMSTRONG, A. M., HALPERN, J. and HIGGINSON, W. C. E., *J. phys. Chem.*, **60**, 1661 (1956)
41 ASHURST, K. G. and HIGGINSON, W. C. E., *J. chem. Soc.*, 3044 (1953)
42 ASHURST, K. G. and HIGGINSON, W. C. E., *J. chem. Soc.*, 343 (1956)
43 ASPRAY, M. J., ROSSEINSKY, D. R. and SHAW, G. B., *Chemy Ind.*, 911 (1963)
44 AST, T., TERWILLIGER, D. T., BEYNON, J. H. and COOKS, R. G., *J. chem. Phys.*, **62**, 3855 (1975)
45 ATHERTON, N. M., *Chem. Phys. Lett.*, **23**, 454 (1973)
46 ATKINSON, L. and DAY, P., *J. chem. Soc., A*, 2423 (1969)
47 ATKINSON, L. and DAY, P., *J. chem. Soc., A*, 2432 (1969)
48 AULICH, H., BARON, B. and DELAHAY, P., *J. chem. Phys.*, **58**, 603 (1973)
49 AULICH, H., BARON, B., DELAHAY, P. and LUGO, R., *J. chem. Phys.*, **58**, 4439 (1973)
50 AULICH, H., DELAHAY, P. and NEMEC, L., *J. chem. Phys.*, **59**, 2354 (1973)
51 AULICH, H., NEMEC, L. and DELAHAY, P., *J. chem. Phys.*, **61**, 4235 (1974)
52 AUSTIN, I. G. and MOTT, N. F., *Adv. Phys.*, **18**, 41 (1969)
53 AUSTIN, I. G. and MOTT, N. F., *Science, N. Y.*, **168**, 71 (1970)
54 AUSTIN, I. G., SPRINGTHORPE, A. J. and SMITH, B. A., *Phys. Lett.*, **21**, 20 (1966)
55 AUSTIN, I. G., SPRINGTHORPE, A. J., SMITH, B. A. and TURNER, C. E., *Proc. phys. Soc.*, **90**, 157 (1967)
56 BAADSGAARD, H. and TREADWELL, W. D., *Helv. chim. Acta*, **38**, 1669 (1955)
57 BAEDE, A. P. M., *Adv. chem. Phys.*, **30**, 463 (1975)
58 BAEDE, A. P. M. and LOS, J., *Physica*, **52**, 422 (1971)
59 BAEDE, A. P. M., MOUTINHO, A. M. C., De VRIES, A. E. and LOS, J., *Chem. Phys. Lett.*, **3**, 530 (1969)
60 BAER, M., *Chem. Phys. Lett.*, **35**, 112 (1975)
61 BAES, C. F., *J. phys. Chem.*, **60**, 805 (1956)
62 BAIZER, M. M. (Ed.), *Organic Electrochemistry*, Marcel Dekker: New York (1973)
63 BAKAČ, A., HAND, T. D. and SYKES, A. G., *Inorg. Chem.*, **14**, 2540 (1975)
64 BAKER, B. R., ORHANOVIC, M. and SUTIN, N., *J. Am. chem. Soc.*, **89**, 722 (1967)
65 BAKER, F. B., BREWER, W.D. and NEWTON, T. W., *Inorg. Chem.*, **5**, 1294 (1966)
66 BAKER, F. B., NEWTON, T.W. and KAHN, M., *J. phys. Chem.*, **64**, 109 (1968)
67 BALAHURA, R. J. and LEWIS, N. A., *Can. J. Chem.*, **53**, 1154 (1975)
68 BALAHURA, R. J. and PURCELL, W. L., *Inorg. Chem.*, **14**, 1469 (1975)
69 BALL, D. L. and KING, E. L., *J. Am. chem. Soc.*, **80**, 1091 (1958)
70 BALLARD, R. E. and GRIFFITHS, G., *Chem. Commun.*, 1472 (1971)
70a BALLARD, R. E. and GRIFFITHS, G., *J. chem. Soc., A*, 1960 (1971)
71 BALZANI, V. and CARASSITI, V., *Photochemistry of Coordination Compounds*, Academic Press: London (1970)
72 BARI, R. A., ADLER, D. and LANGE, R. V., *Phys. Rev. B*, **2**, 2898 (1970)
73 BARNES, J. C., DUNCAN, C. S. and PEACOCK, R. D., *J. C. S. Dalton*, 1875 (1972)
74 BARON, B., CHARTIER, P., DELAHAY, P., and LUGO, R., *J. chem. Phys.*, **51**, 2562 (1969)
75 BARON, B., DELAHAY, P. and LUGO, R., *J. chem. phys.*, **53**, 1399 (1970)
76 BARON, B., DELAHAY, P. and LUGO, R., *J. chem. Phys.*, **55**, 4180 (1971)
76a BARRETT, M. B., *Diss. Abstr.*, **29B**, 2333 (1969)
77 BERRY, R. S., *Chem. Rev.*, **69**, 533 (1969)
78 BASATO, M. and PELOSO, A., *Gazz. chim. ital.*, **102**, 893 (1972)
79 BASOLO, F., MESSING, A. F., WILKS, P. H., WILKINS, R. G. and PEARSON, R. G., *J. inorg. nucl. Chem.*, **8**, 203 (1958)
80 BASOLO, F., MORRIS, M. L. and PEARSON, R. G., *Discuss. Faraday Soc.*, **29**, 80 (1960)
81 BASOLO, F. and PEARSON, R. G., *Mechanisms of Inorganic Reactions*, p. 306, John Wiley: New York (1958)

82 BASOLO, F. and PEARSON, R. G., *Mechanisms of Inorganic Reactions*, 2nd edn, p. 129 ff John Wiley: New York (1967)
83 BASOLO, F. and PEARSON, R. G., *Prog. inorg. Chem.*, **4**, 381–453 (see in particular p. 435 ff.) (1962)
84 BASOLO, F., WILKS, P. H., PEARSON, R. G. and WILKINS, R. G., *J. inorg. nucl. Chem.*, **6**, 161 (1958)
85 BATES, D. R., *Proc. R. Soc. A*, **247**, 294 (1958)
86 BATES, D. R., *Proc. R. Soc. A*, **257**, 22 (1960)
87 BATES, D. R., 'Theoretical Treatment of Collisions between Atomic Systems'. In *Atomic and Molecular Processes*, Ed. by D. R. Bates, chap. 14, Academic Press: New York (1962)
88 BATES, D. R. and BOYD, T. J. M., *Proc. phys. Soc.*, **A69**, 910 (1956)
89 BATES, D. R., LEDSHAM, K. and STEWART, A. L., *Phil. Trans. R. Soc.*, **A246**, 215 (1953)
90 BATES, D. R. and LEWIS, J. T., *Proc. phys. Soc.*, **A68**, 173 (1955)
91 BATES, D. R. and LYNN, N., *Proc. R. Soc., A*, **253**, 141 (1959)
92 BATES, D. R. and McCARROLL, R., *Adv. Phys.*, **11**, 39 (1962)
93 BATES, D. R. and MASSEY, H. S. W., *Phil. Mag. (series 7)*, **45**, 111 (1954)
94 BATES, D. R. and MOISEIWITSCH, B. L., *Proc. phys. Soc.*, **A67**, 805 (1954)
95 BATLOGG, B., KALDIS, E. and WACHTER, P., *J. Magnetism and Magnetic Materials*, **3**, 96 (1976)
96 BAXENDALE, J. H., *Symposium on Relaxation Kinetics*, Buffalo N. Y., (1965), cited in reference 542
97 BAYFIELD, J. E., 'Electron Transfer in Simple Atomic Collisions'. In *Atomic Physics*, Ed. by G. zu Pulitz, E. W. Weber and A. Winnacker, Vol. 5, p 397, Plenum Press: New York (1975)
98 BAYLISS, N. S., *J. chem. Phys.*, **18**, 292 (1950)
99 BEARCROFT, D. J., SEBERA, D., ZWICKEL, A. and TAUBE, H., unpublished data quoted in reference 1128
100 BEARMAN, G. H., EARL, J. D., PIEPER, R. J., HARRIS, H. H. and LEVENTHAL, J. J., *Phys. Rev. A*, **13**, 1734 (1976)
101 BEATTIE, J. K. and BASOLO, F., *Inorg. Chem.*, **6**, 2069 (1967)
102 BEATTIE, J. K. and BASOLO, F., *Inorg. Chem.*, **10**, 486 (1971)
103 BEATTIE, J. K., HUSH, N. S. and TAYLOR, P. R., *Inorg. Chem.*, **15**, 992 (1976)
104 BECK, M. T., *Coord. Chem. Rev.*, **3**, 91 (1968)
105 BECK, M. T., SERES, I. and BARDI, I., *Magy. kém. Foly.*, **69**, 46 (1963); *CA* **59**, 7152a
106 BELL, R. P., *The Proton in Chemistry*, 2nd edn, Chapman and Hall: London, (1973)
107 BRENET, J. and TRAORE, K., *Transfer Coefficients in Electrochemical Kinetics*, Academic Press: London (1971)
108 BEN-NAIM, A., *J. chem. Phys.*, **54**, 1387 (1971)
109 BENNETT, L. E. and TAUBE, H., *Inorg. Chem.*, **7**, 254 (1968)
110 BENSON, P. and HAIM, A., *J. Am. chem. Soc.*, **87**, 3826 (1965)
111 BENSON, S. W., *The Foundations of Chemical Kinetics*, McGraw-Hill: London, (1960)
112 BERGH, A. A. and HAIGHT, G. P., *Inorg. Chem.*, **1**, 688 (1962)
113 BERGH, A. A. and HAIGHT, G. P., *Inorg. Chem.*, **8**, 189 (1969)
114 BERKOWITZ, J., CHUPKA, W. A. and GUTMAN, D., *J. chem. Phys.*, **55**, 2733 (1971)
115 BERNAL, J. R., DASGUPTA, D. R. and MACKAY, A. L., *Clay Miner. Bull.*, **4**, 15 (1959); cited in reference 3
116 BERTA, M. A. and KOSKI, W. S., *J. Am. chem. Soc.*, **86**, 5098 (1964)
117 BERTRAND, J. A. and HIGHTOWER, T. C., *Inorg. Chem.*, **12**, 206 (1973)
118 BETTS, R. H., *Can. J. Chem.*, **33**, 1780 (1955)
119 BHATTACHARYYA, S. S. and RAI DASTIDAR, T. K., *J. Phys. B*, **8**, 1522 (1975)
120 BIRK, J. P., *J. Am. chem. Soc.*, **91**, 3189 (1969)
121 BIRK, J. P., *Inorg. Chem.*, **9**, 125 (1970)
122 BIRK, J. P., *Inorg. Chem.*, **14**, 1724 (1975)
123 BIRK, J. P. and GASIEWSKI, J. W., *Inorg. Chem.*, **10**, 1586 (1971)
124 BIRK, J. P. and LOGAN, T. P., *Inorg. Chem.*, **12**, 580 (1973)
125 BIRK, J. P. and WEAVER, S. V., *Inorg. Chem.*, **11**, 95 (1972)
126 BIRKE, G., LATSCHA, H. P. and PRITZKOW, H., *Z. Naturf.*, **31B**, 1285 (1976)
127 BIRADAR, N. S., STRANKS, D. R. and VAIDYA, M. S., *Trans. Faraday Soc.*, **58**, 2421 (1962)
128 BJERRUM, N., *K. danske Vidensk. Selsk. Skr (Math. fys. Medd.)*, **7**, 1 (1926) p. 9

129 BLOOM, M. B. D., RAYNOR, J. B. and SYMONS, M. C. R., *J. chem. Soc., A*, 3843 (1971)
130 BLUMENTHAL, R. N. and HOFMAIER, R. L., *J. electrochem. Soc.*, **121**, 126 (1974)
131 BLUMENTHAL, R. N. and SHARMA, R. K., *J. Solid State Chem.*, **13**, 360 (1975)
132 BOCKRIS, J. O'M., KHAN, S. U. M. and MATTHEWS, D. B., *J. Res. Inst. Catalysis, Hokkaido Univ.*, **22**, 1 (1974)
133 BODEK, I. and DAVIES, G., *Coord. Chem. Rev.*, **14**, 269 (1974)
134 BOHME, D. K., HASTED, J. B. and ONG, P. P., *Chem. Phys. Lett.*, **1**, 259 (1967)
135 BOHME, D. K., HASTED, J. B. and ONG, P. P., *J. Phys. B*, **1**, 879 (1968)
136 BOK, L. D. C., LEIPOLDT, J. G. and BASSON, S. S., *J. inorg. nucl. Chem.*, **37**, 2151 (1975)
137 BOMCHIL, G., DELAHAY, F. and LEVIN, I., *J. chem. Phys.*, **56**, 5194 (1972)
138 BONNER, N. A., *J. Am. chem. Soc.*, **71**, 3909 (1949)
139 BONNER, N. A. and HUNT, J. P., *J. Am. chem. Soc.*, **82**, 3826 (1960)
140 BOOTH, R. J., STARKIE, H. C. and SYMONS, M. C. R., *J. phys. Chem.*, **76**, 141 (1972)
141 BOOTH, R. J., STARKIE, H. C., SYMONS, M. C. R. and EACHUS, R. S., *J. C. S. Dalton*, 2233 (1973)
142 BORN, M., *Z. Physik*, **1**, 45 (1920)
143 BORN, M. and HUANG, K., *Dynamical Theory of Crystal Lattices*, Oxford University Press: London (1954)
144 BOSMAN, A. J. and CREVECOEUR, C., *Phys. Rev.*, **144**, 763 (1966)
145 BOSMAN, A. J. and CREVECOEUR, C., *J. Phys. Chem. Solids*, **29**, 109 (1968)
146 BOTTCHER, C., ALLISON, A. C. and DALGARNO, A., *Chem. Phys. Lett.* **11**, 307 (1971)
147 BOWERS, M. T. and ELLEMAN, D. D., *Chem. Phys. Lett.*, **16**, 486 (1972)
148 BOWERS, M. T. and LAUDENSLAGER, J. B., *J. chem. Phys.*, **56**, 4711 (1972)
149 BOWERS, M. T. and SU, T., *Adv. Electronics Electron Phys.*, **34**, 223 (1973)
150 BOWLES, J. C. and HALL, D., *Chem. Commun.*, 1523 (1971)
151 BOYD, T. J. M. and MOISEIWITSCH, B. L., *Proc. Phys. Soc. A*, **70**, 809 (1957)
152 BRADDOCK, J. N., CRAMER, J. L. and MEYER, T. J., *J. Am. chem. Soc.*, **97**, 1972 (1975)
153 BRADDOCK, J. N. and MEYER, T. J., *J. Am. chem. Soc.*, **95**, 3158 (1973)
154 BRANSDEN, B. H., *Rep. Prog. Phys.*, **35**, 949 (1972)
155 BRANSKY, I. and TALLAN, N. M., *J. chem. Phys.*, **49**, 1243 (1968)
156 BRANSKY, I., TALLAN, N. M. and HED, A. Z., *J. appl. Phys.*, **41**, 1787 (1970)
157 BRATERMAN, P. S., PHIPPS, P. B. P. and WILLIAMS, R. J. P., *J. chem. Soc.*, 6164 (1965)
158 BRAY, W. C. and GORIN, M. H., *J. Am. chem. Soc.*, **54**, 2124 (1932)
159 BREGMAN-REISLER, H., ROSENBERG, A., and AMIEL, S., *J. chem. Phys.*, **59**, 5404 (1973)
160 BRIDGES, K. L., MUKHERJEE, S. K. and GORDON, G., *Inorg. Chem.*, **11**, 2494 (1972)
161 BRIEGLEB, G., *Elektronen-Donator-Acceptor Komplexe*, Springer-Verlag: W. Berlin (1961)
162 BRION, C. E., McDOWELL, C. A. and STEWART, W. B., *Chem. Phys. Lett.*, **13**, 79 (1972)
163 BROWN, D. M. and DAINTON, F. S., *Trans. Faraday Soc.*, **62**, 1139 (1966)
164 BROWN, G. M., CALLAHAN, R. W., JOHNSON, E. C., MEYER, T. J. and WEAVER, T. R., in reference 585, p. 66
165 BROWN, G. M., MEYER, T. J., COWAN, D. O., Le VANDA, C., KAUFMAN, F., ROLING, P. V. and RAUSCH, M. D., *Inorg. Chem.*, **14**, 506 (1975)
166 BROWNE, C. I., CRAIG, R. P. and DAVIDSON, N., *J. Am. chem. Soc.*, **73**, 1946 (1951)
167 BRUBAKER, C. H. and ANDRADE, C., *J. Am. chem. Soc.*, **81**, 5282 (1959)
168 BRUBAKER, C. H. and COURT, A. J., *J. Am. chem. Soc.*, **78**, 5530 (1956)
169 BRUBAKER, C. H. and MICKEL, J. P., *J. inorg. nucl. Chem.*, **4**, 55 (1957)
170 BRUNDLE, C. R., ROBIN, M. B. and BASCH, H., *J. chem. Phys.*, **53**, 2196 (1970)
171 BRUNING, W. and WEISSMAN, S. I., *J. Am. chem. Soc.*, **88**, 373 (1966)
172 BUCKLEY, R. C. and WARDESKA, J. G., *Inorg. Chem.*, **11**, 1723 (1972)
173 BÜHLER, R. E., *J. phys. Chem.*, **76**, 3220 (1972)
174 BUNN, D., DAINTON, F. S. and DUCKWORTH, S., *Trans. Faraday Soc.*, **57**, 1131 (1961)
175 BURKHART, M. J. and NEWTON, T. W., *J. phys. Chem.*, **73**, 1741 (1969)
176 BURNETT, M. G., CONNOLLY, P. J. and KEMBALL, C., *J. chem. Soc., A*, 800 (1967)

177 BURROUGHS, P., HAMMETT, A. and ORCHARD, A. F., *J. C. S. Dalton*, 565 (1974)
178 BUSER, H. J., LUDI, A., FISCHER, P., STUDACH, T. and DALE, B. W., *Z. phys. Chem. Frankf. Ausg.*, **92**, 354 (1974)
179 BUSER, H. J., LUDI, A., PETTER, W. and SCHWARZENBACH, D., *J. C. S. Chem. Commun.*, (1972) 1299
179a BUSER, H. J., SCHWARZENBACH, D., PETTER, W. and LUDI, A., *Inorg. chem.*, **16**, 2704 (1977)
180 BUTLER, R. D. and TAUBE, H., *J. Am. chem. Soc.*, **87**, 5597 (1965)
181 CADOGAN, J. I. G., *Quart. Rev.*, **16**, 208 (1962)
182 CALDIN, E. F. H., *Fast Reactions in Solution*, Blackwell: Oxford (1964)
183 CALDIN, E. F. and GOLD, V., *Proton-Transfer Reactions*, Chapman and Hall: London, (1975)
184 CALLAHAN, R. W., BROWN, G. M. and MEYER, T. J., *J. Am. chem. Soc.*, **96**, 7829 (1974)
185 CALLAHAN, R. W., BROWN, G. M. and MEYER, T. J., *Inorg. Chem.*, **14**, 1443 (1975)
186 CALLAHAN, R. W. and MEYER, T. J., *Chem. Phys. Lett.*, **39**, 82 (1976)
187 CAMPION, R. J., DECK, C. F., KING, P. and WAHL, A. C., *Inorg. Chem.*, **6**, 672 (1967)
188 CAMPION, R. J., PURDIE, N. and SUTIN, N., *J. Am. chem. Soc.*, **85**, 3528 (1963)
189 CAMPION, R. J., PURDIE, N. and SUTIN, N., *Inorg. Chem.*, **3**, 1091 (1964)
190 CANDLIN, J. P. and HALPERN, J., *Inorg. Chem.*, **4**, 766 (1965)
191 CANDLIN, J. P. and HALPERN, J., *Inorg. Chem.*, **4**, 1086 (1965)
192 CANDLIN, J. P., HALPERN, J. and NAKAMURA, S., *J. Am. chem. Soc.*, **85**, 2517 (1963)
193 CANDLIN, J. P., HALPERN, J. and TRIMM, D. L., *J. Am. chem. Soc.*, **86**, 1019 (1964)
194 CANNON, R. D., *J. Chem. Soc., A*, 1098 (1968)
195 CANNON, R. D., *Nature, Lond.*, **228**, 644 (1970)
196 CANNON, R. D., *J. inorg. nucl. Chem.*, **38**, 1222 (1976)
197 CANNON, R. D., *Adv. inorg. Chem. Radiochem.*, **21**, 179 (1978)
198 CANNON, R. D., *Chem. Phys. Lett.*, **49**, 299 (1977)
199 CANNON, R. D., unpublished calculations
200 CANNON, R. D. and EARLEY, J. E., *J. Am. chem. Soc.*, **88**, 1872 (1966)
201 CANNON, R. D. and GARDINER, J., *J. Am. chem. Soc.*, **92**, 3800 (1970); CANNON, R. D. and GARDINER, J., *Inorg. Chem.*, **13**, 390 (1974)
202 CANNON, R. D. and GARDINER, J., *J.C.S. Dalton*, 622 (1976)
203 CANNON, R. D. and STILLMAN, J. S., *Inorg. Chem.*, **14**, 2202 (1975)
204 CANNON, R. D. and STILLMAN, J. S., *Inorg. Chem.*, **14**, 2207 (1975)
205 CANNON, R. D. and STILLMAN, J. S., *J.C.S. Dalton*, 428 (1976)
206 CANNON, R. D., STILLMAN, J. S., SARAWEK, K. and POWELL, D. B., *J.C.S. Chem. Comm.*, 31 (1976)
207 CARLYLE, D. W. and ESPENSON, J. H., *J. Am. chem. Soc.*, **90**, 2272 (1968)
208 CARLYLE, D. W. and ESPENSON, J. H., *Inorg. Chem.*, **8**, 575 (1969)
209 CARLYLE, D. W. and ESPENSON, J. H., *J. Am. chem. Soc.*, **91**, 599 (1969)
210 CARPENTER, G. B., *Acta crystallogr.*, **8**, 852 (1955)
211 CASABO, J., RIBAS, J. and ALVAREZ, S., *Inorg. Chem. Acta*, **16**, 45 (1976)
212 CAVALIERE, P., FERRANTE, G. and MONTES, B. M., *Chem. Phys. Lett.*, **36**, 583 (1975)
213 CERCEK, B., EBERT, M. and SWALLOW, A. J., *J. Chem. Soc., A*, 612 (1966)
214 CELSI, S., SECCOT, F. and VENTURINI, M., *J.C.S. Dalton*, 793 (1974)
215 CHAFFEE, E. and EDWARDS, J. O., *Prog. inorg. Chem.*, **13**, 205 (1970)
216 CHALLENGER, G. E. and MASTERS, B. J., *J. Am. chem. Soc.*, **78**, 3012 (1956)
217 CHAPMAN, N. B. and SHORTER, J. (Eds), *Advances in Linear Free Energy Relationships*, Plenum Press: London (1972)
218 CHAPMAN, S. and PRESTON, R. K., *J. chem. Phys.*, **60**, 650 (1974)
219 CHAU, M. and BOWERS, M. T., *Chem. Phys. Lett.*, **44**, 490 (1976)
220 CHEN, J. C. Y., *Case Studies in Atomic Physics*, **3**, 305 (1973); CA **79**, 129128h
221 CHEN, J. C. and GOULD, E. S., *J. Am. chem. Soc.*, **95**, 5539 (1973)
222 CHEN, A. C. and HAHN, Y., *Phys. Rev. A*, **12**, 823 (1975)
223 CHEN, Y. H., JOHNSON, R. E., HUMPHRIS, R. R., SIEGEL, M. W. and BORING, J. W., *J. Phys. B*, **8**, 1527 (1975)
224 CHIA, L., NEMEC, L. and DELAHAY, P., *Chem. Phys. Lett.*, **32**, 90 (1975)
225 CHILD, M. S., *Molecular Collision Theory*, Academic Press: London (1974)

226 CHRISTENSEN, R. J. and ESPENSON, J. H., *Chem. Commun.*, 756 (1970)
227 CHRISTENSEN, R. J., ESPENSON, J. H. and BUTCHER, A. B., *Inorg. Chem.*, **12**, 564 (1973)
228 CHRISTOV, S. G., *Ber. Bunsenges. Phys. Chem.*, **76**, 507 (1972)
229 CHRISTOV, S. G., *Ber. Bunsenges. Phys. Chem.*, **78**, 537 (1974)
230 CHRISTOV, S. G., *Ber. Bunsenges. Phys. Chem.*, **79**, 357 (1975)
231 CHRISTOV, S. G., *J. Electrochem. Soc.*, **124**, 69 (1977)
232 CHUPKA, W. A., 'Ion-Molecule Reactions by Photoionisation Techniques'. In *Ion-Molecule Reactions*, Edited by J. L. Franklin Vol. 1, pp. 33–76, Plenum Press: New York (1972)
233 CHUPKA, W. A., RUSSELL, M. E. and REFAEY, K., *J. chem. Phys.*, **48**, 1518 (1968)
234 CLARK, M. G., DiSALVO, F. J., GLASS, A. M. and PETERSON, G. E., *J. chem. Phys.*, **59**, 6209 (1973)
235 CITRIN, P. H., *J. Am. chem. Soc.*, **95**, 6472 (1973)
236 COHEN, A. O. and MARCUS, R. A., *J. phys. Chem.*, **72**, 4249 (1968)
237 COHEN, D., AMIS, E. S., SULLIVAN, J. C. and HINDMAN, J. C., *J. phys. Chem.*, **60**, 701 (1956)
238 COHEN, D., SULLIVAN, J. C., AMIS, E. S. and HINDMAN, J. C., *J. Am. chem. Soc.*, **78**, 1543 (1956)
239 COHEN, D., SULLIVAN, J. C. and HINDMAN, J. C., *J. Am. chem. Soc.*, **76**, 352 (1954)
240 COHEN, H. and MEYERSTEIN, D., *J.C.S. Dalton*, 2477 (1975)
241 COHEN, I. A., JUNG, C. and GOVERNO, T., *J. Am. chem. Soc.*, **94**, 3003 (1972)
242 COHEN, J. S., EVANS, S. A. and LANE, N. F., *Phys. Rev. A*, **4**, 2248 (1971)
243 COLEMAN, J. P. and McDOWELL, M. R. C., *Introduction to the Theory of Ion-Atom Collisions*, North-Holland: Amsterdam (1970)
244 COLVILLE, P. A., ERNST, W. G. and GILBERT, M. C., *Amer. Miner.*, **51**, 1727 (1966)
245 CONANT, J. B. and CUTTER, H. B., *J. Am. chem. Soc.*, **44**, 2651 (1922)
246 CONNICK, R. E. and McVEY, W. H., *J. Am. chem. Soc.*, **75**, 474 (1953)
247 CONOCCHIOLI, T. J., HAMILTON, E. J. and SUTIN, N., *J. Am. chem. Soc.*, **87**, 926 (1965)
248 CONOCCHIOLI, T. J., NANCOLLAS, G. H. and SUTIN, N., *J. Am. chem. Soc.*, **86**, 1453 (1964)
249 CONOCCHIOLI, T. J., NANCOLLAS, G. H. and SUTIN, N., *Inorg. Chem.*, **5**, 1 (1966)
250 COOKE, N. E., KUWANA, T. and ESPENSON, J., *Inorg. Chem.*, **10**, 1081 (1971)
251 COPE, V. W., MILLER, R. G. and FRASER, R. T. M., *J. chem. Soc., A*, 301 (1967)
252 COSBY, P. C., MORAN, T. F. and FLANNERY, M. R., *J. chem. Phys.*, **61**, 1259 (1974)
253 COSBY, P. C., MORAN, T. F., HORNSTEIN, J. V. and FLANNERY, M. R., *Chem. Phys. Lett.*, **24**, 431 (1974); and references cited therein
254 COSGROVE, J. G., COLLINS, R. L. and MURTY, D. S., *J. Am. chem. Soc.*, **95**, 1083 (1973)
255 COULSON, C. A., *J. chem. Soc.*, **778** (1956)
256 COULSON, C. A., *Valence*, 2nd ed., Clarendon Press: Oxford (1963)
257 COWAN, D. O., Le VANDA, C., PARK, J. and KAUFMAN, F., *Accts Chem. Res.*, **6**, 1 (1973)
258 COX, L. T., COLLINS, S. B. and MARTIN, D. S., *J. inorg. nucl. Chem.*, **17**, 383 (1961)
259 CRAFT, R. W. AND GAUNDER, R. G., *Inorg. Chem.*, **14**, 1283 (1975)
260 CRAIG, R. P. and DAVIDSON, N., *J. Am. chem. Soc.*, **73**, 1951 (1951)
261 CRAMER, J. L. and MEYER, T. J., *Inorg. Chem.*, **13**, 1250 (1974)
262 KREEVOY, M. M. and KONASEWICH, D. E., *Adv. chem. Phys.*, **21**, 243 (1971)
263 CREUTZ, C., *Ph.D. Thesis, Stanford University* (1971)
264 CREUTZ, C., GOOD, M. L. and CHANDRA, S., *Inorg. Nucl. Chem. Lett.*, **9**, 171 (1973)
265 CREUTZ, C. and TAUBE, H., *J. Am. chem. Soc.*, **91**, 3988 (1969)
266 CREUTZ, C. and TAUBE, H., *J. Am. chem. Soc.*, **95**, 1086 (1973)
267 CULPIN, D., DAY, P., EDWARDS, P. R. and WILLIAMS, R. J. P., *J. chem. Soc., A*, 1838 (1968)
268 DAINTON, F. S., *Chem. Soc. Rev.*, **4**, 323 (1975)
269 DALGARNO, A., *Proc. phys. Soc., A*, **67**, 1010 (1954)
270 DALGARNO, A. In *Ion-Molecule Reactions* Edited by E. W. MacDaniel, V. Cermak, A. Dalgarno, E. E. Ferguson and L. Friedman, chap. 3, Wiley-Interscience: New York (1970)
271 DALGARNO, A., BOTTCHER, C. and VICTOR, G. A., *Chem. Phys. Lett.*, **7**, 265 (1970)
272 DALGARNO, A. and YADAV, H. N., *Proc. phys. Soc., A*, **66**, 173 (1953)

273 DARLOW, F. S., *Acta crystallogr.*, **14,** 1257 (1961)
274 DARLOW, F. S. and COCHRAN, W., *Acta crystallogr.*, **14,** 1250 (1961)
275 DASH, A. C. and NANDA, R. K., *Inorg. Chem.*, **12,** 2024 (1973)
276 DAUGHERTY, N. A., *J. Am. chem. Soc.*, **87,** 5026 (1965)
277 DAUGHERTY, N. A. and ERBACHER, J. K., *Inorg. Chem.*, **14,** 683 (1975)
278 DAUGHERTY, N. A. and NEWTON, T. W., *J. phys. Chem.*, **68,** 612 (1964)
279 DAUGHERTY, N. A. and SCHIEFELBEIN, B., *Inorg. Chem.*, **9,** 1716 (1970)
280 DAVIDSON, N., *J. Am. chem. Soc.*, **73,** 2361 (1951)
281 DAVIES, G., *Coord. Chem. Rev.*, **4,** 199 (1969)
282 DAVIES, G., *Coord. Chem. Rev.*, **14,** 287 (1974)
283 DAVIES, G. and WARNQVIST, B., *Coord Chem. Rev.*, **5,** 349 (1970)
284 DAVIES, G. and WATKINS, K. O., *J. phys. Chem.*, **74,** 3388 (1970)
285 DAVIES, G. and WATKINS, K. O., *Inorg. Chem.*, **9,** 2735 (1970)
286 DAVIES, K. M. and ESPENSON, J. H., *J. Am. chem. Soc.*, **91,** 3093 (1969)
287 DAVIES, K. M. and ESPENSON, J. H., *J. Am. chem. Soc.*, **92,** 1884 (1970)
288 DAVIES, K. M. and ESPENSON, J. H., *J. Am. chem. Soc.*, **92,** 1889 (1970)
289 DAVIES, M. and EVANS, F. P., *Trans. Faraday Soc.*, **52,** 74 (1956)
290 DAVIES, R. and JORDAN, R. B., *Inorg. Chem.*, **10,** 2432 (1971)
291 DAVIES, R., KIPLING, B. and SYKES, A. G., *J. Am. chem. Soc.*, **95,** 7250 (1973)
292 DAVIES, R., STEVENSON, M. B. and SYKES, A. G., *J. chem. Soc., A*, 1261 (1970)
293 DAVIES, R. and SYKES, A. G., *J. chem. Soc., A*, 2237 (1968)
293a DAY, P., *Inorg. Chem.*, **2,** 452 (1963)
294 DAY, P., *J. chem. Soc., A*, 1835 (1968)
295 DAY, P., *Endeavour*, **29,** 45 (1970)
296 DAY, P. and HALL, I. D., *J. chem. Soc., A*, 2679 (1970)
297 de CASTELLÓ, R. A., MAC-COLL, C. P., EGEN, N. B. and HAIM, A., *Inorg. Chem.*, **8,** 699 (1969)
298 de CASTELLÓ, R. A., MAC-COLL, C. P. and HAIM, A., *Inorg. Chem.*, **10,** 203 (1971)
299 de CHANT, M. J. and HUNT, J. B., *J. Am. chem. Soc.*, **90,** 3695 (1968)
300 de VRIES, B., *J. Catal.*, **1,** 489 (1962)
301 DEBYE, P., *Trans. Electrochem. Soc.*, **82,** 265 (1942)
302 DEBYE, P., and HÜCKEL, E., *Phys. Z.*, **24,** 185 (1923)
303 DELAHAY, P., *J. chem. Phys.*, **55,** 4188 (1971)
304 DELAHAY, P. and NEMEC, L., *J. chem. Phys.*, **57,** 2135 (1972)
305 DELAHAY, P., 'Quasi-free electrons in Polar Liquids'. In *Electrons in Fluids* Edited by J. Jortner and N. R. Kestner, Springer: New York (1973) pp. 131–138
306 DELAHAY, P., private communication cited in reference 197
307 DELAHAY, P., CHARTIER, P. and NEMEC, L., *J. chem. Phys.*, **53,** 3126 (1970)
308 DENISON, J. T. and RAMSEY, J. R., *J. Am. chem. Soc.*, **77,** 2615 (1955)
309 DEUTSCH, E., SULLIVAN, J. C. and WATKINS, K. O., *Inorg. Chem.*, **14,** 550 (1975)
310 DEUTSCH, E. and TAUBE, H., *Inorg. Chem.*, **7,** 1532 (1968)
311 DIAZ, H. and TAUBE, H., *Inorg. Chem.*, **9,** 1304 (1970)
312 DICKINSON, W. L. and JOHNSON, R. L., *Inorg. Chem.*, **12,** 2048 (1973)
313 DIEBLER, H., DODEL, P. H. and TAUBE, H., *Inorg. Chem.*, **5,** 1688 (1966)
314 DIEBLER, H. and SUTIN, N., *J. phys. Chem.*, **68,** 174 (1964)
315 DIEBLER, H. and TAUBE, H., *Inorg. Chem.*, **4,** 1029 (1965)
316 DIXON, M. and WEBB, E. C., *Nature, Lond.*, **184,** 1298 (1959)
317 DOCKAL, E. R., EVERHART, E. T. and GOULD, E. S., *J. Am. chem. Soc.*, **93,** 5661 (1971)
318 DOCKAL, E. R. and GOULD, E. S., *J. Am. chem. Soc.*, **94,** 6673 (1972)
318a DODEL, P. H. and TAUBE, H., *Z. phys. Chem. (Frankfurt)*, **44,** 92 (1965)
319 DODSON, R. W., *J. Am. chem. Soc.*, **75,** 1795 (1953)
320 DODSON, R. W. and SCHWARZ, H. A., *J. phys. Chem.*, **78,** 892 (1974)
321 DOGONADZE, R. R., *Doklady Chem.*, **124,** 9 (1959)
322 DOGONADZE, R. R., *Doklady Chem.*, **133,** 765 (1960)
323 DOGONADZE, R. R., *Doklady Chem.*, **142,** 156 (1962)
324 DOGONADZE, R. R., KRISHTALIK, L. I. and PLESKOV, Yu. V., *Sov. Electrochem.*, **10,** 489 (1974)
325 DOGONADZE, R. R. and KUZNETSOV, A. M., *Teor. Eksp. Khim.*, **6,** 298 (1970); *CA* **74,** 46043x

326 DOGONADZE, R. R. and KUZNETSOV, A. M., *Sov. Electrochem.*, **7**, 735 (1971). See also references cited in reference 329, and discussion in references 329, 921, 922
327 DOGONADZE, R. R. and KUZNETSOV, A. M., *Zh. vses. Khim. Obshch.*, **19**, 242 (1974); *CA*, **81**, 111518s
328 DOGONADZE, R. R. and KUZNETSOV, A. M., *Proc. Symp. Electrocatal.*, 195 (1974); *CA* **82**, 49108
329 DOGONADZE, R. R. and KUZNETSOV, A. M., *Electrochim. Acta*, **19**, 961 (1974)
330 DOGONADZE, R. R. and KUZNETSOV, A. M., *J. electroanal. Chem.*, **65**, 545 (1975)
331 DOGONADZE, R. R., KUZNETSOV, A. M. and CHERNENKO, A. A., *Russ. chem. Rev.*, **34**, 759 (1965)
332 DOGONADZE, R. R., KUZNETSOV, A. M. and VOROTYNTZEV, M. A., *Z. phys. Chem. (N.F.)*, **100**, 1 (1976)
333 DOGONADZE, R. R., ULSTRUP, J. and KHARKATS, Yu. I., *J. C. S. Faraday II*, **68**, 744, (1972)
334 DOGONADZE, R. R., ULSTRUP, J. and KHARKATS, Yu. I., *J. electroanal. Chem.*, **39**, 47 (1972)
335 DOGONADZE, R. R., ULSTRUP, J. and KHARKATS, Yu. I., *J. theor. Biol.*, **40**, 259 (1973)
336 DOGONADZE, R. R., ULSTRUP, J. and KHARKATS, Yu. I., *J. theor. Biol.*, **40**, 279 (1973)
337 DOLCETTI, G., PELOSO, A. and TOBE, M. L., *J. chem. Soc.*, 5196 (1965)
338 DOLPHIN, D., NIEM, T., FELTON, R. H. and FUJITA, I., *J. Am. chem. Soc.*, **97**, 5288 (1975)
339 DORFMAN, M. K. and GRYDER, J. W., *Inorg. Chem.*, **1**, 799 (1962)
340 DOTY, L. F. and CAVELL, R. G., *Inorg. Chem.*, **13**, 2722 (1974)
340a DOYLE, J. and SYKES, A. G., *J. chem. Soc., A*, 795 (1967)
341 DOYLE, J. and SYKES, A. G., *J. chem. Soc., A*, 215 (1968)
342 DOYLE, J., SYKES, A. G. and ADIN, A., *J. chem. Soc., A*, 1314 (1968)
343 DREYER, R., KÖNIG, K. and SCHMIDT, H., *Z. phys. Chem.*, **227**, 257 (1964)
344 DRYE, D. J., HIGGINSON, W. C. E. and KNOWLES, P., *J. chem. Soc.*, 1137 (1962)
345 DUCHART, B. S., FLUENDY, M. A. D. and LAWLEY, K. P., *Chem. Phys. Lett.*, **14**, 129 (1972)
346 DUGAN, J. V. and PALMER, R. W., *Chem. Phys. Lett.*, **13**, 144 (1972)
347 DUKE, F. R. and PARCHEN, F. R., *J. Am. chem. Soc.*, **78**, 1540 (1956)
348 DULZ, G. and SUTIN, N., *J. Am. chem. Soc.*, **86**, 829 (1964)
349 DUNKIN, D. B., FEHSENFELD, F. C. and FERGUSON, E. E., *Chem. Phys. Lett.*, **15**, 257 (1972)
350 DURUP, J. and DURUP, M., *J. chim. Phys.*, **64**, 386 (1967)
351 DWYER, F. P. and GYARFAS, E. C., *Nature, Lond.*, **166**, 481 (1950)
352 DYKE, R. and HIGGINSON, W. C. E., *J. chem. Soc.*, 2802 (1963)
353 EARLEY, J. E., *Prog. inorg. Chem.*, **13**, 243 (1970)
354 EARLEY, J. E. and CANNON, R. D., *Transit. Metal Chem. (New York)*, **1**, 33 (1965); see also p. 38, and Appendix, Table 22
355 EBERSON, L. and SCHÄFER, H., 'Organic Electrochemistry', *Topics in Current Chemistry, No. 21*, Springer Verlag: Berlin (1971)
356 EDWARDS, J. O., *Inorganic Reaction Mechanisms*, W. A. Benjamin: New York (1964)
357 EFRIMA, S. and BIXON, M., *Chem. phys. Lett.*, **25**, 34 (1974)
358 EIGEN, M., KRUSE, W., MAASS, G. and DeMAEYER, L., *Progr. React. Kinet.*, **2**, 285 (1964)
359 EIGEN, M. and WILKINS, R. G., *Adv. Chem. Ser.*, **49**, 55 (1965)
360 EIGENMANN, K. and GÜNTHARD, Hs. H., *Chem. Phys. Lett.*, **13**, 58 (1972)
361 EKSTROM, A., *Inorg. Chem.*, **13**, 2237 (1974)
362 EKSTROM, A., McLAREN, A. B. and SMYTHE, L. E., *Inorg. Chem.*, **14**, 2899 (1975)
363 EKSTROM, A. and FARRAR, Y., *Inorg. Chem.*, **11**, 2610 (1972)
364 EKSTROM, A. and JOHNSON, D. A., *J. inorg. nucl. Chem.*, **36**, 2557 (1974)
365 EKSTROM, A., McLAREN, A. B. and SMYTHE, L. E., *Inorg. Chem.*, **14**, 2899 (1975)
366 ELEY, D. D. and EVANS, M. G., *Trans. Faraday Soc.*, **34**, 1093 (1938)
367 ELIAS, J. E. and DRAGO, R. S., *Inorg. Chem.*, **11**, 415 (1972)
368 ELLIS, J. D. and SYKES, A. G., *J. C. S. Dalton*, 2553 (1973)
369 ELLIS, J. D. and SYKES, A. G., *J. C. S. Dalton*, 537 (1973)
370 EL-TANTAWY, Y. A. and ABU-SHADY, A. I., *Z. phys. Chem. Frankf. Ausg.*, **78**, 317 (1972); **88**, 141 (1974)

371 ELVING, P. J. and TEITELBAUM, C., *J. Am. chem. Soc.*, **71**, 3916 (1949)
372 ENDICOTT, J. F. and TAUBE, H., *J. Am. chem. Soc.*, **84**, 4985 (1962)
373 ENDICOTT, J. F. and TAUBE, H., *J. Am. chem. Soc.*, **86**, 1686 (1964)
374 ENDICOTT, J. F. and TAUBE, H., *Inorg. Chem.*, **4**, 437 (1965)
375 ENDO, K., YAMAUCHI, S., FUEKI, K. and MUKAIBO, T., *J. Solid St. Chem.*, **19**, 13 (1976)
376 ESPENSON, J. H., *J. Am. chem. Soc.*, **86**, 1883 (1964); **86**, 5101 (1964)
377 ESPENSON, J. H., *Inorg. Chem.*, **4**, 121 (1965)
378 ESPENSON, J. H., *Inorg. Chem.*, **4**, 1025 (1965)
379 ESPENSON, J. H., *Inorg. Chem.*, **4**, 1533 (1965)
380 ESPENSON, J. H., *J. Am. chem. Soc.*, **89**, 1276 (1967)
381 ESPENSON, J. H., *Inorg. Chem.*, **7**, 631 (1968)
382 ESPENSON, J. H., *Accts. chem. Res.*, **3**, 347 (1970)
383 ESPENSON, J. H., *J. Am. chem. Soc.*, **92**, 1880 (1970)
384 ESPENSON, J. H. and BIRK, J. P., *J. Am. chem. Soc.*, **87**, 3280 (1965)
385 ESPENSON, J. H. and BOONE, D. J., *Inorg. Chem.*, **7**, 636 (1968)
386 ESPENSON, J. H. and CARLYLE, D. W., *Inorg. Chem.*, **5**, 586 (1966)
387 ESPENSON, J. H. and CHRISTENSEN, R. J., *J. Am. chem. Soc.*, **91**, 3769 (1969)
388 ESPENSON, J. H. and KING, E. L., *J. Am. chem. Soc.*, **85**, 3328 (1963)
389 ESPENSON, J. H. and KINNEY, R. J., *Inorg. Chem.*, **10**, 376 (1971)
390 ESPENSON, J. H. and KRUG, L. A., *Inorg. Chem.*, **8**, 2633 (1969)
391 ESPENSON, J. H. and McCARLEY, R. E., *J. Am. chem. Soc.*, **88**, 1063 (1966)
392 ESPENSON, J. H., SHAW, K. and PARKER, O. J., *J. Am. chem. Soc.*, **89**, 5730 (1967)
393 ESPENSON, J. H. and WANG, R. T., *Chem. Commun.*, 207 (1970)
394 ESPENSON, J. H. and WANG, R. T., *Inorg. Chem.*, **11**, 955 (1972)
395 EVANS, S. A., COHEN, J. S. and LANE, N. F., *Phys. Rev. A*, **4**, 2235 (1971)
396 EYRING, H., HIRSCHFELDER, J. O. and TAYLOR, H. S., *J. chem. Phys.*, **4**, 479 (1936)
397 EYRING, H., WALTER, J. and KIMBALL, G. E., *Quantum Chemistry*, John Wiley: New York (1944)
398 FALCINELLA, B., FELGATE, P. D. and LAURENCE, G. S., *J. C. S. Dalton*, 1367 (1974)
399 FALCINELLA, B., FELGATE, P. D. and LAURENCE, G. S., *J. C. S. Dalton*, 1 (1975)
400 FAN, F-R. F. and GOULD, E. S., *Inorg. Chem.*, **13**, 2639 (1974)
401 FAN, F-R. F. and GOULD, E. S., *Inorg. Chem.*, **13**, 2647 (1974)
402 FARR, J. K., HULETT, L. G., LANE, R. H. and HURST, J. K., *J. Am. chem. Soc.*, **97**, 2654 (1975)
403 FAY, D. P. and SUTIN, N., *Inorg. Chem.*, **9**, 1291 (1970)
404 FEHSENFELD, F. C. and FERGUSON, E. E., *J. chem. Phys.*, **56**, 3066 (1972)
404a FELIX, F. and LUDI, A., *Inorg. Chem.*, **17**, 1782 (1978)
405 FERGUSON, A. F., *Proc. R. Soc. A*, **264**, 540 (1961)
406 FERGUSON, E. E., *Ann. Rev. phys. Chem.*, **26**, 17 (1975)
407 FERGUSON, E. E., DUNKIN, D. B. and FEHSENFELD, F. C., *J. chem. Phys.*, **57**, 1459 (1972)
408 FERGUSON, E. E., FEHSENFELD, F. C., and SCHMELTEKOPF, A. L., *J. chem. Phys.*, **47**, 3085 (1967)
409 FERGUSON, E. E., FEHSENFELD, F. C. and SCHMELTEKOPF, A. L., *Adv. at. & mol. Phys.*, **5**, 1 (1969)
410 FETISOV, I. K. and FIRSOV, O. B., *Soviet Phys. JETP*, **10**, 67 (1960)
411 FIELD, G. B. and STEIGMAN, G., *Astrophys. J*, **166**, 59 (1971)
412 FIELDING, P. E. and MELLOR, D. P., *J. chem. Phys.*, **22**, 1155 (1954)
413 FIRSOV, O. B., *Soviet Phys. JETP*, **15**, 906 (1962)
414 FITE, W. L., BRACKMANN, R. T. and SNOW, W. R., *Phys. Rev.*, **112**, 1161 (1958)
415 FITE, W. L., STEBBINGS, R. F., HUMMER, D. G. and BRACKMANN, R. T., *Phys. Rev.*, **119**, 663 (1960)
416 FLAKS, I. P. and SOLOV'EV, E. S., *Soviet Phys. tech. Phys.*, **3**, 564 (1958), cited in reference 745
417 FLEISCHER, E. B., KRISHNAMURTHY, M. and CHEUNG, S. K., *J. Am. chem. Soc.*, **97**, 3873 (1975)
418 FLORENCE, T. M. and SHIRVINGTON, B. J., *Analyt. Chem.*, **37**, 950 (1965)
419 FLUENDY, M. A. D. and LAWLEY, K. P., *Chemical Application of Molecular Beam Scattering*, Chapman and Hall: London (1973), see Chapter 8

420 FORD, P. C., *Coord. Chem. Rev.*, **5**, 75 (1970)
421 FORD, P. C., KUEMPEL, J. R. and TAUBE, H., *Inorg. Chem.*, **7**, 1976 (1968)
422 FORD, P., RUDD, DE F. P., GAUNDER, R. and TAUBE, H., *J. Am. Chem. Soc.*, **90**, 1187 (1968)
423 FORD, P. C. and SUTTON, C., *Inorg. Chem.*, **8**, 1544 (1969)
424 FORCHEIMER, O. L. and EPPLE, R. P., *J. Am. chem. Soc.*, **74**, 5772 (1952)
425 FOSTER, R., *Organic Charge Transfer Complexes*, Academic Press: London (1969)
426 FOUST, R. D. and FORD, P. C., *J. inorg. nucl. Chem.*, **36**, 930 (1974)
427 FRANCK, J. and PLATZMAN, R. In *Radiation Biology*, Vol. 1, chap. 2, McGraw-Hill Book Co.: New York (1953)
428 FRANCK, J. and SCHEIBE, G., *Z. phys. Chem.*, **A139**, 22 (1929)
429 FRANK, H. S., *J. chem. Phys.*, **23**, 2023 (1955)
430 FRANKLIN, J. L., *Ion–Molecule Reactions*, Vol. 1, Plenum Press: New York (1972)
431 FRASER, R. T. M:, *Proc. chem. Soc.*, 317 (1960)
432 FRASER, R. T. M., *J. Am. chem. Soc.*, **87**, 564 (1961)
433 FRASER, R. T. M., *Inorg. Chem.*, **3**, 255 (1964)
434 FRASER, R. T. M., SEBERA, D. K. and TAUBE, H., *J. Am. chem. Soc.*, **81**, 2906 (1959)
435 FRASER, R. T. M. and TAUBE, H., *J. Am. chem. Soc.*, **81**, 5000 (1959)
436 FRASER, R. T. M. and TAUBE, H., *J. Am. chem. Soc.*, **81**, 5514 (1959)
437 FRASER, R. T. M. and TAUBE, H., *J. Am. chem. Soc.*, **83**, 2239 (1961)
438 FRASER, R. T. M. and TAUBE, H., *J. Am. chem. Soc.*, **83**, 2242 (1961)
439 FREDIN, L. and NELANDER, B., *J. Am. chem. Soc.*, **96**, 1672 (1974)
440 FRIEDMAN, H. L. and ZELTMANN, A. H., *J. chem. Phys.*, **28**, 878 (1958)
441 FRÖHLICH, H., *Adv. Phys.*, **3**, 325 (1954)
442 FRÖHLICH, H., *The Theory of Dielectrics*, 2nd edn., chap. 4, Clarendon Press: Oxford (1958)
443 FRONCZEK, F. R. and SCHAEFER, W. P., *Inorg. Chem.*, **13**, 727 (1974)
444 FROST, A. A., *J. Am. chem. Soc.*, **73**, 2680 (1951)
445 FROST, A. A. and PEARSON, R. G., *Kinetics and Mechanism*, 2nd edn., John Wiley: London (1961)
446 FRUCHART, J. M., HERVE, G., LAUNAY, J. P. and MASSART, R. *J. inorg. nucl. Chem.*, **38**, 1627 (1976)
447 FULTON, R. B. and NEWTON, T. W., *J. phys. Chem.*, **74**, 1661 (1970)
448 FUOSS, R. M., *J. Am. chem. Soc.*, **80**, 5059 (1958); *see also* reference 15, pp. 551–2
449 FUSCHILLO, N., LALEVIC, B. and LEUNG, B., *J. appl. Phys.*, **46**, 310 (1975)
450 FUTRELL, J. H. and TIERNAN, T. O. 'Ion-Molecule Reactions'. In P. Ausloos (Ed.) *Fundamental Processes in Radiation Chemistry*, p. 171, Wiley-Interscience: New York (1968)
451 GASIEWSKI, J. W. and BIRK, J. P. unpublished experiments cited in reference 689
452 GASWICK, D. and HAIM, A., *J. Am. chem. Soc.*, **93**, 7347 (1971)
453 GASWICK, D. and HAIM, A., *J. Am. chem. Soc.*, **96**, 7845 (1974)
454 GAUNDER, R. G. and TAUBE, H. *Inorg. Chem.*, **9**, 2627 (1970)
455 GEORGE, P. and GRIFFITH, G. 'Electron Transfer and Enzyme Catalysis'. In P. D. BOYER, H. LARDY and K. MYRBÄCK (Eds.) *The Enzymes*, 2nd edn., pp. 347–389 Academic Press: New York (1959)
456 GEORGE, P. and IRVINE, D. H., *J. chem. Soc.*, 587 (1954)
457 GERMAN, É. D., DVALI, V. G., DOGONADZE, R. R. and KUZNETSOV. A. M. *Sov. Electrochem.*, **12**, 639 (1976)
458 GHOSE, S., *Acta Crystallog.*, **14**, 622 (1961)
459 GIESE, C. F., *Adv. Mass. Spectrom.*, **3**, 321 (1966)
460 GIESE, C. F. and GENTRY, W. R., *Phys. Rev. A*, **10**, 2156 (1974)
461 GILKS, S. W. and WAIND, G. M., *Discuss. Faraday Soc.* **29**, 102 (1960)
462 GILLEN, K. T., RULIS, A. M. and BERNSTEIN, R. B., *J. chem. Phys.*, **54**, 2831 (1971)
463 GIOUMOUSIS, G. and STEVENSON, D. P., *J. chem. Phys.*, **29**, 294 (1958)
464 GIULIANO, C. R. and McCONNELL, H. M., *J. inorg. nucl. Chem.*, **9**, 171 (1959)
465 GJERTSEN, L. and WAHL, A. C., *J. Am. chem. Soc.*, **81**, 1572 (1959)
466 GLASSTONE, S., LAIDLER, K. J. and EYRING, H., *The Theory of Rate Processes*, McGraw-Hill Book Co.: New York (1941)
466a GLAUSER, R., HAUSER, U., HERREN, F., LUDI, A., RODER, P., SCHMIDT, E., SIEGENTHALER, H. and WENK, F., *J. Am. chem. Soc.*, **95**, 8457 (1973)
467 GLIKMAN, T. S., ZABRODA, O. V. and ZAVGORODNYAYA, L. N. *Dok (Akd. Nauk SSSR) Phys. Chem.*, **177**, 810 (1967)

468 GOODENOUGH, J. B., *Mater. Res. Bull.*, **5**, 621 (1970); *CA* **73** 124659y
469 GOMM, P. S. and UNDERHILL, A. E., *Inorg. Nucl. Chem. Lett.*, **10**, 309 (1974)
470 GORDON, B. M., WILLIAMS, L. L. and SUTIN, N., *J. Am. chem. Soc.*, **83**, 2061 (1961)
471 GORDON, G., *Inorg. Chem.*, **2**, 1277 (1963)
472 GORDON, G and TAUBE, H., *J. inorg. nucl. Chem.*, **16**, 272 (1961)
473 GORDON, G. and TAUBE, H., *Inorg. Chem.*, **1**, 69 (1962)
474 GOULD, E. S. and TAUBE, H., *J. Am. chem. Soc.*, **86**, 1318 (1964)
475 GOULD, E. S., *J. Am. chem. Soc.*, **87**, 4730 (1965)
476 GOULD, E. S., *J. Am. chem. Soc.*, **88**, 2983 (1966)
477 GOULD, E. S., *J. Am. chem. Soc.*, **89**, 5792 (1967)
478 GOULD, E. S., *J. Am. chem. Soc.*, **94**, 4360 (1972)
479 GOULD, E. S., *J. Am. chem. Soc.*, **96**, 2373 (1974)
479a GOULD, E. S. and TAUBE, H., *J. Am. chem. Soc.*, **85**, 3706 (1963)
480 GOULD, E. S. and TAUBE, H., *J. Am. chem. Soc.* **86**, 1318 (1964)
481 GREEN, M., HIGGINSON, W. C. E., STEAD, J. B. and SYKES, A. G., *J. chem. Soc.*, A, 3068 (1971)
482 GREEN, M., SCHUG, K. and TAUBE, H., *Inorg. Chem.*, **4**, 1184 (1965)
483 GREEN, M. and SYKES, A. G., *J. chem. Soc.*, A, 3221 (1970); 3067 (1971)
484 GROSSMAN, B. and HAIM, A., *J. Am. chem. Soc.*, **92**, 4835 (1970)
485 GROSSMAN, B. and HAIM, A., *J. Am. chem. Soc.*, **93**, 6490 (1971)
486 GRYDER, J. W. and DODSON, R. W., *J. Am. chem. Soc.*, **73**, 2890 (1951)
487 GRYDER, J. W. and DORFMAN, M. C., *J. Am. chem. Soc.*, **83**, 1254 (1961)
488 GUENTHER, P. R. and LINCK, R. G., *J. Am. chem. Soc.*, **91**, 3769 (1969)
489 GUGGENHEIM, E. A., *Thermodynamics, an advanced treatment for Chemists and Physicists*, 5th edn.; p. 300, North-Holland Publishing Co.: Amsterdam (1967)
490 GUPTA, K. S., *Indian J. Chem.*, **12**, 990 (1974)
491 GUPTA, K. S., *J. inorg. nucl. Chem.*, **36**, 3879 (1974)
492 GURNEE, E. F. and MAGEE, J. L., *J. chem. Phys.*, **26**, 1237 (1957)
493 HAIGHT, G. P., personal communication quoted in reference 797, note 15
494 HAIGHT, G. P., HUANG T. J. and PLATT, H., *J. Am. chem. Soc.*, **96**, 3137 (1974)
495 HAIM, A., *J. Am. chem. Soc.*, **88**, 2324 (1966)
496 HAIM, A., *Inorg. Chem.*, **5**, 2081 (1966)
497 HAIM, A., *Inorg. Chem.*, **7**, 1475 (1968)
498 HAIM, A. and SUTIN, N., *J. Am. chem. Soc.*, **87**, 4210 (1965); **88**, 434 (1966)
499 HAIM, A. and SUTIN, N., *J. Am. chem. Soc.*, **88**, 434 (1966)
500 HAIM, A. and SUTIN, N., *J. Am. chem. Soc.*, **88**, 5343 (1966)
502 HAIM, A. and SUTIN, N., *Inorg. Chem.*, **15**, 476 (1976)
503 HAIM, A. and WILMARTH, W. K, *J. Am. chem. Soc.*, **83**, 509 (1961)
504 HALDANE, J. B. S., *Nature, Lond.*, **179**, 832 (1957)
505 HALPERN, J., *Can. J. Chem.*, **37**, 148 (1959)
506 HALPERN, J., *Quart. Rev.*, **15**, 207 (1961)
507 HALPERN, J., LEGARE, R. J. and LUMRY, R., *J. Am. chem. Soc.*, **85**, 680 (1963)
508 HALPERN, J. and NAKAMURA, S., *J. Am. chem. Soc.*, **87**, 3002 (1965)
509 HALPERN, J. and ORGEL, L. E., *Discuss. Faraday Soc.*, **29**, 32 (1960)
510 HALPERN, J. and PRIBANIĆ, M., *J. Am. chem. Soc.*, **90**, 5942 (1968)
510a HALPERN, J. and SMITH, J. G. *Can. J. Chem.*, **34**, 1419 (1956)
511 HAND, T. P., HYDE, M. R. and SYKES, A. G., *Inorg. Chem.*, **14**, 1720 (1975)
512 HANSON, J. R., *Synthesis*, **1**, 1 (1974); *CA* **80**, 59014n
513 HANUŠ, V. and BRDIČKA, R., *Chemicke Listy*, **44**, 291 (1950); *CA* **45**, 3735d
514 HARA, Y. and MINOMURA, S., *J. Chem. Phys.*, **61**, 5339 (1974); cf. also HARA, Y. and MINOMURA, S., *Proc. Int. Conf. High Pressure*, 4th, pp. 355–60 (1974); *CA* **83**, 90046k
515 HARA, Y., SHIROTANI, I. and MINOMURA, S., *Inorg. Chem.*, **14**, 1834 (1975)
516 HARBOTTLE, G. and DODSON, R. W., *J. Am. chem. Soc.*, **73**, 2442 (1951)
517 HARKNESS, A. C. and HALPERN, J., *J. Am. chem. Soc.*, **81**, 3526 (1959)
518 HARLAND, P. W. and RYAN, K. R., *Int. J. Mass Spectrom. and Ion Phys.*, **18**, 215 (1975); *CA*, **84**, 9172z
518a HARRIMAN, J. E. and MAKI, A. H., *J. chem. Phys.*, **39**, 778 (1963)
519 HASAN, F. and ROČEK, J., *J. Am. chem. Soc.*, **96**, 6802 (1974) and references cited therein
520 HASTED, J. B., *Proc. R. Soc. A*, **205**, 421 (1951); HASTED, J. B., *Proc. R. Soc. A*, **212**, 235 (1952)
521 HASTED, J. B., *J. appl. Phys.*, **30**, 25 (1959)
522 HASTED, J. B., *Adv. Electronics Electron Phys.*, **13**, 1 (1960)

523 HASTED, J. B., 'Charge Transfer and Collisional Detachment'. In D. R. Bates (Ed.) *Atomic and Molecular Processes*, chap. 18, Academic Press: New York (1962)
524 HASTED, J. B., *Adv. A. & Mol. Phys.*, **4**, 237 (1968)
525 HASTED, J. B., *Physics of Atomic Collisions*, Butterworths: London (1972)
526 HASTED, J. B. and CHONG, A. Y. J., *Proc. Phys. Soc. A*, **80**, 441 (1962)
527 HAVEMANN, U., ZÜLICKE, L., NIKITIN, E. E. and ZEMBEKOV, A. A., *Chem. Phys. Lett.*, **25**, 487 (1974)
528 HAYDEN, H. C. and AMME, R. C., *Phys. Rev.*, **172**, 104 (1968)
529 HAYON, E. and SIMIC, M., *J. Am. chem. Soc.*, **95**, 2433 (1973)
530 HEGEDUS, L. S. and HAIM, A., *Inorg. Chem.*, **6**, 664 (1967)
531 HEIKES, R. R. and JOHNSTON, W. D., *J. chem. Phys.*, **26**, 582 (1957)
532 HEIKES, R. R., MARADUDIN, A. A. and MILLER, R. C., *Ann. phys. (Paris)*, **8**, 733 (1963)
533 HEIKES, R. R. and URE, R. W., (Eds.), *Thermoelectricity: science and engineering*, chap. 4, John Wiley, Interscience: New York (1961)
534 HEMMES, P., *J. Am. chem. Soc.*, **94**, 75 (1972)
535 HENCHMAN, M., 'Rate Constants and Cross-Sections'. In J. L. Franklin (Ed.) *Ion-Molecule Reactions*, Vol. 1, p. 101–259, and in particular p. 183, Plenum Press: New York (1972)
536 HENGLEIN, A. and MUCCINI, G. A., *Z. Naturf.*, **17A**, 452 (1962); **18A**, 753 (1963)
537 HERBISON-EVANS, D., PHIPPS, P. B. P. and WILLIAMS, R. J. P., *J. chem. Soc.*, 6170 (1965)
538 HERBST, E., PATTERSON, T. A. and LINEBERGER, W. C., *J. chem. Phys.*, **61**, 1300 (1974)
539 HERSCHBACH, D. R., KWEI, G. H. and NORRIS, J. A., *J. chem. Phys.*, **34**, 1842 (1961)
540 HERY, M. and WIEGHARDT, K., *J. C. S. Dalton*, 1536 (1976)
541 HICKS, K. W., *J. inorg. nucl. Chem.*, **38**, 1381 (1976)
542 HICKS, K. W. and SUTTER, J. R., *J. phys. Chem.*, **75**, 1107 (1971)
543 HICKS, K. W., TOPPEN, D. L. and LINCK, R. G., *Inorg. Chem.*, **11**, 310 (1972)
544 HIGGINSON, W. C. E., *Discuss. Faraday Soc.*, **29**, 135 (1960) (Commenting on reference 461)
545 HIGGINSON, W. C. E., ROSSEINSKY, D. R., STEAD, J. B. and SYKES, A. G., *Discuss. Faraday Soc.*, **29**, 49 (1960)
546 HILSCH, R. and POHL, R. W., *Z. Physik.*, **57**, 145 (1929); and references cited in references 428, 872
547 HINDMAN, J. C., SULLIVAN, J. C. and COHEN, D., *J. Am. chem. Soc.*, **81**, 2316 (1959)
548 HODGKINSON, D. P. and BRIGGS, J. S., *J. Phys. B, (Atm. and Molec. Phys.)* **9**, 255 (1976)
549 HOFFMAN, A. B. and TAUBE, H., *Inorg. Chem.*, **7**, 903 (1968)
550 HOFFMAN, M. Z. and SIMIC, M., *J. Am. chem. Soc.*, **94**, 1757 (1972)
551 HOFSTEE, B. H. J., *Nature, Lond.*, **184**, 1296 (1959)
552 HOLM, R. H., AVERILL, B. A., HERSKOVITZ, T., FRANKEL, R. B., GRAY, H. B., SIIMAN, O. and GRUNTHANER, F. J., *J. Am. chem. Soc.*, **96**, 2644 (1974)
553 HOLSTEIN, T., *J. phys. Chem.*, **56**, 832 (1952)
554 HOMER, J. B., LEHRLE, R. S., ROBB, J. C. and THOMAS, D. W., *Adv. Mass. Spectrom.*, **3**, 415 (1966)
555 HONIG, J. M., *J. chem. Educ.*, **43**, 76 (1966)
556 HOPPENJANS, D. W., GORDON, G. and HUNT, J. B., *Inorg. Chem.*, **10**, 754 (1971)
557 HORNIG, H. C., ZIMMERMAN, G. L. and LIBBY, W. F., *J. Am. chem. Soc.*, **72**, 3808 (1950)
558 HUCHITAL, D. H. and HODGES, R. J., *Inorg. Chem.*, **12**, 998 (1973)
559 HUCHITAL, D. H. and HODGES, R. J., *Inorg. Chem.*, **12**, 1004 (1973)
560 HUCHITAL, D. H., SUTIN, N. and WARNQVIST, B., *Inorg. Chem.*, **6**, 838 (1967)
561 HUCHITAL, D. H. and WILKINS, R. G., *Inorg. Chem.*, **6**, 1022 (1967)
562 HUGHES, B. M., LIFSHITZ, C. and TIERNAN, T. O., *J. chem. Phys.*, **59**, 3162 (1973)
563 HUIZENGA, J. R. and MAGNUSSON, L. B., *J. Am. chem. Soc.*, **73**, 3202 (1951)
564 HUNT, J. B. and EARLEY, J. E., *J. Am. chem. Soc.*, **82**, 5312 (1960)
565 HUNT, J. P., unpublished work, cited in reference 1013
566 HURST, J. K. and LANE, R. H., *J. Am. chem. Soc.*, **95**, 1703 (1973)
567 HURST, J. K. and TAUBE, H., *J. Am. chem. Soc.*, **90**, 1178 (1968)
568 HUSH, N. S., *Z. Elektrochem.*, **61**, 734 (1957)
569 HUSH, N. S., *J. chem. Phys.*, **28**, 962 (1958)

570 HUSH, N. S., *Trans. Faraday Soc.*, **57**, 557 (1961)
571 HUSH, N. S., *Prog. inorg. Chem.*, **8**, 391 (1967)
572 HUSH, N. S., *Electrochim. Acta*, **13**, 1005 (1968)
573 HUSH, N. S., *Chem. Phys.*, **10**, 361 (1975)
574 HWANG, C. and HAIM, A., *Inorg. Chem.*, **9**, 500 (1970)
576 HYDE, F. J., *Semiconductors*, Macdonald: London (1965)
577 HYDE, M. R. Ph.D. Thesis, University of Leeds (1973) quoted in reference 511
578 HYDE, M. R., DAVIES, R. and SYKES, A. G., *J. C. S. Dalton*, 1838 (1972)
579 HYDE, M. R., TAYLOR, R. S. and SYKES, A. G., *J. C. S. Dalton*, 2730 (1973)
580 HYDE, M. R., SCOTT, K. L., WIEGHARDT, K. and SYKES, A. G., *J. C. S. Dalton*, 690 (1976)
581 HYDE, M. R. WIEGHARDT, K. and SYKES, A. G. *J. C. S. Dalton*, 690 (1976)
582 ICE, G. E. and OLSON, R. E., *Phys. Rev. A*, **11**, 111 (1975)
583 IMAI, H., *Bull. chem. Soc. Japan*, **30**, 873, (1957)
584 INDELLI, A. and AMIS, E. S., *J. Am. chem. Soc.*, **81**, 4180 (1959)
585 INTERRANTE, L. V., *Extended Interactions between Metal Ions in Transition Metal Complexes (A. C. S. Symposium series, 5)*, Washington D.C. (1974)
586 INTERRANTE, L. V. and BROWALL, K. W., *Inorg. Chem.*, **13**, 1162 (1974)
587 INTERRANTE, L. V., BROWALL, K. W. and BUNDY, F. P., *Inorg. Chem.*, **13**, 1158 (1974)
588 IRVINE, D. H., *J. Chem. Soc.*, 1841 (1957)
589 ISHIDA, S., ISHIGAMI, T. and KOHMURA, T., *Progr. theor. Phys., Osaka* **55**, 710 (1976)
590 ISIED, S. S. and TAUBE, H., *J. Am. chem. Soc.*, **95**, 8198 (1973)
591 ITZKOWITZ, M. M. and NORDMEYER, F. R., *Inorg. Chem.*, **14**, 2124 (1975)
592 IVANOV, G. K., *Optika Spektrosk*, **37**, 636 (1974); CA **82**, 35198
593 IWASAKI, M., EDA, B. and TORIYAMA, K., *J. Am. chem. Soc.*, **92**, 3211 (1970)
594 JACKS, C. A. and BENNETT, L. E., *Inorg. Chem.*, **13**, 2035 (1974)
595 JAFFÉ, H. H. and ORCHIN, M., *Theory and Applications of Ultraviolet Spectroscopy*, Chap. 9, J. Wiley, Inc.: New York (1962)
596 JAMES, R. V. and KING, E. L., *Inorg. Chem.*, **9**, 1301 (1970)
597 JANSEN, P. H. and ULSTRUP, J., *Acta chem. scand.*, **23**, 1822 (1969)
598 JOHNSON, C. E., *J. Am. chem. Soc.*, **74**, 959 (1952)
599 JONES, F. A. and AMIS, E. S., *J. inorg. nucl. Chem.*, **26**, 1045 (1964)
600 JONES, J. R., *The Ionisation of Carbon Acids*, Academic Press: London (1973)
601 JONKER, G. H., *Physics, Chem. Solids*, **9**, 165 (1959)
602 JØRGENSEN, C. K., *Absorption Spectra and Chemical Bonding in Complexes* Chap. 9, Pergamon Press: Oxford (1960)
603 JUNGST, R. and STUCKY, G., *Inorg. Chem.*, **13**, 2404 (1974)
604 JWO, J.-J. and HAIM, A., *J. Am. chem. Soc.*, **98**, 1172 (1976)
605 KALLEN, T. W. and EARLEY J. E., *Chem. Commun.*, 851 (1970)
606 KALLEN, T. W. and EARLEY, J. E., *Inorg. Chem.*, **10**, 1152 (1971)
607 KANE-MAGUIRE, N. A. P., TOLLISON, R. M. and RICHARDSON, D. E., *Inorg. Chem.*, **15**, 499 (1976)
608 KANE-MAGUIRE, N. A. P. and LANGFORD, C. H., *J.C.S. Chem. Commun.*, 351 (1973)
609 KATZIN, L. I., *J. chem. Phys.*, **23**, 2055 (1955)
610 KAUZMANN, W., *Quantum Chemistry. An Introduction*, Academic Press: New York (1957)
611 KAWAMURA, H., SHIROTANI, I. and INOKUCHI, H., *Chem. Phys. Lett.*, **24**, 549 (1974)
612 KEMPTER, V., KNESER, Th. and SCHLIER, Ch., *J. chem. Phys.*, **52**, 5851 (1970)
612a KEMPTER, V., MECKLENBRAUCK, W., MENZINGER, M., SCHULLER, G., HERSCHBACH, D. R. and SCHLIER, Ch., *Chem. Phys. Lett.*, **6**, 97 (1970)
613 KESTNER, N. R., LOGAN, J. and JORTNER, J., *J. Phys. Chem.*, **78**, 2148 (1974)
614 KHARKATS, Yu. I., *Sov. Electrochem.*, **8**, 1266 (1972)
615 KHARKATS, Yu. I., *Sov. Electrochem.*, **9**, 845 (1973)
616 KHARKATS, Yu. I., *Sov. Electrochem.*, **10**, 588 (1974)
617 KHARKATS, Yu. I., *Sov. Electrochem.*, **10**, 1083 (1974)
618 KHARKATS, Yu. I., MADUMAROV, A. K. and VOROTYNTSEV, M. A., *J. C. S. Faraday 2*, **70**, 1578 (1974)

619 KHARKATS, Yu. I. and CHONISHVILI, G. M., *Sov. Electrochem.*, **11**, 161 (1975)
620 KHARKATS, Yu. I. and CHONISHVILI, G. M., *Sov. Electrochem.*, **11**, 164 (1975)
622 KINSEY, J. L., 'Survey of Reactive Scattering Results'. In J. C. POLANYI (Ed.), *M.T.P. Int. Rev. Sci., Phys. Chem., Series One*, Vol. 9, pp. 173–212, Med. Tech. Publishing Co.: London (1972)
623 KIRKWOOD, J. G., *J. chem. Phys.*, **2**, 351 (1934)
624 KIRKWOOD, J. G. and WESTHEIMER, F. H., *J. chem. Phys.*, **6**, 506 (1938)
625 KIRWIN, J. B., PROLL, P. J. and SUTCLIFFE, L. H., *Trans. Faraday Soc.*, **60**, 119 (1964)
626 KLINGER, H., MÜLLER, A. and SALZBORN, E., *J. Phys. B.: (Atm Molec. Phys.)*, **8**, 230 (1975)
627 KOEPPL, G. W. and KRESGE, A. J., *J. C. S. Chem. Common.*, 371 (1973)
628 KOIDE, S., *J. phys. Soc. Japan*, **20**, 123 (1965)
629 KOSOWER, E. M., *J. Am. chem. Soc.*, **80**, 3253 (1958)
630 KOSOWER, E. M., *J. Chim. Phys.*, **61**, 230 (1964)
631 KOWALSKY, A., *Biochemistry*, **4**, 2382 (1965)
632 KRENTZIEN, H. and TAUBE, H., *J. Am. chem. Soc.*, **98**, 6379 (1976)
633 KRISHNAMURTY, K. V. and WAHL, A. C., *J. Am. chem. Soc.*, **80**, 5921 (1958)
634 KRUSE, W. and TAUBE, H., *J. Am chem. Soc.*, **82**, 526 (1960)
635 KSENDZOV, Ya. M., ANSELIM', L. M., VASIL'EVA, L. L. and LATYSHEVA, V. M., *Soviet Phys. solid St.*, **5**, 1116 (1963)
636 KUBO, R. and TOYOZAWA, Y., *Prog. theor. Phys. Osaka*, **13**, 160 (1955)
637 KUDRYASHOV, I. V. and KOCHETKOV, V. L., *Kinet. Katal.*, **9**, 68 (1968); *CA* **69**, 76229b
638 KURIEN, K. C. and BURTON, M., *Summary of Proc. 4th Informal Conf., Radiation Chem. of Water, Notre Dame, Indiana*, p. 14 (1961); cited in reference 788, note 30
639 KURZ, J. L. and KURZ, L. C., *J. Am. chem. Soc.*, **94**, 4451 (1972)
640 KŮTA, J. and YEAGER, E., *J. electroanal. Chem.*, **31**, 119 (1971)
640a KUZNETSOV, K. E. and LETNIKOV, F. A., *Russ. J. phys. Chem.*, **47**, 1707 (1973)
641 LA MAR, G. N. and VAN HECKE, G. R., *J. Am. chem. Soc.*, **94**, 9042, 9049 (1972)
642 LA MAR, G. N., *J. Am. chem. Soc.*, **94**, 9055 (1972)
643 LABUDDE, R. A., KUNTZ, P. J., BERNSTEIN, R. B. and LEVINE, R. D., *Chem. Phys. Lett.*, **19**, 7 (1973), and references cited therein
644 LAIDLER, K. J., *Can. J. Chem.*, **37**, 138 (1959)
645 LAM, S. K., DOVERSPIKE, L. D. and CHAMPION, R. L., *Phys. Rev., A*, **7**, 1595 (1973)
646 LANDAU, L., *Phys. Z. Sowjun.*, **2**, 46 (1932); *CA* **27**, 224
647 LANDAU, L., *Phys. Z. Sowjun.*, **3**, 664 (1933)
648 *LANDOLT-BÖRNSTEIN ZAHLENWERTE und FUNKTIONEN aus PHYSIK, CHEMIE* [etc.], 6 Auflage, II Band, 8 Teil, 'Optische Konstanten', Springer-Verlag: Berlin (1962)
649 LANGEVIN, P., *Annls Chim. phys.*, **5**, 245 (1905)
650 LANGFORD, C. H. and CAREY, J. H., *Can. J. Chem.*, **53**, 2430 (1975)
651 LANGFORD, C. H., VUIK, C. P. J. and KANE-MAGUIRE, N. A. P., *Inorg. nucl. chem. Lett.*, **11**, 377 (1975)
652 LARSSON, R., *Acta chem. scand.*, **21**, 257 (1967)
653 LATIMER, W. M., PITZER, K. S. and SLANSKY, C. M., *J. chem. Phys.*, **7**, 108 (1939)
654 LAUDENSLAGER, J. B., HUNTRESS, W. T. and BOWERS, M. T., *J. chem. Phys.*, **61**, 4600 (1974)
655 LAURENCE, G. S. and ELLIS, K. J., *J.C.S. Dalton*, 2229 (1972)
656 LAVALLEE, C. and NEWTON, T. W., *Inorg. Chem.*, **11**, 2616 (1972)
656a LAVALLEE, D. K., LAVALLEE, C., SULLIVAN, J. C., and DEUTSCH, E., *Inorg. Chem.*, **12**, 570 (1973)
657 LAWANI, S. A., *J. phys. Chem.*, **80**, 105 (1976)
657a LAWTON, S. L. and JACOBSON, R. A., *J. Am. chem. Soc.*, **88**, 616 (1967)
658 LE VANDA, C., COWAN, D. O., LEITCH, C. and BECHGAARD, K., *J. Am. chem. Soc.*, **96**, 6788 (1974)
659 LEFFERT, C. B., JACKSON, W. M. and ROTHE, E. W., *J. chem. Phys.*, **58**, 5801 (1973)
660 LEFFLER, J. E. and GRUNWALD, E., *Rates and Equilibria of Organic Reactions*, Wiley: New York (1963)
661 LEVENTHAL, J. J., EARL, J. D. and HARRIS, H. H., *Phys. review Lett.*, **35**, 719 (1975)
662 LEVICH, V. G., *Adv. Electrochem. & electrochem. Eng.*, **4**, 249 (1966)
663 LEVICH, V. G. and DOGONADZE, R. R., *Colln Czech. chem. Commun., Engl. edn* **26**, 193 (1961)

664 LEVICH, V. G. and KUZNETSOV, A. M., *Teor. Eksp. Khim.*, **6,** 291 (1970)
665 LEVINE, R. D. and BERNSTEIN, R. B., *Molecular Reaction Dynamics*, Oxford University Press: London (1974)
666 LEWIS, W. B. and CORYELL, C. D., *Brookhaven Conference Report BNL-C8*, p. 131 (1948); *CA* **46,** 1343c
667 LEWIS, W. B., CORYELL, C. D. and IRVINE, J. W., *J. chem. Soc.*, S386 (149)
668 LESZCZYNSKI, Z., KUBICA, J. and RYBACKI, L., *Przem. Chem.*, **42,** 92 (1963); *CA* **59,** 5013f
669 LIANG, A. and GOULD, E. S., *J. Am. chem. Soc.*, **92,** 6791 (1970)
670 LIANG, A. and GOULD, E. S., *Inorg. Chem.*, **12,** 12 (1973)
671 LIBBY, W. F., *Abstract Physical Inorganic Section, 115th Meeting of the American Chemical Society, San Francisco, Calif., U.S.A.*, March-April (1949)
672 LIBBY, W. F., *J. phys. Chem.*, **56,** 863 (1952); with discussion, p. 866
673 LIBBY, W. F., *J. chem. Phys.*, **38,** 420 (1963)
674 LIBBY, W. F., *Topics Mod. Phys.*, 205–18 (1971); *CA* **75,** 156965c
675 LICHTEN, W., *Phys. Rev.*, **131,** 229 (1963)
676 LICHTEN, W., *At. Phys.*, **4,** 249 (1975)
677 LINCK, R. G., *Inorg. Chem.*, **7,** 2394 (1968)
678 LINCK, R. G., *M.T.P. (Med. Tech. Publishing Co.) Int. Rev. Sci.: Inorg. Chem., Ser. 1*, **9,** 303 Butterworths: London (1972)
679 LINCK, R. G., *M. T. P. Int. Rev. Sci.: Inorg. Chem., Ser. 2*, **9,** 173 Butterworths: London (1974)
680 LINCK, R. G., unpublished observations, quoted in reference 678
681 LINHARD, M., *Z. Elektrochem.*, **50,** 224 (1944)
682 LIPELES, M., *J. chem. Phys.*, **51,** 1252 (1969)
683 LITEPLO, M. P. and ENDICOTT, J. F., *J. Am. chem. Soc.*, **91,** 3982 (1969)
684 LITEPLO, M. P. and ENDICOTT, J. F., *Inorg. Chem.*, **10,** 1420 (1971)
685 LITTLER, J. G. F. and WILLIAMS, R. J. P., *J. chem. Soc.*, 6368 (1965)
686 LITVAK, H. E., URENA, A. G. and BERNSTEIN, R. B., *J. chem. Phys.*, **61,** 4091 (1974)
687 LOCKWOOD, G. J. and EVERHART, E., *Phys. Rev.*, **125,** 567 (1962)
688 LOEWENSTEIN, A. and RON, G., *Inorg. Chem.*, **6,** 1604 (1967)
689 LOGAN, T. P. and BIRK, J. P., *Inorg. Chem.*, **12,** 2464 (1973)
690 LORRAIN, P. and CORSON, D. R., *Electromagnetic Fields and Waves*, 2nd edn, Freeman: San Francisco (1970)
691 LOVE, C. M., QUINN, L. P. and BRUBAKER, C. H., *J. inorg. nucl. Chem.*, **27,** 2183 (1965)
692 LUDI, A., personal communication quoted in reference 747
693 LUTHER, R. and RUTTER, T. F., *Z. anorg. Chem.*, **54,** 1 (1907)
694 McCANN, K. J., FLANNERY, M. R., HORNSTEIN, J. V. and MORAN, T. F., *J. Chem. Phys.*, **63,** 4998 (1975)
695 McCARROLL, R., *Proc. R. Soc., A* **264,** 547 (1961)
695a McCONNELL, H. M., *J. chem. Phys.*, **35,** 508 (1961)
696 McCONNELL, H. M., HAM, J. S. and PLATT, J. R., *J. chem. Phys.*, **21,** 66 (1953)
697 McCONNELL, H. M. and BERGER, S. B., *J. chem. Phys.*, **27,** 230 (1957)
698 McCONNELL, H. M. and DAVIDSON, N., *J. Am. chem. Soc.*, **71,** 3845 (1949)
699 McCONNELL, H. M. and DAVIDSON, N., *J. Am. chem. Soc.*, **72,** 3168 (1950)
700 McCONNELL, H. M. and DAVIDSON, N., *J. Am. chem. Soc.*, **72,** 5557 (1950)
701 McCONNELL, H. M. and WEAVER, H. E., *J. chem. Phys.*, **25,** 307 (1956)
702 McCURDY, W. H. and GUILBAULT, G. C., *J. phys. Chem.*, **64,** 1825 (1960)
703 McGLYNN, S. P., quoted in reference 778, p. 110
704 McKELLAR, J. R. and WEST, B. O., *J.C.S. Dalton*, 796 (1974)
705 McKNIGHT, G. F. and HAIGHT, G. P., *Inorg. Chem.*, **12,** 1619 (1973)
706 McKNIGHT, G. F. and HAIGHT, G. P., *Inorg. Chem.*, **12,** 1934 (1973)
707 McNEIL, D. A. C., RAYNOR, J. B. and SYMONS, M. C. R., *J. chem. Soc.*, 410 (1965)
708 McRAE, E. G., *J. phys. Chem.*, **61,** 562 (1957)
709 MAGEE, J. L., *J. chem. Phys.*, **8,** 687 (1940)
710 MAHAN, B. H., *J. chem. Phys.*, **55,** 1436 (1971)
711 MAHAN, B. H., *Accts Chem. Res.*, **8,** 55 (1975)
712 MAHMOOD, A. J. and MUTTALIB, M. A., *Dacca Univ. Stud.*, **22,** 67 (1974); *CA* **83,** 048905
713 MAIER, W. B., *J. chem. Phys.*, **60,** 3588 (1974)

714 MAIRANOVSKII, S. G. and LISHCHETA, L. I., *Colln Czech. chem. Commun., Engl. Edn*, **25**, 3025 (1960)
715 MAIRANOVSKII, S. G. and LISHCHETA, L. I., *Izv. Akad. Nauk SSSR Otd. Khim. Nauk*, 1749 (1961); *CA* **58**, 1934g
716 MAIRANOVSKII, S. G. and LISHCHETA, L. I., *Izv. Akad. Nauk SSSR Otd. Khim. Nauk*, 1984 (1962); *CA* **58**, 7610e
717 MALIK, W. U. and ALI, S. I., *Talanta*, **8**, 737 (1961); *CA* **56**, 4352a
718 MALLIARIS, A. and KATAKIS, D., *J. Am. chem. Soc.*, **87**, 3077 (1965)
719 MANNING, P. V. and JARNAGIN, R. C., *J. phys. Chem.*, **67**, 2884 (1963)
720 MARCUS, R. A., *J. chem. Phys.*, **24**, 966 (1956)
721 MARCUS, R. A., *J. chem. Phys.*, **24**, 979 (1956)
722 MARCUS, R. A., *J. chem. Phys.*, **26**, 867 (1957)
723 MARCUS, R. A., *J. chem. Phys.*, **26**, 872 (1957)
724 MARCUS, R. A., *Discuss. Faraday Soc.*, **29**, 21 (1960)
725 MARCUS, R. A., *J. phys. Chem.*, **67**, 853 (1963)
726 MARCUS, R. A., *J. phys. Chem.*, **67**, 853 (1963)
727 MARCUS, R. A., *Ann. Rev. phys. Chem.*, **15**, 155 (1964)
728 MARCUS, R. A., *J. chem. Phys.*, **43**, 679 (1965)
729 MARCUS, R. A., *Electrochim. Acta*, **13**, 995 (1968)
730 MARCUS, R. A., *J. phys. Chem.*, **72**, 891 (1968)
731 MARCUS, R. A. and SUTIN, N., *Inorg. Chem.*, **14**, 213 (1975)
732 MARCUS, R. J., ZWOLINSKI, B. J. and EYRING, H., *J. phys. Chem.*, **58**, 432 (1954)
733 MARTIN, A. H. and GOULD, E. S., *Inorg. Chem.*, **14**, 873 (1975)
734 MASON, W. R., *Coord. Chem. Rev.*, **7**, 241 (1971)
735 MASON, W. R. and JOHNSON, R. C., *Inorg. Chem.*,.**4**, 1258 (1965)
736 MASSEY, H. S. W., *Rep. Prog. Phys.*, **12**, 248 (1949)
737 MASSEY, H. S. W. and BURHOP, E. H. S., *Electronic and Ionic Impact Phenomena*, Clarendon Press: Oxford (1952)
738 MASSEY, H. S. W., BURHOP, E. H. S. and GILBODY, H. B., *Electronic and Ionic Impact Resonance*, 2nd edn (5 volumes) Clarendon Press: Oxford (1969)
739 MASSEY, H. S. W. and SMITH, R. A., *Proc. R. Soc., A* **142**, 142 (1933)
740 MASTERS, B. J. and SCHWARTZ, L. L., *J. Am. chem. Soc.*, **83**, 2620 (1961)
741 MATSUZAWA, M., *J. Phys. B: (Atm. Molec. phys)*, **8**, 2114 (1975), and references cited therein
742 MATHEWS, D. M., HEFLEY, J. D. and AMIS, E. S., *J. phys. Chem.*, **63**, 1236 (1959)
743 MATTHEWS, B. A., TURNER, J. V. and WATTS, D. W., *Aust. J. Chem.*, **29**, 551 (1976)
744 MATVEENKO, A. V. and PONOMAREV, L. I., *Soviet Phys. JETP*, **30**, 1131 (1970)
745 MATVEENKO, A. V. and PONOMAREV, L. I., *Soviet Phys. JETP*, **41**, 456 (1975)
746 MAXWELL, J. C., *A treatise on Electricity and Magnetism*, 3rd edn Vol. 1, p. 272, Clarendon Press: Oxford (1904)
746a MAYOH, B. and DAY, P., *J. Am. chem. Soc.*, **94**, 2885 (1972)
747 MAYOH, B. and DAY, P., *J.C.S. Dalton*, 846 (1974)
748 MEIER, D. J. and GARNER, C. S., *J. phys. Chem.*, **56**, 853 (1952)
749 MEITES, L. and ZUMAN, P., *Electrochemical Data, Part I: Organic, Organometallic and Biochemical Substances*, John Wiley Interscience: London (1974)
750 MELIUS, C. F., and GODDARD, W. A., *Chem. Phys. Lett.*, **15**, 524 (1972) See further references cited in reference 35
751 MELIUS, C. F. and GODDARD, W. A., *Phys. Rev. A*, **10**, 1541 (1974)
752 MERIDETH, C. W., Ph.D. Thesis, University of California (UCRL-11704) (1965)
753 MICHAELIS, L. and MENTEN, M. L., *Biochem. Z.*, **49**, 333 (1913)
754 MICIC, O. I. and CERCEK, B., *J. phys. Chem.*, **78**, 285 (1974)
755 See reference 10
756 MILLER, J. D. and PRINCE, R. H., *J. chem. Soc., A*, 1048 (1966)
757 MITEVA, M., BONCHEV, P., MALINOVSKI, A. and KABASONOV, K., *Dokl. bulg. Akad. Nauk*, **29**, 81 (1976); *CA* **85**, 114661
758 MITOFF, S. P., *J. chem. Phys.*, **35**, 882 (1961)
759 MOISEIWITSCH, B. L., *Proc. phys. Soc. A*, **69**, 653 (1956)
760 MOISEIWITSCH, B. L., *J. atmos. Terr. Phys.* (Special Suppl., 'Meteors', ed. T. R. Kaiser) London: Pergamon Press, Vol. 2, p. 23, quoted in reference 526
761 MONACELLI, F., BASOLO, F. and PEARSON, R. G., *J. inorg. nucl. Chem.*, **24**, 1241 (1962)

762 MOORE, F. M. and HICKS, K. W., *Inorg. Chem.*, **14**, 413 (1975)
763 MOORE, G. E., *J. opt. soc. Am.*, **43**, 1045 (1953)
764 MOORE, R. L., *J. Am. chem. Soc.*, **77**, 1504 (1955)
765 MORAN, T. F., FLANNERY, M. R. and ALBRITTON, D. L., *J. chem. Phys.*, **62**, 2869 (1975)
766 MORIN, F. J., *Phys. Rev.*, **83**, 1005 (1951)
767 MORRISON, W. H. and HENDRICKSON, D. N., *Inorg. Chem.*, **14**, 2331 (1975); and references cited therein
768 MORRISON, W. H., KROGSRUD, S. and HENDRICKSON, D. N., *Inorg. Chem.*, **12**, 1998 (1973)
769 MOTT, N. F. and MASSEY, H. S. W., *The Theory of Atomic Collisions*, chap. 19, Clarendon Press: Oxford (1965)
770 MOVIUS, W. G. and LINCK, R. G., *J. Am. chem. Soc.*, **91**, 5394 (1969)
771 MOVIUS, W. G. and LINCK, R. G., *J. Am. chem. Soc.*, **92**, 2677 (1970)
771a MÜLLER, A., KLINGER, H. and SALZBORN, E., *J. Phys. B: (Atm. Molec. Phys)*, **9**, 291 (1976)
772 MÜLLER, E. (Ed.), *Methoden der organischen Chemie (Houben-Weyl)*, Band III, Teil 2, Thieme Verlag: Stuttgart (1955)
773 MULLIKEN, R. S., *J. chem. Phys.*, **7**, 14 and 20 (1939)
774 MULLIKEN, R. S., *J. chem. Phys.*, **8**, 234 (1940)
775 MULLIKEN, R. S., *J. chem. Phys.*, **8**, 382 (1940)
776 MULLIKEN, R. S. and RIEKE, C. A., *Rep. Prog. Phys.*, **8**, 231 (1941)
777 MULLIKEN, R. S., *J. Am. chem. Soc.*, **74**, 811 (1952)
778 MULLIKEN, R. S. and PERSON, W. B., *A. Rev. phys. Chem.*, **13**, 107 (1962)
779 MURMANN, R. K., TAUBE, H. and POSEY, F. A., *J. Am. chem. Soc.*, **79**, 262 (1957)
780 MURMANN, R. K. and SULLIVAN, J. C., *Inorg. Chem.*, **6**, 892 (1967)
781 MURRAY-RUST, P., *Acta Crystallogr.*, **B31**, 978 (1975)
782 MURRELL, J. N., *The Theory of the Electronic Spectra of Organic Molecules*, Methuen: London (1963)
783 NAKAGAWA, I. and SCHIMANOUCHI, T., *Spectrochim. Acta*, **22**, 759 (1966)
784 NALLEY, S. J., COMPTON, R. N., SCHWEINLER, H. S. and ANDERSON, V. E., *J. chem. Phys.*, **59**, 4125 (1973)
785 NANDA, R. K. and DASH, A. C., *J. inorg. nucl. Chem.*, **36**, 1595 (1974)
786 NAVON, G. and MEYERSTEIN, D., *J. phys. Chem.*, **74**, 4067 (1970)
787 NAVON, G. and STEIN, G., *J. phys. Chem.*, **69**, 1384 (1965)
788 NAVON, G. and STEIN, G., *J. phys. Chem.*, **69**, 1390 (1965)
789 NEMEC, L., *J. chem. Phys.*, **59**, 6092 (1973)
790 NEMEC, L., BARON, B. and DELAHAY, P., *Chem. Phys. Lett.*, **16**, 278 (1972)
791 NEUMANN, H. M. and BROWN, H., *J. Am. chem. Soc.*, **78**, 1846 (1956)
792 NEWTON, R. G., *Scattering Theory of Waves and Particles*, McGraw-Hill: New York (1966)
793 NEWTON, T. W., *J. phys. Chem.*, **62**, 943 (1958)
794 NEWTON, T. W., *J. phys. Chem.*, **63**, 1493 (1959)
795 NEWTON, T. W., *J. chem. Educ.*, **45**, 571 (1968)
796 NEWTON, T. W., *J. phys. Chem.*, **74**, 1655 (1970)
797 NEWTON, T. W., *Inorg. Chem.*, **14**, 2394 (1975)
798 NEWTON, T. W. and BAKER, F. B., *Inorg. Chem.*, **1**, 368 (1962)
799 NEWTON, T. W. and BAKER, F. B., *J. phys. Chem.*, **67**, 1425 (1963)
800 NEWTON, T. W. and BAKER, F. B., *Inorg. Chem.*, **3**, 569 (1964)
801 NEWTON, T. W. and BAKER, F. B., *Adv. Chem. Ser.*, **71**, 268 (1967)
802 NEWTON, T. W. and BAKER, F. B., *Inorg. Chem.*, **4**, 1166 (1965)
803 NEWTON, T. W. and BAKER, F. B., *J. phys. Chem.*, **69**, 176 (1965)
804 NEWTON, T. W. and BAKER, F. B., *J. phys. Chem.*, **70**, 1943 (1966)
805 NEWTON, T. W. and BURKHART, M. J., *Inorg. Chem.*, **10**, 2323 (1971)
806 NEWTON, T. W. and FULTON, R. B., *J. phys. Chem.*, **74**, 2797 (1970)
807 NEWTON, T. W., McCRARY, G. E. and CLARK, W. G., *J. phys. Chem.*, **72**, 4333 (1968)
808 NEWTON, T. W. and RABIDEAU, S. W., *J. phys. Chem.*, **63**, 365 (1959)
809 NEYNABER, R. H., *Adv. at. & mol. Phys.*, **5**, 57 (1969)
810 NORDMEYER, F. R. and TAUBE, H., *J. Am. chem. Soc.*, **88**, 4295 (1966)
811 NORDMEYER, F. R. and TAUBE, H., *J. Am. chem. Soc.*, **90**, 1162 (1968)
812 NORRIS, A. R. and TOBE, M. L., *Inorg. Chim. Acta*, **1**, 41 (1967)

813 NORRIS, C. and NORDMEYER, F. R., *Inorg. Chem.*, **10**, 1235 (1971)
814 NORRIS, C. and NORDMEYER, F. R., *J. Am. chem. Soc.*, **93**, 4044 (1971)
815 NOYES, Z., *Z. phys. Chem.*, **16**, 576 (1895)
816 NOYES, R. M., *J. Am. chem. Soc.*, **86**, 971 (1964)
817 OGARD, A. E. and TAUBE, H., *J. Am. chem. Soc.*, **80**, 1084 (1958)
818 OGINO, H., TAKAHASHI, M. and TANAKA, N., *Bull. Chem. Soc. Japan*, **47**, 1426 (1974)
819 OGINO, H. and TANAKA, N., *Bull. chem. Soc. Japan*, **41**, 1622 (1968)
820 OHASHI, K., *Bull. chem. Soc. Japan*, **45**, 3093 (1972); **46**, 1880 (1973)
821 OLSON, M. V. and TAUBE, H., *Inorg. Chem.*, **9**, 2072 (1970)
822 OMIDVAR, K., *Phys. Rev. A*, **12**, 911 (1975)
822a OPPENHEIMER, M., BOTTCHER, C. and DALGARNO, A., *Chem. Phys. Lett.*, **15**, 24 (1972)
822b *Organic Electronic Spectral Data* vols. 1–6 Wiley Interscience: New York (1960–70)
823 ORGEL, L. E., *Quart. Rev.*, **8**, 422 (1954)
824 ORGEL, L. E., in *'Quelques Problemes de Chimie Minerale'*, Report of the 10th Solvay Council, Bruxelles, p. 289 (1956); *CA* **52**, 19654b
825 PADOVA, J., *Electrochim. Acta*, **12**, 1227 (1967)
826 PADOVA, J. I., In D. F. Conway and J. O'M. Bockris (Eds.), *Modern Aspects of Electrochemistry*, No. 7, Butterworths: London (1971)
827 PALMER, K. F. and WILLIAMS, D., *J. opt. Soc. America*, **64**, 1107 (1974)
828 PARKER, O. J. and ESPENSON, J. H., *Inorg. Chem.*, **8**, 185 (1969)
829 PARKER, O. J. and ESPENSON, J. H., *J. Am. chem. Soc.*, **91**, 1968 (1969)
830 PARSEGIAN, V. A. and NINHAM, B. W., *J. colloid and interface Sci.*, **37**, 332 (1971)
831 PATEL, R. C., BALL, R. C., ENDICOTT, J. F. and HUGHES, R. G., *Inorg. Chem.*, **9**, 23 (1970)
832 PATEL, R. C. and ENDICOTT, J. F., *J. Am. chem. Soc.*, **90**, 6364 (1968)
833 PAULSON, J. F., In *Ion-Molecule Reactions in Gases*, Adv. Chem. Series, No. 58, P. J. Ausloos (Ed.), p. 28
834 PEARSON, R. G., *J. Am. chem. Soc.*, **85**, 3533 (1963)
835 PEKKARINEN, L., *Ann. Acad. Sci. Fennicae, Ser. A*, **II**, No. 62 (1954); No. 85 (1957); *CA* **49**, 15401c
836 PELOSO, A., *Coord. Chem. Rev.*, **10**, 123 (1973)
837 PELOSO, A., *Coord. Chem. Rev.*, **16**, 95 (1975)
838 PELOSO, A. and BASATO, M., *J. chem. Soc., A*, 725 (1971)
839 PELOSO, A. and BASATO, M., *J.C.S. Dalton*, 2040 (1972)
840 PELOSO, A. and BASATO, M., *Coord. Chem. Rev.*, **8**, 111 (1972)
841 PENNINGTON, D. E. and HAIM, A., *J. Am. chem. Soc.*, **88**, 3450 (1966)
842 PEREL, J., DALEY, H. L., PEEK, J. M. and GREEN, T. A., *Phys. Rev. Lett.*, **23**, 677 (1969)
843 PEREL, J., DALEY, H. L. and SMITH, F. J., *Phys. Rev. A*, **1**, 1626 (1970)
844 PEREL, J., VERNON, R. H. and DALEY, H. L., *Phys. Rev. A*, **138**, 937 (1965)
845 PHILLIPS, C. S. G. and WILLIAMS, R. J. P., *Inorganic Chemistry*, 2 vols., Clarendon Press: Oxford (1965–6)
846 PLADZIEWICZ, J. R. and ESPENSON, J. H., *Inorg. Chem.*, **10**, 634 (1971)
847 PLADZIEWICZ, J. R. and ESPENSON, J. H., *J. phys. Chem.*, **75**, 3381 (1971)
848 PLADZIEWICZ, J. R. and ESPENSON, J. H., *J. Am. chem. Soc.*, **95**, 56 (1973)
849 PLADZIEWICZ, J. R., MEYER, T. J., BROOMHEAD, J. A. and TAUBE, H., *Inorg. Chem.*, **12**, 639 (1973)
850 PLATZMAN, R. L. and FRANCK, J. In *L. Farkas Memorial Volume*, A. Farkas and E. P. Wigner, (Eds.) Research Council of Israel: Jerusalem; Special Publication No. 1, pp. 21–36 (1952); *CA*, **49**, 10738b
851 PLATZMAN, R. L. and FRANCK, J., *Z. Phys.*, **138**, 411 (1954)
852 PO, H. N., WONG, W.-K. and CHEN, K. D., *J. inorg. nucl. Chem.*, **36**, 3872 (1974)
853 POLISSAR, M. J., *J. Am. chem. Soc.*, **58**, 1372 (1936)
854 POSTMUS, C. and KING, E. L., *J. phys. Chem.*, **59**, 1216 (1955)
855 POSPÍŠIL, L. and KUTA, J., *Colln. Czech. chem. Commun. Engl. Edn.*, **34**, 742 (1969)
856 POWELL, R. E. and LATIMER, W. M., *J. chem. Phys.*, **19**, 1139 (1951)
857 POWERS, M. J., CALLAHAN, R. W. SALMON, D. J. and MEYER, T. J., *Inorg. Chem.*, **15**, 894 (1976)
858 PRESENT, R. D., *Kinetic Theory of Gases*, McGraw-Hill: New York (1958)

859 PRESTWOOD, R. J. and WAHL, A.C., *J. Am. chem. Soc.*, **71**, 3137 (1949)
860 PRICE, H. J. and TAUBE, H., *Inorg. Chem.*, **7**, 1 (1968)
861 PRINCE, R. H. and SEGAL, M. G., *J.C.S. Dalton*, 330 (1975)
862 PRINCE, R. H. and SEGAL, M. G., *J.C.S. Dalton*, 1245 (1975)
863 Ibid, *J.C.S. Chem. Commun.*, **3**, 100 (1976)
864 PRZYSTAS, T. J. and SUTIN, N., *J. Am. chem. Soc.*, **95**, 5545 (1973)
865 PTITSYN, B. V., ZEMSKOV, S. V. and NIKOLAEV, A. V., *Dokl. (Acad Nank SSSR) Chem.*, **167**, 304 (1966)
866 RABIDEAU, S. W., *J. Am. chem. Soc.*, **79**, 6350 (1957)
867 RABIDEAU, S. W., *J. phys. Chem.*, **62**, 414 (1958)
868 RABIDEAU, S. W. and KLINE, R. J., *J. phys. Chem.*, **62**, 617 (1958)
869 RABIDEAU, S. W. and KLINE, R. J., *J. phys. Chem.*, **63**, 1502 (1959)
870 RABIDEAU, S. W. and KLINE, R. J., *J. inorg. nucl. Chem.*, **14**, 91 (1960)
871 RABIDEAU, S. W. and MASTERS, B. J., *J. phys. Chem.*, **65**, 1256 (1961)
872 RABINOWITCH, E., *Rev. mod. Phys.*, **14**, 112 (1942)
873 RAO, P. S. and HAYON, E., *J. phys. Chem.*, **77**, 2753 (1973)
874 RAPP, D. and FRANCIS, W. E. *J. chem. Phys.*, **37**, 2631 (1962)
875 RAWOOF, R. A. and SUTTER, J. R., *J. phys. Chem.*, **71**, 2767 (1967)
876 RAY, P. R., *Quart. J. Indian. chem. Soc.*, **4**, 327 (1927); *CA* **22**, 191
877 RAY, P. R. and DUTT, N. K., *Z. anorg. allg. Chem.*, **234**, 65 (1937)
878 REHM, D. and WELLER, A., *Israel J. Chem.*, **8**, 259 (1970)
879 REICHEL, T. L. and SAWYER, D. T., *Inorg. Chem.*, **14**, 1869 (1975)
880 REICHARDT, C., *Angew. Chem. Inte. Ed. Engl.*, **4**, 29 (1965)
881 REINSCHMIEDT, K., SULLIVAN, J. C. and WOODS, M., *Inorg. Chem.*, **12**, 1639 (1973)
882 RESTIVO, R. J., FERGUSON, G. and BALAHURA, R. J., *Inorg. Chem.*, **16**, 167 (1977)
883 REUBEN, B., *Chem. Br.*, **10**, 434 (1974)
884 REUTER, B., JASKOWSKY, J., and RIEDEL, E., *Z. Elektrochem.*, **63**, 937 (1959)
885 REYNOLDS, W. L., and LUMRY, R. W., *J. chem. Phys.*, **23**, 2460 (1955)
886 REYNOLDS, W. L. and LUMRY, R. W., *Mechanisms of Electron Transfer*, Ronald Press: New York (1966)
887 RICH, R. L. and TAUBE, H., *J. Am. chem. Soc.*, **76**, 2608 (1954)
888 RICH, R. L. and TAUBE, H., *J. phys. Chem.*, **58**, 6 (1954)
889 RICHARDSON, J. H., STEPHENSON, L. M. and BRAUMAN, J. I., *Chem. Phys. Lett.*, **25**, 318 (1974)
890 RIGG, T., STEIN, G. and WEISS, J., *Proc. R. Soc. A*, **211**, 375 (1952)
891 RILLEMA, D. P. and ENDICOTT, J. F., *Inorg. Chem.*, **11**, 2361 (1972)
892 RILLEMA, D. P. and ENDICOTT, J. F., *J. Am. chem. Soc.*, **94**, 8711 (1972)
893 RILLEMA, D. P., ENDICOTT, J. F. and PATEL, R. C., *J. Am. chem. Soc.*, **94**, 394 (1972)
894 ROBBINS, D. W. and STRENS, R. G. J., *Min. Mag.*, **38**, 551 (1972); quoted in reference 747
895 ROBIN, M. B., *Inorg. Chem.*, **1**, 337 (1962)
896 ROBIN, M. B. and DAY, P., *Adv. inorg. Chem. Radiochem.*, **10**, 247 (1967)
897 ROBINSON, R. A. and STOKES, R. H., *Electrolyte Solutions*, 2nd edn., Butterworths: London (1959)
898 RODBERG, L. S. and THALER, R. M., *Introduction to the Quantum Theory of Scattering*, Academic Press: New York (1967)
898a RODD, E. H., *The Chemistry of Carbon Compounds*, Vol. I, p. 993, Elsevier: London (1952)
899 RONA, E., *J. Am. chem. Soc.*, **72**, 4339 (1950)
900 ROOKSBY, H. P., *Acta Crystallogr.*, **1**, 226 (1948)
901 ROSENHEIN, L., SPEISER, D. and HAIM, A., *Inorg. Chem.*, **13**, 1571 (1974)
902 ROSSEINSKY, D. R., *J. Chem. Soc.*, 1181 (1963)
903 ROSSEINSKY, D. R., *Electrochim. Acta*, **16**, 19 (1971)
904 ROSSEINSKY, D. R., *Chem. Rev.*, **72**, 215 (1972)
905 ROSSEINSKY, D. R., *J. chem. Educ.*, **53**, 617 (1976)
906 ROSSEINSKY, D. R. and HIGGINSON, W. C. E., *J. Chem. Soc.*, 31 (1960)
907 ROSSEINSKY, D. R. and HILL, R. J., *J. C. S. Dalton*, 715 (1972)
908 ROSSEINSKY, D. R. and NICOL, M. J., *Trans. Faraday Soc.*, **61**, 2718 (1965)
909 ROSSEINSKY, D. R. and NICOL, M. J., *Electrochim. Acta*, **11**, 1069 (1966)
910 ROSSEINSKY, D. R. and NICOL, M. J., *J. Chem. Soc. A*, 2887 (1969)

911 ROSSEINSKY, D. R. and NICOL, M. J., *J. Chem. Soc. A*, 1196 (1970)
912 ROSSOTTI, F. J. C., *J. inorg. nucl. Chem.*, **1**, 159 (1955)
913 ROTHWELL, H. L., VAN ZYL, B. and AMME, R. C., *J. chem. Phys.*, **61**, 3851 (1974)
914 RUBINSHTEĬN, A. M., *Compt. Rend. acad. sci. U.R.S.S.*, **28**, 55 (1940); *CA* **35**, 2401/5 and other references cited in reference 80
915 RUFF, I., *Quart. Rev.*, **22**, 199 (1968)
916 RYKOV, A. G. and BLOKIN, N. B. *Radiokhimiya*, **16**, 799 (1974); *CA* **82**, 160752t
917 RYKOV, A. G. and FROLOV, A. A., *Radiokhimiya*, **16**, 810 (1974); *CA* **82**, 90572
918 SACHER, E. and LAIDLER, K. J., *Trans. Faraday Soc.*, **59**, 396 (1963)
919 SAJI, T., YAMADA, T. and AOYAGUI, S., *J. electroanal. Chem.*, **61**, 147 (1975)
920 SCHMICKLER, W., *J. C. S. Faraday II*, **72**, 307 (1976)
921 SCHMICKLER, W. and VIELSTICH, W., *Electrochim. Acta*, **18**, 883 (1973)
922 SCHMICKLER, W. and VIELSTICH, W., *Electrochim. Acta*, **19**, 963 (1974)
923 SCHMIDT, P. P., *Aust. J. Chem.*, **23**, 1287 (1970)
924 SCHMIDT, P. P., *J. chem. Phys.*, **57**, 3749 (1972)
925 SCHMIDT, P. P., *J. chem. Phys.*, **58**, 4384 (1973)
926 SCHMIDT, P. P., *J. phys. Chem.*, **77**, 488 (1973)
927 SCHMIDT, P. P., *J. phys. Chem.*, **78**, 1684 (1974)
928 SCHMIDT, P. P., *Electrochemistry*, (Chemical Society Specialist Periodical Report) **5**, 21 (1975)
929 SCHMIDT, P. P. and MARK, H. B., *J. chem. Phys.*, **58**, 4290 (1973)
930 SCHWARZ, H. A., COMSTOCK, D., YANDELL, J. K., and DODSON, R. W., *J. phys. Chem.*, **78**, 488 (1974)
931 SCOTT, K. L., personal communication, quoted in reference 511
932 SCOTT, K. L. and SYKES, A. G., *J. C. S. Dalton*, 1832 (1972)
933 SCOTT, K. L., WIEGHARDT, K. and SYKES, A. G., *Inorg. Chem.*, **12**, 655 (1973)
934 SCOTT, P. D., GLASSER, D. and NICOL, M. J. *J. C. S. Faraday I*, **71**, 1413 (1975)
935 SEBERA, D. K. and TAUBE, H., *J. Am chem. Soc.*, **83**, 1785 (1961)
936 SECCO, F., CELSI, S. and GRATI, C., *J. C. S. Dalton*, 1675 (1972)
937 SEEWALD, D., SUTIN, N. and WATKINS, K. O., *J. Am. chem. Soc.*, **91**, 7307 (1969)
938 SEN, R. K., YEAGER, E. and O'GRADY, W. E., *Ann. Rev. Phys.*, **26**, 287 (1975)
939 ŠEVČÍK, P. and TREINDL, L'., *Coll. Czech. chem. Commun. Engl. Edn.*, **37**, 2725 (1972)
940 SHAFFER, P. A., *J. Am. chem. Soc.*, **55**, 2169 (1933)
941 SHAFFER, P. A., *J. phys. Chem.*, **40**, 1021 (1936)
942 SHEA, C., and HAIM, A., *J. Am. chem. Soc.*, **93**, 3055 (1971)
944 SHEPPARD, J. C., *J. phys. Chem.*, **68**, 1190 (1964)
945 SHEPPARD, J. C. and BROWN, L. C., *J. phys. Chem.*, **67**, 1025 (1963)
946 SHEPPARD, J. C. and WAHL, A. C., *J. Am. chem. Soc.*, **79**, 1020 (1957)
947 SHIMADA, K., MOSHUK, G., CONNOR, H. D., CALUWE, P. and SZWARC, M., *Chem. Phys. Lett.*, **14**, 396 (1972)
948 SHPORER, M., RON, G., LOEWENSTEIN, A. and NAVON, G., *Inorg. Chem.*, **4**, 361 (1965)
949 SILLÉN, L. G., *Quart. Rev.*, **13**, 146 (1959)
950 SILLÉN, L. G. and MARTELL, A. E., (eds.), *Stability Constants* (Chemical Society Special Publication No. 17), (1964); *Supplement No. 1* (Chemical Society Special Publication No. 23), (1971)
951 SILVERMAN, J. and DODSON, R. W., *J. phys. Chem.*, **56**, 846 (1952)
952 SIMÁNDI, L. and NAGY, F., *Acta chim. hung.*, **46**, 101 (1965)
953 SMIRNOV, B. M., *Soviet Phys. JETP*, **19**, 692 (1964)
954 SMIRNOV, B. M., *Soviet Phys. JETP*, **20**, 345 (1965)
955 SMITH, C. P., *Dielectric Behaviour and Structure*, McGraw-Hill: New York (1955)
956 SMITH, D. L. and FUTRELL, J. H., *J. chem. Phys.*, **59**, 463 (1973)
957 SMITH, F. J., *Phys. Lett.*, **20**, 271 (1966)
958 SMITH, F. T., *Phys. Rev.*, **179**, 111 (1969)
959 SMOLUCHOWSKI, M. V., *Phys. Z.*, **17**, 557, 585 (1916); *Z. phys. Chem.*, **92**, 129 (1917)
960 SNELLGROVE, R. and KING, E. L., *J. Am. chem. Soc.*, **84**, 4609 (1962)
961 SNELLGROVE, R. and KING, E. L., *Inorg. Chem.*, **3**, 288 (1964)
962 SØNDERGAARD, N. C., ULSTRUP, J. and JORTNER, J., *Chem. Phys.*, **17**, 417 (1976)
962a SPIECKER, H., and WIEGHARDT, K., *Inorg. Chem.*, **16**, 1290 (1977)
963 SPINNER, T. and HARRIS, G. M., *Inorg. Chem.*, **11**, 1067 (1972)
964 SRINIVASAN, V. and ROČEK, J., *J. Am. chem. Soc.*, **96**, 127 (1974)

965 STEARN, A. E. and EYRING, H., *J. chem. Phys.*, **3**, 778 (1935) (cf. reference 466, p. 151)
966 STEBBINGS, R. F., *Adv. chem. Phys.*, **10**, 195 (1966)
967 STEIN, G. and TREININ, A., *Trans. Faraday Soc.*, **55**, 1091 (1959)
968 STOCKER, R. N. and NEUMANN, H., *J. chem. Phys.*, **61**, 3852 (1974)
969 STORM, D. and RAPP, D., *J. chem. Phys.*, **53**, 1333 (1970)
970 STORM, D. and RAPP, D., *J. chem. Phys.*, **57**, 4278 (1972)
970a STRANKS, D. R., *Pure appl. Chem.*, **38**, 304 (1974)
972 STRATTON, J. A., *Electromagnetic Theory*, pp. 149–151, McGraw-Hill: New York (1941)
973 STREKAS, T. C. and SPIRO, T. G., *Inorg. Chem.*, **15**, 974 (1976)
974 STRITAR, J. A. and TAUBE, H., *Inorg. Chem.*, **8**, 2281 (1969)
975 STUECKELBERG, E. C. G., *Helv. phys. Acta*, **5**, 369 (1932); *CA* **27**, 2072
976 SULFAB, Y., AL-OBADIE, M. S. and AL-SALEM, N. A., *Z. phys. Chem.*, **94**, 77 (1975)
977 SULLIVAN, J. C., *J. Am. chem. Soc.*, **84**, 4256 (1962)
978 SULLIVAN, J. C., *Inorg. Chem.*, **3**, 315 (1964)
979 SULLIVAN, J. C., *J. Am. chem. Soc.*, **87**, 1495 (1965)
980 SULLIVAN, J. C., COHEN, D. and HINDMAN, J. C., *J. Am. chem. Soc.*, **76**, 4275 (1954)
981 SULLIVAN, J. C., COHEN, D. and HINDMAN, J. C., *J. Am. chem. Soc.*, **79**, 4029 (1957)
982 SULLIVAN, J. C., ZIELEN, A. J. and HINDMAN, J. C., *J. Am. chem. Soc.*, **82**, 5288 (1960)
983 SULLIVAN, J. C. and THOMPSON, R. C., *Inorg. Chem.*, **6**, 1795 (1967)
984 SULLIVAN, J. C., ZIELEN, A. J. and HINDMAN J. C., *J. Am. chem. Soc.*, **82**, 5288 (1960)
985 SUTIN, N., *Ann. Rev. nucl. Sci.*, **12**, 285 (1962); *CA*, **58**, 6375h
986 SUTIN, N., *Electrochim. Acta*, **13**, 1175 (1968)
987 SUTIN, N., *Accts. Chem. Res.*, **1**, 225 (1968)
988 SUTIN, N., *Chem. Br.*, **8**, 148 (1972)
989 SUTIN, N. In *Inorganic Biochemistry*, (Ed. G. L. Eichhorn), **2**, 611–53 Elsevier: Amsterdam (1973)
990 SUTIN, N. and FORMAN, A., *J. Am. chem. Soc.*, **93**, 5274 (1971)
990a SUTIN, N., ROWLEY, J. K. and DODSON, R. W., *J. phys. Chem.*, **65**, 1248 (1961)
991 SUTTER, J. H. and HUNT, J. B., *J. Am. chem. Soc.*, **91**, 3107 (1969)
992 SUTTER, J. H., COLQUITT, K. and SUTTER, J. R., *Inorg. Chem.*, **13**, 1444 (1974)
993 SVATOS, G. and TAUBE, H., *J. Am. chem. Soc.*, **83**, 4172 (1961)
994 SYKES, A. G., *J. Chem. Soc.*, 5549 (1961)
995 SKYES, A. G., *Chem. Commun.*, 442 (1965)
996 SYKES, A. G., *Adv. inorg. Chem. Radiochem.*, **10**, 153–245 (1967)
997 SYKES, A. G., *Chem. Br.*, **6**, 159 (1970)
998 SYKES, A. G., private communication
999 SYKES, A.G. and THORNELY, R. N. F., *J. Chem. Soc. A*, 232 (1970)
1000 SYKES, A. G. and WEIL, J. A., *Prog. inorg. Chem.*, **13**, 1–106 (1970)
1001 SYMONS, M. C. R., WEST, D. X. and WILKINSON, J. G., *J. C. S. Chem. Commun.*, 917 (1973)
1002 TAKAKI, G. T. and FRASER, R. T. M., *Proc. Chem. Soc.*, 116 (1964)
1003 TAKEDA, M. and GREENWOOD, N. N., *J. C. S. Dalton*, 2207 (1975)
1004 TAL' ROZE, V. L., *Pure appl. Chem.*, **5**, 455 (1962)
1005 TAL' ROZE, V. L. and KARACHEVTSEV, G. V., *Adv. Mass. Spectrom.*, Vol. 3, p. 211, Institute of Petroleum; London (1966); *CA*, **69**, 3880h
1006 TAUBE, H., *Chem. Rev.* **50**, 69 (1952)
1007 TAUBE, H., *J. Am. chem. Soc.*, **77**, 4481 (1955)
1008 TAUBE, H., *Proc. Int. Conf. Coordination Chem. London (1959)*, Chem. Soc. Special Publication No. 13, pp. 57–71, The Chemical Society: London (1959)
1009 TAUBE, H., *Can. J. Chem.*, **37**, 129–137 (1959)
1010 TAUBE, H., *Adv. inorg. Chem. Radiochem.*, **1**, 1 (1959)
1011 TAUBE, H., *Adv. Chem. Ser.*, **49**, 107–125
1012 TAUBE, H., *Proc. R. A. Welch Foundation*, **6**, 7–29, 30–43 (1962)
1013 TAUBE, H., *Pure appl. Chem.*, **24**, 289 (1970)
1014 TAUBE, H., *Ber. Bunsenges. phys. Chem.*, **76**, 964 (1972)
1015 TAUBE, H. and GOULD, E. S., *Accts. Chem. Res.*, **2**, 321 (1969)
1015a TAUBE, H. and KING, E. L., *J. Am. chem. Soc.*, **76**, 4053 (1954)
1016 TAUBE, H. and MYERS, H., *J. Am. chem. Soc.*, **76**, 2103 (1954)
1017 TAUBE, H., MYERS, H. and RICH, R. L., *J. Am. chem. Soc.*, **75**, 4118 (1953)

1018 TAYLOR, R. S. and SYKES, A. G., *J. Chem. Soc. A*, 1628 (1971)
1019 TAYLOR, R. S. and SYKES, A. G., *J. Chem. Soc. A*, 2419 (1969)
1020 TENDLER, Y. and FARAGGI, M., *J. chem. Phys.*, **57**, 1358 (1972)
1021 TEWES, H. A., RAMSAY, J. B. and GARNER, C. S., *J. Am. chem. Soc.*, **72**, 2422 (1950)
1022 THAMBURAJ, P. K. and GOULD, E. S., *Inorg. Chem.*, **14**, 15 (1975); and references cited therein
1023 THAMBURAJ, P. K. and GOULD, E. S., *Inorg. Chem.*, **14**, 15 (1975)
1024 THOMAS, J. K., GORDAN, S. and HART, E. J., *J. phys. Chem.*, **68**, 1524 (1964)
1025 THOMAS, L, and HICKS, K. W., *Inorg. Chem.*, **13**, 749 (1974)
1026 THOMAS, T. W. and UNDERHILL, A. E., *Chem. Commun.*, 1344 (1969)
1027 THOMAS, T. W. and UNDERHILL, A. E., *J. Chem. Soc. A*, 512 (1971)
1028 THOMPSON, G. A. K. and SYKES, A. G. *Inorg. Chem.*, **15**, 638 (1976)
1029 THOMPSON, M. A., SULLIVAN, J. C. and DEUTSCH, E., *J. Am. chem. Soc.*, **93**, 5667 (1971)
1030 THOMPSON, R. C., *J. Am. chem. Soc.*, **70**, 1045 (1948)
1031 THOMPSON, R. C., *Inorg. Chem.*, **15**, 1080 (1976)
1032 THOMPSON, R. C. and SULLIVAN, J. C., *J. Am. chem. Soc.*, **92**, 3028 (1970)
1033 THORNELEY, R. N. F. and SYKES, A. G., *J. Chem. Soc. A*, 1036 (1970)
1034 THORNELEY, R. N. F. and SYKES, A. G., *Chem. Commun.*, 331 (1969)
1035 THORNELEY, R. N. F. and SYKES, A. G., *J. Chem. Soc. A*, 862 (1970)
1036 THORNTON, A. T., WIEGHARDT, K. and SYKES, A. G., *J.C.S. Dalton*, 147 (1976)
1037 TOM, G. M. and TAUBE, H., *J. Am. chem. Soc.*, **97**, 5310 (1975)
1038 TOM, G. M., CREUTZ, C. and TAUBE, H., *J. Am. chem. Soc.*, **96**, 7827 (1974)
1039 TOMA, H. E., *J. inorg. nucl. Chem.*, **38**, 431 (1976)
1040 TOMA, H. E. and MALIN, J. M., *J. Am. chem. Soc.*, **97**, 288 (1975)
1041 TOMIYASU, H. and FUKUTOMI, H., *Bull. chem. Soc. Japan*, **48**, 13 (1975)
1042 TONG, J. Y-P. and KING, E. L., *J. Am. chem. Soc.*, **82**, 3805 (1960)
1043 TOPPEN, D. L. and LINCK, R. G., *Inorg. Chem.*, **10**, 2635 (1971)
1044 TORIKAI, A., SUZUKI, T., MIYAZAKI, T., FUEKI, K. and KURI, Z., *J. phys. Chem.*, **75**, 482 (1971)
1044a TROTTER, P. J., *J. Am. chem. Soc.*, **88**, 5721 (1966)
1045 TULLY, F. P., LEE Y. T. and BERRY, R. S., *Chem. Phys. Lett.*, **9**, 80 (1971)
1046 ULSTRUP, J., *Acta. chem. scand.*, **23**, 3091 (1969)
1047 ULSTRUP, J. and JORTNER, J., *J. chem. Phys.*, **63**, 4358 (1975)
1048 ULSTRUP, J., *Trans. Faraday Soc.*, **67**, 2645 (1971)
1049 ULSTRUP, J., *J. C. S. Faraday I*, **71**, 435 (1975)
1050 VAN DUYNE, R. P. and FISCHER, S. F., *Chem. Phys.*, **5**, 183 (1974)
1051 VAN HOUTEN, S., *Physics Chem. Solids*, **17**, 7 (1960)
1052 VAUGHAN, W. E., 'Tables of Dielectric Constants, Dipole Moments and Dielectric Relaxation Times', *Digest of Literature on Dielectrics*, **32**, 22 (1968); **33**, 21, (1969); **34**, 17 (1970). *CA*, **73**, 29781y; **75**, 113597s; **76** 159671k
1053 VERWEY, E. J. W. and De BOER, J. H., *Proc. phys. Soc.*, (Extra Part) **49**, 59 (1937)
1054 VERWEY, E. J. W., quoted in reference 295
1055 VETTER, K. J., *Electrochemical Kinetics*, Academic Press: London (1967)
1056 VOEVODSKIĬ, V. V., SOLODOVNIKOV, S. P., and CHIBSIKIN, V. M., *Dokl. Akad. Nauk SSSR*, **129**, 1082 (1959); *CA* **55**, 25482a
1057 VOGLER, A. and KUNKELY, H., *Ber Bunsenges. Phys. Chem.*, **79**, 83 (1975)
1058 VOGLER, A. and KUNKELY, H., *Ber. Bunsenges. Phys. Chem.*, **79**, 301 (1975)
1059 VOGT, D., *Int. J. Mass Spectrom. Ion Phys.*, **3**, 81 (1969)
1060 VOGT, E. and WANNIER, G. H., *Phys. Rev.*, **95**, 1190 (1954)
1061 VOL'KENSHTEIN, M. V., DOGONADZE, R. R., MADUMAROV, A. K. and KHARKATS, Yu. I., *Dokl. (Akad, Nauk. SSSR) Phys. Chem.*, **199**, 569 (1971)
1062 VOLTZ, S. E., *J. phys. Chem.*, **56**, 867 (1952) (Discussion of reference 672)
1063 VON BUSCH, F., HORMES, J. and LIESEN, H. D., *Chem. Phys. Lett.*, **35**, 372 (1975)
1064 VON STACKELBERG, M., *Electrochemical Potentials of Organic Substances*. In *Methoden der Organischen Chemie (Houben-Weyl)*, Ed. E. Müller, Band III, Teil 2, p. 254, Thieme verlag: Stuttgart (1955)
1065 VON STACKELBERG, 'Polarography of Organic Substances'. In *Methoden der Organischen Chemie (Houben-Weyl)*, Ed. E. Müller, Band III, Teil 2, p. 295, Thieme verlag: Stuttgart (1955)
1066 VRACHNOU-ASTRA, E. and KATAKIS, D., *J. Am. chem. Soc.*, **89**, 6772 (1967)

1067 VRACHNOU-ASTRA, E., SAKELLARIDIS, P. and KATAKIS, D., *J. Am. chem. Soc.*, **92**, 811 (1970)
1068 VRACHNOU-ASTRA, E., SAKELLARIDIS, P. and KATAKIS, D., *J. Am. chem. Soc.* **92**, 3936 (1970)
1069 WADA, G. and TAMAKI, K., *Bull. chem. Soc. Japan*, **47**, 1422 (1974)
1070 WALTON, E. G., CORVAN, P. J., BROWN, D. B. and DAY, P., *Inorg. Chem.*, **15**, 1737 (1976)
1071 WANG, B.-C., SCHAEFER, W. P. and MARSH, R. F., *Inorg. Chem.*, **10**, 1492 (1971)
1072 WANG, R. T. and ESPENSON, J. H., *J. Am. chem. Soc.*, **93**, 380 (1971)
1073 WANG, R. T. and ESPENSON, J. H., *J. Am. chem. Soc.*, **93**, 1629 (1971)
1073a WARD, J. R. and HAIM, A., *J. Am. chem. Soc.*, **92**, 475 (1970)
1073b WARNOCK, T. T., BERNSTEIN, R. B. and GROSSER, A. E., *J. chem. Phys.*, **46**, 1685 (1967)
1074 WARNQVIST, B. and DODSON, R. W., *Inorg. Chem.*, **10**, 2624 (1974)
1075 WATKINS, K. O., SULLIVAN, J. C. and DEUTSCH, E., *Inorg. Chem.*, **13**, 1712 (1974)
1076 WATSON, W. D., *Astrophys. J.*, **188**, 35 (1974)
1077 WEAR, J. O., *J. Chem. Soc.*, 5596 (1965)
1078 WEAVER, M. J. and ANSON, F. C., *J. phys. Chem.*, **80**, 1861 (1976)
1079 WEBER, E. W. and VETTER, J., *Phys. Lett.*, **56A**, 446 (1976)
1080 WEISER, H. B., MILLIGAN, W. O. and BATES, J. B., *J. phys. Chem.*, **46**, 99 (1942)
1081 WEISS, J., *J. Chem. Soc.*, 309 (1944)
1082 WEISS, J., *Proc. R. Soc. A*, **222**, 128 (1954)
1083 WEISSMAN, S. I., *J. Am. chem. Soc.*, **80**, 6462 (1958)
1084 WELBER, B., *J. chem. Phys.*, **42**, 4262 (1965)
1084a WELLS, A. F., *Structural Inorganic Chemistry* 4th edn., p. 556, Oxford University Press: London (1975)
1085 WELLS, P. R., *Linear Free Energy Relationships*, Academic Press: London (1968)
1086 WESCHLER, C. J. and DEUTSCH, E., *Inorg. Chem.*, **15**, 139 (1976)
1087 WESTHEIMER, F. H. and KIRKWOOD, J. G., *J. chem. Phys.*, **6**, 513 (1938)
1088 WESTHEIMER, F. H., *Chem. Rev.*, **45**, 419 (1949)
1089 WETTON, E. A. M. and HIGGINSON, W. C. E., *J. chem. Soc.*, 5890 (1965)
1090 WHARTON, R. K. VON and WIEGHARDT, K., *Z. anorg. allg. Chem.*, **425**, 145 (1976)
1091 WHARTON, R. K., OJO, J. F. and SYKES, A. G., *J. C. S. Dalton*, 1526 (1975)
1092 WHITNEY, J. E. and DAVIDSON, N., *J. Am. chem. Soc.*, **69**, 2076 (1947)
1093 WHITNEY, J. E. and DAVIDSON, N., *J. Am. chem. Soc.*, **71**, 3809 (1949)
1094 WHITTAKER, E. J. W., *Acta Crystallogr*, **2**, 312 (1949)
1095 WIEGHARDT, K. and SYKES, A. G., *J. C. S. Dalton*, 651 (1974)
1096 WILES, D. R., *Can. J. Chem.*, **36**, 167 (1958)
1097 WILETS, L. and GALLAHER, D. F., *Phys. Rev.*, **147**, 13 (1966)
1098 WILKINS, R. G. and YELIN,-R., *J. Am. chem. Soc.*, **89**, 5496 (1967)
1099 WILLIAMS, D. J., GOEDDE, A. O. and PEARSON, J. M., *J. Am. chem. Soc.*, **94**, 7580 (1972)
1099a WILSON, C. R., and TAUBE, H., *Inorg. Chem.*, **14**, 2276 (1975)
1100 WILSON, S. T., BONDURANT, R. F., MEYER, T. J. and SALMON, D. J., *J. Am. chem. Soc.*, **97**, 2285 (1975)
1101 WINOGRAD, N. and KUWANA, T., *J. chem. Soc.*, **93**, 4343 (1971)
1102 WOLFGANG, R., *Accts. Chem. Res.*, **2**, 248 (1969)
1103 WOLFGANG, R., *Accts. Chem. Res.*, **3**, 48 (1970)
1104 WOLFGANG, R. L. and DODSON, R. W., *J. phys. Chem.*, **56**, 872 (1952)
1105 WOOD, D. L. and REMEIKA, R. P., *J. appl. Phys.*, **37**, 1232 (1966)
1106 WOODS, M. and SULLIVAN, J. C., *Inorg. Chem.*, **13**, 2774 (1974)
1107 WOODWARD, R. B. and HOFFMAN, R., *The Conservation of Orbital Symmetry*, Verlag Chemie; Weinheim (1970)
1108 WREN, D. J. and MENZINGER, M., *J. chem. Phys.*, **63**, 4557 (1975)
1109 WREN, D. J. and MENZINGER, M., *Chem. Phys. Lett.*, **25**, 378 (1974); **27**, 572 (1974); and references cited therein
1110 WRIGHT, R. C. and LAURENCE, G. S., *J. C. S. Chem. Commun.*, 132 (1972)
1111 WU, T-Y. and OHMURA, T., *Quantum Theory of Scattering*, Prentice Hall: New York (1962)
1112 YAMADA, S. and TSUCHIDA, R., *Bull. chem. Soc. Japan*, **29**, 421 (1956)
1113 YAMADA, S. and TSUCHIDA, R., *Bull. chem. Soc. Japan*, **29**, 894 (1956)

1114 YANDELL, J. K., FAY, D. P. and SUTIN, N., *J. Am. chem. Soc.*, **95**, 1131 (1973)
1115 YATSIMIRSKIĬ, K. B. and FEDEROVA, T. I., *Izv. vyssh. ucheb. zaved*, 40 (1958); *CA* **53**, 1977 (1959)
1116 YATSIMIRSKIĬ, K. B., TIKHONOVA, L. P. and SVARKOVSKAYA, I. P., *Russ. J. inorg. Chem.*, **14**, 1572 (1969)
1117 YUAN, Y-M. and MICHA, D. A., *J. chem. Phys.*, **65**, 4876 (1976); and references cited therein
1118 ZANELLA, A. and TAUBE, H., *J. Am. chem. Soc.*, **94**, 6403 (1972)
1119 ZEMSKOV, S. V., PTITSYN, B. V., LYUBIMOV, V. N. and MALAKHOV, V. F., *Russ. J. inorg. Chem.*, **12**, 648 (1967)
1120 ZENER, C., *Proc. R. Soc. A*, **137**, 696 (1932)
1121 ZHDANOV, V. P., *Zh. tekh. Fiz.*, **46**, 204 (1976); *CA* **84**, 95741
1122 ZHUZE, V. P. and SHELYKH, A. I., *Sov. Phys. solid St.*, **5**, 1278 (1963)
1123 ZIOLO, R. F., GAUGHAN, A. P., DORI, Z., PIERPONT, C. G. and EISENBERG, R., *J. Am. chem. Soc.*, **92**, 738 (1970); *Inorg. Chem.*, **10**, 1289 (1971)
1124 ZUMAN, P., *Colln. Czech. Chem. Commun. Engl. Edn.*, **15**, 1107 (1950)
1125 ZUMAN, P., *The Elucidation of Organic Electrode Processes*, Academic Press: London (1969)
1126 ZUMAN, P., *Prog. Polarogr.*, **3**, 73 (1972)
1127 ZUMAN, P. and PERRIN, C. L., *Organic Polarography*, John Wiley Interscience: London (1969)
1128 ZWICKEL, A. and TAUBE, H., *J. Am. chem. Soc.*, **81**, 1288 (1959)
1129 ZWICKEL, A. and TAUBE, H., *J. Am. chem. Soc.*, **81**, 2915 (1959)
1130 ZWICKEL, A. and TAUBE, H., *J. Am. chem. Soc.*, **83**, 793 (1961)
1131 ZWOLINSKI, B. J., MARCUS, R. J. and EYRING, H., *Chem. Rev.*, **55**, 157 (1955)

Index

For oxidants, reductants and electron transfer intermediates, the reader is directed to the specific indexes.

(a) and (b) character *see also* Hard acid, Soft acid, 260, 262, 264, 266
Abnormal region *see* Anomalous region
Accidental resonance, 23
Acid catalysis, 255
Acoustic vibrations, 203
Activation, volume of, 154
Activation energy, 177, 184, 186, 188, 204, 205, 224, 266, 274, 275, 277, 299, 301, 312
Activation enthalpies, 106–108
 low or negative, 107
 outer-sphere reactions, in, 107
Activation entropy, 104, 154, 217, 281
Activation free energy, 82, 103, 107, 195, 218, 226
 relation with overall free energy change, 149
 variation in, 238
Activation parameters, 115, 215, 244
Activation process, 51
 bond stretching, 186
 derivation, 194
 electrostatic considerations, 191
 solvent as dielectric continuum, 188
Activity coefficient, 115, 117
Adiabatic collision, Massey's criteria for, 18, 33
Adiabatic electron transfer, 13, 17, 18, 33, 42, 215, 219, 225, 226, 300
 chemical mechanism, 231
 distinction from non-adiabatic transfer, 219
 energy curves, 184
Adiabatic parameter, 19
Adjacent attack, 134
Anharmonicity, 211
Anomalous region, 215, 216, 233, 235, 237, 238
Arrhenius activation energy, 299
Arrhenius' law, 313
Asymmetrical transitions *see* Intervalence charge transfer

Atom–atom reactions, 33
Atomic polarisation, 194
Atoms, unlike, electron transfer between, 15–23
Atom transfer reaction, 26–32, 40
Average-valency complex *see also* Mixed-valence, 60, 270, 288

Binuclear intermediates, 97
 break-up of, 141
 definitions, 98
 free energy profiles, 121
 historical note, 97
 inert, 99, 155, 287
 isolation of, 137, 139
 kinetic evidence of, 116
 labile, 108, 123–131
  chelation, and, 127
  direct observation, 124
  examples, 115
  indeterminate rate laws, 112
  ion-pair complexes, 126
  symmetry, 114
 mechanisms, 98
 optical electron transfer, in, 285
 rate determining steps, 104
 rate of formation, 98
 role of, 98
 structure of, 139
Bjerrum, N, 101
Bohr radius, 8
Boltzmann distribution law, 183, 276, 277
Bond stretching model, 186, 222, 228, 277
Born–Oppenheimer approximation, 5, 17, 28
Bound polaron hopping *see* Hopping
Bridged binuclear species, 133
Bridged complexes, 285

339

Bridging groups, 135, 136, 224
  inner sphere mechanism, in, 209
  reducibility of, 247
  transfer of, 141, 152
    inert reactants, 141
    inorganic systems, 155
    labile reactants, 145
    organic systems, 159
    remote attack, in, 155
Bridging ligands, 134, 153, 296
Bridging mechanisms *see also* Double-bridged mechanisms, 128, 133, 154, 223–266, 285, 286
  chemical mechanism, 230
    evidence for, 238
  detection of intermediate, 238
  evidence of, 152
  free energy relationships, 241
  halide ions, by, 257
  kinetic isotope effect, 247
  models, 223–226
  organic ligands, by, 247
  protons in, 246
  reducibility of group, 247
  resonance transfer, 228
  single exchange, 227
  superexchange, 229
  theory, 223–237
Broadening *see* Energy broadening
de Broglie, 4

Cage complex, 87, 102
Catalysis, 149, 255
  inner- and outer-sphere mechanisms and, 149, 166, 167
Cation–cation reactions, 213
Ceiling unimolecular rate constant, 102
Charge carrier, 299–302
Charge exchange reactions, 26, 27
Charge transfer absorption, 267, 275
Charge transfer complexes, 292
Charge transfer to solvent, 283
Chelate complexes, 127
Chelation *see also under* Mechanisms, 108, 153, 247
Chemical mechanisms, 225, 230–247, 248, 251, 297
  definition, 230
  energy profile, 233
  evidence for, 238
  free energy relationships, 241
  inner-sphere, 238
  kinetic isotope effect, 247
  outer-sphere, 239
Collision complexes, 28, 40
Complementary reactions, 52, 68, 76, 84, 88, 89
  two-electron, 88
Comproportionation, 287

Conduction (electrical), 223, 268, 298, 299
Conductivity, 262, 264, 265, 298, 299
Continuum model, 189, 222, 227, 281, 282, 295
Controlled-valency semiconductors, 303–304
Coulombic interactions, 209
Coulomb's law, 20
Coupling, 15
Critical reaction distance, 19
Crocidolite, 306, 309–311
Cross-reaction, 44, 205, 217
Cross relation, 205, 206, 207, 209
  usefulness of, 207
Cross section, 7, 8, 9, 10, 11, 17, 177
  decrease in, 32
  non-adiabatic reactions, of, 21
Crystals, 203, 299
Current density, 298
Current (electric), 298
Curvature term, 212
Cyclic transition states, 108
Cytochrome *c* molecule, 169

Debye equation, 100, 213
Debye–Huckel equation, 101–102
Diabatic curves, 21
Dielectric constant, 191, 192, 193
Dielectric saturation, 191
Diffusion, 100, 222, 301
Diffusion-controlled mechanism,
  evidence for, 104
  examples, 105
  Smoluchowski model, 100
  theory of, 100
Diffusion-controlled rate, 68, 99, 100–105, 154, 207, 213–214, 222, 234, 237, 238, 301
Dipole–dipole forces, 102, 103
Dipole–dipole interactions, 209
Dipole strength, 276
Direct electron transfer, 134, 135
Direct exchange, 229
Displacement, 191
Double electron transfer, 84–91
Double exchange mechanism, 107, 224, 229
'Dry' electrons, 189, 283

Einstein, Albert, 102
Electrical conduction, 223, 268, 298, 300
Electricity, 190
Electrochemical data, 251
Electrochemical kinetics, 220
Electrochemical rate constant, 220
Electrocyclic reactions, 44
Electron affinity, 36
Electronic polarisation, 194
Electronic state correlation, 40
Electron mobility, 223
Electron spin resonance, 55, 169, 238, 245, 270
Electron transfer *see under* Mechanisms and individual mechanisms

Electron tunnelling, 180, 243
Electrostatic considerations, 191
Electrostatic potential, 302
Elementary reaction, 49, 50
Ellipsoidal complex, 202
Energy broadening, 185
Enthalpy, 218, 277
Entropy, 101, 217, 218, 280, 281, 303
  activation, 104, 154, 217, 281
e.s.r. *see* Electron spin resonance
Exchange reactions, 217, 220
Excited states, 277, 279
Exothermic charge transfer, 33
Expanded coordination, 171
Extinction coefficient, 276, 292
Extrinsic semiconductors, 299
Eyring, 28, 300

Facial attack, 166, 170
Field, 194
Field strength, 299
Flash photolysis, 89, 90
Fluendy and Lawley's classification, 28, 30, 32
Force constants, 188
Fractionation factor, 154
Franck–Condon barrier *see* Reorganisation energy
Franck–Condon energy *see* Reorganisation energy
Franck–Condon factor, 35, 36, 38, 39, 40, 279
Franck–Condon principle, 34–40, 177, 180, 193, 194, 267, 280, 281, 282, 312
  calculation of parameters, 36
  historical notes, 176
  internal energy of reactants, 37
  kinetic energy of reactants, 36
  rate comparisons governed by, 39
Free-electron models, 223
Free-energy change,
  correlation with rate law, 82
  relation with activation free energy, 149
Free-energy profiles, 80–83, 121–123, 232–238, 245
Free-energy relationships, 241
Free small polarons, 313
Fuess equation, 101, 102

Gases, kinetic theory of, 9, 222
Gas phase, electron transfer in, 3
Gibbs free energy, 190
Gioumousis, 24
Grazing collision, 44

Halide ions, bridging by, 257, 305
Hall effect, 302, 313, 314
Halpern, 228
Hard acids, 153, 260
Harmonic approximation, 228

Harmonic oscillator, 182, 184
Harned's rule, 117
Harpoon mechanism, 27, 32
  initial step, 34
Heitler–London approximation, 7–8, 10, 228
Helmholtz free energy, 190
Hofstee plot, 111
Hole transfer, 230, 231, 234, 292
Hopping conductivity, examples of, 306
Hopping mechanism, 298, 300–301, 304, 306, 307, 310, 311, 314
Hush, 184, 185, 192, 195, 197–201, 205, 215, 222, 270, 275–277, 282, 288, 295, 301, 302
Hydrogen bonding, 102, 209
Hydrolysis, 115, 117, 136

Image effects, 192, 193
Image forces, 102, 192, 193
Impact parameter, 7, 8, 10, 11, 23, 28, 33, 44
Impedance, 302
Induced oxidation *see* Induced reactions
Induced reactions, 55, 59, 65, 68, 73, 86
Induction (chemical) *see* Induced reactions
Induction factor, 65, 66, 68
Inert and labile complexes, 133
Inert polarisation, 193
Inner potential, 189
Inner space transfer, substitution controlled, 106
Inner-sphere charge transfer, 268
Inner-sphere ligands, effect on rates, 296
Inner-sphere mechanism, 26, 83, 99, 100, 131, 133, 134–137, 141, 151, 210, 223, 228, 262, 268
  bond model, 189
  bridging groups, 152
  catalysis and, 119, 149
  chemical mechanism, 238
  concentration of ionic charge in, 107
  distinguishing from outer-sphere, 141
  effect of variation of oxidant and reductant, 242
  energy, 217
  energy terms, 204, 281
  experimental evidence, 135, 137
  free energy relations, 212
  isolation of binuclear complex, 137
  isomeric reaction products, 147
  linear free relationships, 149
  proof of, 142
  sub-divisions of, 134
  substitution controlled transfer, 147
  summary of, 134–136
  systematic difference to outer sphere, 154
  transfer of bridging groups, 141, 209
Inner-sphere (reorganisation) energy, 204, 205, 281
Interaction energy *see* Resonance energy
Intermolecular electron transfer, 297

Intervalence charge transfer *see also* Optical charge transfer, 90, 177, 266, 270, 302, 312
Intramolecular electron transfer, 125, 178, 240, 267
Intrinsic barrier *see* Intrinsic free energy barrier
Intrinsic free energy barrier *see also* Reorganisational barrier, 124, 182, 197, 200, 206, 210, 213, 214, 222, 233, 234
Intrinsic semiconductors, 299
Ion–dipole forces, 34
Ion–dipole interaction, 24, 102, 205, 217
Ionic atmosphere *see also* Medium effects, 101
Ion–molecule reactions, 27
Ion pair, 100, 101, 103, 126
  complexes, 126
  rates of formation and dissociation, 103
Ions of multiple valency, 78
Isomeric reaction products, 147
Isotopic effects, 32, 106, 247
Isotopic exchange, 132
Isotopic fractionation change, 154

Kinetic isotope effects *see* Isotope effects
Kinetic theory of gases, 9, 222

Labile and inert complexes, 133
Labile intermediates,
  chelate complexes, 127
  direct observation, 124
  examples, 115
  free energy profiles, 121–123
  indeterminate rate laws, 112
  ion pairs, 126
  symmetry, 114
Labile polarisation, 193
Landau expression, 20, 177
Langevin model, 24, 25, 33, 38
Large polaron, 312
Lattice relaxation term, 303
Lawley, classification, 28, 30
LCAO approximation, 7–8, 10, 228, 271
Lennard–Jones potential, 101
Lewis acid, 260
Libby's mechanism, 133
Libration, 277
Ligand reducibility, 255
Linear free energy relationships, 135, 149, 150, 151, 210–213, 294
  inner sphere mechanisms, for, 212
  multiple application of, 151
  summary of, 150
Lines of force, 192
Linked electron transfer, 135–136
Low-energy collisions, 14
Low-temperature processes, 300

Magee mechanism, 32
Magnetic exchange, 223

Maleic acid, 245
Marcus, 103, 168, 182–185, 193–200, 205, 210, 211, 220–222, 231, 234, 260, 262, 277, 282, 301
Marcus cross relation, 131, 205–210
Marcus–Hush approximation *see* Medium overlap
Massey's criterion for adiabatic collision 18, 19, 24, 33, 34
Mass law retardation effect *see* Retardation
Maxwell distribution, 24
Mechanisms of electron transfer, 132–171 *see also* Adiabatic electron transfer, Adjacent attack, Atom transfer, Diffusion control, Facial attack, Grazing collision, Harpoon mechanism, Hopping, Near diabatic, Non-adiabatic, One-electron transfer, Spectator stripping, Stripping, Tunnelling, Two-electron transfer, etc.
  bridging *see also* Double-bridged mechanism, 128, 133, 154, 223–266, 285, 286
    chemical, 230
    detection of intermediate, 238
    evidence of, 152, 238
    free energy relationships, 241
    halide ions, by, 257
    kinetic isotope effect, 247
    models, 223–226
    organic ligands, by, 247
    protons in, 246
    reductibility of group, 247
    resonance transfer, 228
    single exchange, 227
    superexchange, 229
    theory, 223–237
  chemical, 225, 230–247, 248, 251, 297
    definition, 230
    energy profile, 233
    evidence for, 238
    free energy relationships, 241
    inner sphere, 238
    kinetic isotope effect, 247
    outer sphere, 239
  classification, 26, 134, 135
  direct, 134, 135
  double bridged, 108
  double exchange, 107, 224, 229
  expanded coordination, 171
  experimental evidence of, 135
  inner-sphere, 26, 83, 99, 100, 131, 133, 134–137, 141, 151, 210, 223, 228, 262, 268
    bond model, 189
    bridging groups, 141, 152, 209
    catalysis and, 119, 149
    chemical mechanism, 238
    concentration of ionic charge in, 107
    distinguishing from outer-sphere, 141
    effect of variation of oxidant and reductant, 242
    energy, 217

Mechanisms of electron transfer
  inner-sphere (*cont.*)
    energy terms, 204, 281
    experimental evidence, 135, 137
    free energy relations, 212
    isolation of binuclear complex, 137
    isomeric reaction products, 147
    linear free relationships, 149
    proof of, 142
    sub-divisions of, 134
    substitution controlled transfer, 147
    summary of, 134–136
    systematic difference to outer-sphere, 154
    transfer of bridging groups, 141, 209
  linked, 135–136
  models of, 175
  non-bridged, 133, 226, 227
  outer-sphere, 26, 83, 84, 99, 100, 106, 133, 134–136, 141, 143, 151, 161–171, 181, 187, 208, 210, 211, 217, 224, 228, 262, 284, 304
    activation enthalpies in, 107
    bridging, 145, 166, 167, 233
    catalysis and, 149, 166, 167
    chemical, 239
    continuum model, 189
    distinguishing from inner-sphere, 141
    energy terms, 204, 281
    experimental evidence, 135, 137
    facial attack, 170
    general case, 103
    isolation of binuclear complex, 137
    isomeric reaction products, 147
    linear free energy relationships, 149
    resonance energy, 185
    specific orientation of reactants, 169
    specific rates, 208
    sub-divisions, 166
    substitution controlled, 147
    systematic difference to inner-sphere, 154
  push-pull, 230
  quasi-resonant, 33
  resonant (*as opposed to* chemical), 224–225, 228–230, 240, 243, 244, 246, 247, 297
  resonant (between free atoms), 1–15, 22, 27, 29, 38, 44
    alternative reaction pathways, 13
    different ground state orbitals, 13
    extension of theory, 12
    extreme velocity ranges, 14
    highly excited reactants, 13
    low-angle scattering measurements, 10
    oscillatory dependence, 14
    oscillatory effects, 10
    qualitative description, 3
    total cross section measurements, 11
    translational electronic energy, 13
    two-state approximation, 5, 13
  single exchange, 224, 227
  specific orientation, 166, 169

Mechanisms of electron transfer (*cont.*)
  substitution-controlled, 106, 143, 147–148, 154
  superexchange, 224, 225, 229–238, 248, 251, 255
Medium effects *see also* Ionic atmosphere, 117, 257
Medium overlap case, 184, 185, 218, 220
Metal–ligand bond stretching, 186
Michaelis–Menten equation, 111, 124
Microscopic reversibility, 110
Mixed oxide systems, 305
Mixed valency and mixed valence compounds *see also* Average-valence compounds, 60, 99, 266, 268ff, 304, 305
  classification of, 269
Mixing coefficient, 272
Molecular beam methods, 4, 29, 30
Molecules, electron transfer involving, 26–43
  classification of mechanisms, 26
  factors influencing, 33
  followed by atom transfer, 32
  principal mechanisms, 28
  with atom transfer, 29
  without atom transfer, 28
Momentum transfer, 13
Multi-electron reagents, reaction of, 68, 76
  one-electron reagents, with, 68
  three-electron reagent, with, 76
  two-electron reagent, with, 72
Multiphonon model, 211
Multiple electron transfer, 14, 52
Multiple path reactions, 108
Multiple valency, ions of, 78
  reagents of, 68

Near-adiabatic, 18, 205
Net activation process, 50, 51, 262, 263
Net activation rate constants, 99, 109
Non-adiabatic electron transfer, 13, 17, 28, 33, 87, 177, 205, 215, 219, 226, 300
  chemical mechanism, 231
  cross sections, 21
  distinction from adiabatic transfer, 219
  energy curves for, 184
Non-bridged mechanism, 133, 226, 227
Non-bridging ligands, 106, 289
Non-complementary reactions, 52–68, 79, 82, 85, 89, 142
  bridging mechanisms, with, 142
  classification of, 53
  rate laws and mechanisms of, 56, 57, 58
Non-linear free energy relations, 213–214
Non-radiative transitions, 182
Normal modes, 203
N–Q transition, 292
Nuclear tunnelling *see* Tunnelling

Ohm's law, 298, 299

One-electron reagents, reaction of, 68
One-electron transfer, 52, 53, 54, 76–77, 80, 176, 178, 270
  description, 84
  oscillatory dependence of cross section, 14
Optical activity, effect on rates, 169
Optical charge transfer *see also* Intervalance charge transfer, 204, 274–296, 301
Optical charge transfer spectroscopy, 267
Optical electron transfer, 267–297
  bridged complexes, 285
  charge transfer absorption band, 275
  classification, 268
  classification by symmetry, 270
  connection with electron transfer kinetics, 272
  directly bonded systems, 291
  energy compared with thermal transfer, 309
  examples, 282
  photo-induced, 296
  reorganisation energies, 285, 295
  Robin and Day classification, 270
  solvent effects, 281
  structural classification, 268
  symmetrical inert complexes, 287
  theory, 271
  thermodynamic considerations, 277
  two state description, 271
Optical vibrations, 203
Orbital correlation, 40, 41
Orgel, 228
Oscillator strength, 276
Oscillatory effects due to resonance, 10
Outer-sphere charge transfer, 268
Outer-sphere complex, 100
  formation between oppositely charged ions, 126
Outer-sphere mechanism, 26, 83, 84, 99, 100, 106, 133, 134–136, 141, 143, 151, 161–171, 181, 187, 208, 210, 211, 217, 224, 228, 262, 284, 304
  activation enthalpies in, 107
  bridging, 145, 166, 167, 233
  catalysis, 149, 166, 167
  chemical mechanism, 239
  continuum model, 189
  distinguishing from inner-sphere, 141
  energy terms, 204, 281
  experimental evidence, 135, 137
  facial attack, 170
  general case, 103
  isolation of binuclear complex, 137
  isomeric reaction products, 147
  linear free energy relationships, 149
  resonance energy, 185
  specific orientation of reactants, 169
  specific rates, 208
  sub-divisions, 166
  substitution controlled transfer, 147
  systematic difference to inner-sphere, 154

Outer-sphere (reorganisation) energy, 204–205, 227, 281

Pendent groups, 247, 248
Penning ionisation, 36
Perturbation theory, 6, 8
Phonon model, 203, 205, 218, 227
Photo-electron spectrum, 36, 288
Photo-emission spectra, 283
Photo-induced electron transfer, 90, 128, 245, 296, 314
Photo-ionisation, 35
Polanyi, 28
Polarisability, 191, 193, 196, 223
Polarisation, 191, 193, 194, 199, 203
  inert and labile, 193, 204
  non-equilibrium, 193
Polarisation model, 223
Polarons,
  free small, 313
  large, 313
  small, 313
Polaron hopping, 309, 314
Polymerisation, 117
Potential, 298, 302
Potential wells, 181
Precursor complex, 98, 99, 100, 102, 103, 104, 107, 108, 125–128, 135, 137, 155, 178, 180, 181, 182, 196, 199, 240, 264, 267, 268, 270, 271, 277, 280, 300
  definition, 97
  formation and dissociation, 109
  isolation, 137, 155
  long lived, 135
  stability constants, 206
  substitutionally inert, 137, 138
Precursor electronic configuration, 264, 265
Predissociation, 231
Principle of similitude, 123, 200
Proton assisted electron transfer, 200, 214, 219
  examples, 256
Protons, in bridge electron transfers, 246
Pull-push, push-pull mechanism, 230

Quasi-free electron, 189, 283
Quasi-resonant mechanisms, 33

Radiationless transitions, 203
Rate constants, 62, 86
  calculation of, 179
  ceiling unimolecular, 102
  net activation, 109
  variations, 102
Rate expressions, for electron transfer, 112
Rate law, 50, 51, 58, 62, 64, 67, 71, 75
  correlation with free energy charge, 82
  correlation with redox potential, 83

Rate law (*cont.*)
  data required, 117
  distinctive features, 54
  fractional, 94
  indeterminate, 111, 112
  multiple-path reactions, of, 108
  non-complementary reactions, of, 56, 57
  non-integral, 95
  parameters, 51
  third order, 91
  transition state guessed from, 135
  unusual, 92, 93
  zero-order, 95
Rate measurements, 176
Reaction cross section *see* Cross section
Reaction mechanisms, 49
Reaction path, 49–96
  binuclear intermediates, 97–129
  catalysed and uncatalysed, 108, 114
  definitions, 97
  diffusion controlled mechanism, 100
  double electron transfer, 83
  free energy profiles, 80
  kinetics, 49
  labile intermediates,
    direct observation, 123
    kinetic evidence, 108
  non-complementary reactions, 52
  oxidation states, 49–96
  rate-determining steps, 104
  rate laws, 91
  reagents of multiple valence, 68
Reaction probability, 183
Reagents of multiple valence, 68
Rearrangement reactions, 34
Redox potentials, 207, 251, 285, 286
  correlation with rate law, 83
  distinction with electron transfer, 176
Reducibility of bridging group, 247, 251, 255
Refractive index, 193
Relative permittivity, 191
Remote attack, 134, 135, 136, 155–161, 162
  examples of, 156
  experimental evidence of, 155
  inert binuclear complexes in, 155
  organic ligands, 162–165
  primary product inferred from decomposition products, 160
  transfer of bridging group, 155
Reorganisation barrier *see also* Intrinsic free energy barrier, 241
Reorganisation effects, 40
Reorganisation energy, 168, 182, 204, 220, 227, 277, 281, 287, 293, 295, 297, 312
  ellipsoid model, 296
  separate sphere model, 295
Resistance, 298
Resonance, 3, 266, 271
  accidental, 23
Resonance condition, 180, 194
Resonance energy, 17, 22, 179, 184, 185, 265, 272, 294
  estimation of, 34
Resonance integral, 183, 184, 274, 282
Resonance transfer mechanism (*as opposed to* chemical transfer), 224–225, 228–230, 240, 243, 244, 246, 247, 297
Resonance transfer (between free atoms), 1–15, 22, 27, 29, 38, 44
  alternate reaction pathways, 13
  different ground state orbitals, 13
  extension of theory, 12
  extreme velocity ranges, 14
  highly excited reactants, 13
  low-angle scattering measurements, 10
  oscillatory dependence, 14
  oscillatory effects, 10
  qualitative description, 3
  total cross section measurements, 11
  translational electronic energy, 13
  two-state approximation, 5, 13
Retardation, 54, 55, 57, 59, 63, 71, 81, 82
Robin and Day classification, 270, 272

Scattering of reactants, 32
Schrödinger wave equation, 4, 5, 6
Seebeck, 302, 303, 304, 313
Self energy, 191, 193
Self-exchange, 205, 213
Semiconductors, 299, 302
  activation energy, 312
  controlled valency, 303
  extrinsic, 299
  intrinsic, 299
  n-type, 303
  p-type, 303
  temperature and, 302
Similitude, principle of, 123, 200
Single exchange mechanism, 224, 227
Small overlap approximation, 197
Small polaron, 314
Smoluchowski, 100
Soft acid, 153, 260
Solid state, electron transfer in, 299–314
  examples, 305
  theory, 299
Solution, electron transfer in, 49
Solvation energies, 196
Solvent cage, 87, 102
Solvent isotope effect, 154
Solvent polarity scales, 282
Specific orientation, 166, 169
Spectator stripping mechanism, 27, 32
Spherical ions,
  medium-range transfer, 201
  point dipole, 202
  short-range transfer, 201
State correlation, 40
Steady state conditions, 61, 86
Steady state kinetics, 51, 53, 69, 76, 85, 103

Index

Stereoselectivity, 170
Stevenson, 24
Stoicheiometry, 49
Stokes' law, 100
Stripping, 27, 28, 32
Strong overlap case, 84, 185
Stueckelberg, 20
Substitution controlled mechanism, 106, 147–148, 154
Successor complex, 99, 104, 108, 129–131, 135, 137, 178, 181, 196, 199, 267, 270, 271, 280
  decomposition, 130
  dissociation of, 104
  formation and dissociation, 109
  isolation of, 139, 155
  long lived, 135
  stability constants, 206
  substitutionally inert, 140
Successor electronic configuration, 264, 265
Successor state, 186
  definition, 97
Superexchange mechanism, 224, 225, 229, 238, 248, 251, 255
Sykes' equation, 118
Symmetrical exchange process, 99, 143
Symmetrical transitions, 270, 274
Symmetry, 114

Thermal mechanism, 179, 180, 274
  compared with optical, 309
Thexi-state, 279
Three-atom transition state, 60
Three-electron reagents, reaction with multi-electron reagent, 76
Three-electron transfer, 91
Three-state model, 224
Transfer probability, 13
Transition complex, 98
Transition moment, 135, 276, 292
Transition probability, 20, 182–185, 205, 218, 224, 225, 228, 231
Transition state *see also* Eyring, 27, 28, 49, 98, 107, 132, 185, 262, *et seq.*, 272
  cyclic, 108
  free energies of, 80
  guessed from rate law, 135
  theory *see also* Eyring, 50, 102, 103, 185, 196, 218, 219
  thermodynamic properties of, 107
  three-atom, 60

Transition state (*cont.*)
  two-atom, 61
Transition state theory, 28, 50
Transmission coefficient, 185, 218, 220, 227, 265
Transport number, 299
Trapped number, 300
Trapped valencies, 274, 278, 279, 288
Tunnelling,
  electron, 180, 243
  nuclear or system, 179, 215, 219
Two-atom transition state, 61
Two-body collisions, 62
Two-centre description, 271
Two-electron reagents, 52
  reaction with multi-electron reagent, 72
Two-electron transfer, 52, 53, 54, 55, 67, 74, 76–77, 80, 84, 88, 176
  bridging and, 143
  detection of intermediates, 86
  examples, 87
Two-state model, 5, 178, 204, 227, 271

Uncertainty principle, 18, 185

Van der Waals bonding, 294
Vertical electron attachment, 282, 293
Vertical-ionisation, 34, 282, 293
Vertical transitions, 274
Vibrations,
  acoustic, 203
  optical, 203
Virtual charge density, 200
Virtual probability distribution, 200
Virtual state, 224
Volume of activation, 154

Weak-overlap case, 183, 185, 226
Work of charging, 189
Work terms, 206–210, 218, 220, 293, 294, 295
  effects of, 207

X-irradiation, effect on exchange rate, 89, 90

Zener, 20, 177, 183, 185
Zero-order states, 42, 224

Index of oxidants, reductants and electron transfer intermediates

Oxidants

In most cases only the central atom and its formal oxidation state are shown.

Ag^{II} 60, 66–67, 150
Am^{VI} 122–123, 208
Ar^V 14
Ar^{II} 20, 44
Ar^I 9, 28, 33, 34, 39, 40
Au^{III} 56, 88, 90, 143
Au^{II} 91
Ba^{II} 38
Ba^I 38
Br_2 32–33, 34
CH_4^+ 28, 39
CO^+ 33, 38 (See also Organic molecules below)
Ce^{IV} 53, 58, 59, 60, 89, 92, 93, 130, 133, 150, 176, 212, 219
Cl^{VII} 148
Cl 20
$Co^{III}aq$ 70, 94, 104, 105, 135, 148, 149, 150, 212, 213, 214, 219, 280
$Co^{III}(NH_3)_5X)$ with various ligands X, including NH_3)   95, 104, 105, 106, 116, 117, 121, 125, 126, 127, 131, 133, 134, 136, 142, 144, 145, 149, 150, 151, 153, 154, 155, 156, 157, 158, 159–160, 161, 162–165, 166, 169, 170, 171, 219, 221, 228, 238–240, 241, 242, 243, 245, 246–247, 250, 253, 258–266, 297
$Co(en)_3^{3+}$ 125, 129, 154, 166, 167, 168, 176, 219
$Co(phen)_3^{3+}$ 125, 167, 168, 208, 221
Co^{III} (other complexes)   57, 59, 66, 90, 92, 93, 94, 107, 116, 119, 126, 131, 145, 150–151, 167, 175, 212, 213, 239, 258–266
Co^{II} 60, 168
Cr^{VI} 55, 63, 64, 66, 67, 76, 83, 84, 91, 92, 94, 125, 130, 134, 140, 142, 150, 286
Cr^V 55, 64, 66–68, 76, 83
Cr^{IV} 55, 64, 65, 67, 83, 141
Cr^{III} 53, 95, 117, 125, 128, 135, 141, 145, 150, 152, 153, 154, 155, 156, 157, 158, 159, 162–165, 170, 177, 209, 212, 219, 221, 226, 238, 239, 241, 242, 243, 244, 245, 246, 257, 258–266
Cs^I 29, 31, 40
Cu^{II} 286
Cu^I 23, 207
D_2^+ 30, 38
Eu^{III} 133, 176, 221, 262
Fe^{IV} 141
$Fe^{III}aq$ 57, 59, 70, 71, 80, 87, 90, 94, 115, 133, 135, 146, 150, 151, 152, 167, 168, 175, 177, 207, 209, 212, 214, 219, 221, 227, 228, 258–266, 286
$Fe(CN)_6^{3-}$ 105, 130, 131, 140, 148, 166, 167, 176, 177, 219, 221, 226, 280, 287, 297
$Fe(bpy)_3^{3+}$, $Fe(phen)_3^{3+}$ **and related complexes**   60, 105, 127, 167, 169, 208, 209, 212, 216, 218, 221
Fe^{III}(cytochrome-c)   167, 169
Fe^{III} (other complexes)   105, 107–108, 141, 146, 147, 156, 168, 221, 242, 258–266, 314
H^+ 1–9, 10, 11, 12, 22, 25, 134, 170
H 32, 35
H_2^+, HD^+ 30, 32, 33–34, 38, 40, 45 (See also D_2^+)
H_2 94
H_2O^+ see OH_2^+
He^{II} 12, 39
He^I 9, 10, 20, 23, 25, 28, 37, 39, 40
Hg^{II} 59–63, 74–76, 80, 82, 84, 88, 90, 134
Hg^I 9, 61, 74, 75, 82
I^+ 65, 67
I 28
ICH_3 30, 31
IOH, IOH_2^+ 65
In^{III} 56
In^{II} 53
Ir^{IV} 63, 105, 107, 126, 140, 141, 208
Ir^{III} 60, 258–266
Kr^{II} 25

Kr^I 14, 33, 39
Li^I 9, 292
Mn^{VII} 56, 69, 70, 71, 78, 92, 94, 105, 132, 134, 142, 148, 166, 167, 176, 208, 221, 226
Mn^{VI} 69, 70, 71, 78
Mn^V 69, 71
Mn^{IV} 58, 62
Mn^{III} 58, 60, 62, 71, 132, 148, 212, 219
Mo^{VI} 88
Mo^V 88, 105, 148, 208, 286
N^{2+} 28, 33, 38, 39
NO^+ 170
NO 37
NO_2 28, 36–37, 40
N_2O 38
Na^I 9, 20, 23
Nb^V 314
Ne^I 38, 39
Ni^{III} 312
Np^{VII} 60, 72, 78, 116
Np^{VI} 69, 78, 80, 92, 93, 95, 123, 125, 219
Np^V 69, 78, 92, 122
O^{VIII} 12
O^I 25
O 29
OH 58, 89
OH_2^+ 45
O_2^+ 38
O_2 29, 90, 166, 170, 296
Os^{III} 105, 148, 176
Pb^{III} 53
Pt^{IV} 56, 88, 89, 143, 144, 167
Pt^{III} 52, 53, 89

Pu^{VI} 56, 69, 71, 72, 77, 78, 80, 85, 88, 116, 119, 122, 123
Pu^V 53, 71, 72, 77, 78, 85
Pu^{IV} 67, 69, 71, 78
Rb^I 30, 31
Rh^{III} 257, 258–266
$Ru^{III}(NH_3)_5X$ (with various ligands X, including NH_3) 130, 140, 241, 242, 245, 258–266
Ru^{III} (other complexes) 60, 95, 130, 131, 140, 141, 148, 170
SF_6 29
Sb^V 88, 92, 98
Sn^{IV} 56, 59, 81, 88, 90
Sn^{III} 52, 53, 72, 81
$Ta_6Cl_{12}^{4+}$, $Ta_6Br_{12}^{4+}$ 69, 148
$Ta_6Cl_{12}^{3+}$, $Ta_6Br_{12}^{3+}$ 79, 148
Ti^{IV} 53, 70, 71, 287, 303
Tl^{III} 55–58, 73–74, 80, 81, 84, 85, 87, 88, 90, 91, 92, 93, 167, 176
Tl^{II} 53, 55–58, 73–74, 81, 87, 213
U^{VI} 56, 69, 71–72, 77, 78, 80, 88, 92, 98, 115–117, 122, 123, 140, 214, 221
U^V 53, 69, 72, 76, 77, 78, 80
U^{IV} 90
V^{VI} 57
V^V 68, 69, 70, 72, 73, 75, 78, 80, 84, 92, 98, 125, 140, 150
V^{IV} 68, 70, 72, 74, 75, 78, 79, 94, 140, 141, 177, 287
V^{III} 68, 70, 74, 116, 118, 125, 219, 221
W^V 94, 148
Xe^I 39
Organic molecules 30, 31, 44, 167, 169, 221, 246

Reductants

In most cases only the central atom and its formal oxidation state are shown.

Ag^I 64, 66, 69, 82, 83
Ar^I 20, 33, 44
Ar 9, 14, 23, 28, 34, 39, 40
Am^{IV} 122, 123
As^{III} 66
Au^{II} 91
Ba 38
Br^- 94, 148
CH_4 28, 39
CO 33, 38
CO_2^- 166, 169, 240 (See also Organic molecules below)
Cd^I 105

Ce^{III} 53, 56, 58, 64, 69, 81, 83, 89, 92, 93, 133, 176, 219
Ce^{II} 59
Cl^- 20, 148
Co^{II} 56, 57, 68, 69, 81, 82, 94, 116, 130, 131, 140, 141, 142, 149, 153, 154, 155–157, 166, 175, 212, 213, 219, 221, 258–266, 296
Co^I 59
Cr^V 55, 64, 67, 83
Cr^{IV} 64, 67, 76, 83
Cr^{III} 63, 64, 65–66, 83, 92, 94, 286
$Cr^{II}aq$ 31, 53, 55, 56, 60, 67, 68, 69, 70, 82, 84, 93, 95, 104, 105, 106, 107, 116, 117, 118, 119,

125, 129, 130, 131, 133, 134, 135, 136, 140, 141, 143, 144, 145, 146, 147, 149, 150, 151, 152, 153, 154, 155, 156, 157, 158, 159, 161, 162, 164, 165, 167, 168, 169, 171, 177, 207, 209, 211, 212, 213, 219, 221, 226, 230, 238, 239, 241, 242, 243, 244, 245, 246, 247, 249, 250, 256, 257, 258–266, 272

Cr^{II} (other complexes) 56, 92, 93, 94, 106, 149, 150, 153, 154, 167, 170, 221

Cs 29, 30, 31, 36, 40

Cu^{I} 68, 124, 129, 150, 154, 245, 258, 286, 297

D_2 38

e^-aq 212, 239

Eu^{II} 68, 69, 104, 106, 133, 135, 136, 150, 151, 152, 154, 165, 176, 213, 214, 221, 246, 256, 258–266

$Fe^{II}aq$ 56, 57, 59, 60, 64, 65, 68, 69, 70, 71, 73, 81, 82, 83, 89, 92, 94, 98, 104, 106, 107, 108, 116, 119, 124, 125, 126, 127, 133, 141, 145, 147, 150, 151, 153, 167, 168, 169, 175, 176, 177, 208, 209, 212, 216, 218, 219, 221, 242, 258–66, 280, 286, 287, 297

$Fe(bpy)_3^{2+}$, $Fe(phen)_3^{2+}$ and related complexes 56, 64, 68, 69, 70, 81, 83, 92, 105, 107, 140, 208, 209, 212, 221

$Fe(CN)_6^{4-}$ 56, 64, 68, 69, 124, 125, 126, 129, 131, 135, 140, 166, 167, 176, 177, 208, 219, 221, 226, 286

Fe^{II} (cytochrome-c) 169

Fe^{II} (other complexes) 56, 64, 68, 106, 138, 140, 154, 168, 212, 221, 314

H 1–9, 10, 11, 12, 14, 24, 25, 105, 134

H_2 28, 30, 32, 33, 38, 40

HD 34

H_2O see OH_2

H_2O_2 see O_2H_2

H_3^+ 32

HN_3 see N_3^-

HNO_2 see NO_2^-

He^{I} 12

He 9, 10, 12, 20, 22, 25, 28, 37, 39, 40

He_2 39

$(Hg^{I})_2$ 59–63

Hg^{I} 59–63, 75, 82

Hg 9, 59–63, 72, 75, 88, 134, 170

I^- 28, 36, 37, 65, 67, 68, 94

In^{II} 53

Ir^{III} 56, 64, 68, 105, 125, 127, 130, 135

K 30, 32, 33, 34, 36

Kr 14, 25, 33, 39

Li 9, 23, 36

Mn^{VI} 71, 78, 105, 132, 148, 166, 167, 176, 221, 226

Mn^{III} 62

Mn^{II} 56, 58, 62, 69, 81, 82, 92, 94, 106, 132, 219

Mo^{V} 63

Mo^{IV} 64, 69, 105, 140, 208, 286

NCS^- See SCN^-

NH_3 39

NH_3OH^+ 148

NO_2^-, HNO_2 28, 36, 40, 148

NO_3^- 58

N_2 28, 33, 37, 38, 39

$N_2H_5^+$ 148

N_2O 38

N_3^-, HN_3 148

Na 9, 20, 23, 36, 292

Ne 38, 39

Ni^{II} 23, 312

Np^{VI} 56

Np^{V} 64, 69, 79, 82, 83, 125, 208, 214, 219

Np^{IV} 69, 79, 92, 122, 123

Np^{III} 68, 69, 79, 80

O^{VII} 12

O 25

O^- 29

OH^- 89

OH_2 45, 58

O_2H_2 148

O_2 38

O_2^- 29, 314

$O_2[Co(CN)_5]_2^{6-}$ 297

Os^{II} 56, 105, 148, 176, 208

PH_3 39

Pb^{III} 53

Pt^{III} 52, 53, 89

Pt^{II} 88, 89, 143

Pu^{V} 53, 76, 79, 93, 95

Pu^{IV} 64, 76, 77, 79, 83, 122, 123, 140

Pu^{III} 64, 67, 76, 79, 80, 82, 83

Rb 31

$Ru^{II}(NH_3)_5X$ (with various ligands X including NH_3) 68, 107, 142, 150, 151, 154, 166, 170, 212, 287

Ru^{II} (other complexes) 56, 95, 135, 142, 148, 170

S^{IV} 66

SCN^- 144, 148

SF_5 29

Sb^{III} 78, 88, 90, 92

SiH_4 39

Sn^{III} 52, 53, 59, 72, 82

Sn^{II} 59, 72, 73, 80, 84, 85, 88, 90, 91

$Ta_6Cl_{12}^{3+}$, $Ta_6Br_{12}^{3+}$ 69, 78

$Ta_6Cl_{12}^{2+}$, $Ta_6Br_{12}^{2+}$ 64, 78, 92, 148

Ti^{III} 53, 56, 68, 69, 70, 71, 72, 116, 130, 135, 154, 246, 258–266, 303

Ti^{II} 53

Tl^{II} 53, 57, 58, 74, 82, 87, 167, 168, 213

Tl^I 55, 58, 72, 87, 88, 89, 90, 116, 176
U^V 53, 69, 71, 72, 76, 77, 78, 80, 82, 93, 125, 221
U^{IV} 56, 69, 72, 76, 77, 78, 80, 84, 85, 88, 90, 92, 98, 116, 122, 123, 142
U^{III} 78, 80, 92, 104, 105, 116, 213, 214, 258–266
V^{IV} 56, 57, 64, 66, 68, 69, 70, 71, 72, 73, 75, 76, 78, 81, 82, 83, 150, 151
V^{III} 56, 60, 63, 64, 66, 68, 69, 70, 72–74, 75, 76, 77, 78, 82, 83, 115, 116, 126, 177, 219
V^{II} 56, 60, 62, 64, 68, 69, 70, 72, 73, 74, 76, 77, 78, 80, 82, 83, 93, 95, 116, 121, 126, 131, 135, 140, 141, 145, 147, 148, 149, 150, 151, 152, 154, 165, 167, 168, 170, 171, 208, 211, 212, 213, 221, 242, 246, 247, 256
W^{IV} 69, 148
Xe 39
Yb^{II} 150, 258–266
Organic molecules 28, 39, 44, 45, 148, 167, 169, 221, 240

Electron transfer intermediates

This list comprises precursor complexes, successor complexes intervalence charge transfer ground states, and mixed-valence compounds. In most cases only the two central atoms characterising the oxidising and reducing centres, and their formal oxidation states, are shown. In each case the potentially oxidising centre is written first.

Ag^I–Ag^0 269
Ar^I–H_2 42
Au^{III}–Ag^I 307, 308
Au^{III}–Au^I 307
Au^{III}–Cu^{II} 307
Au^{III}–Pd^{II} 307
BF_3–NMe_3 292
Bi^V–Bi^{III} 276
Ce^{IV}–Ce^{III} 306
Co^{III}–Co^{II} 269, 306
Co^{III}–Cr^{II} 106, 116, 120, 121
Co^{III}–Cu^I 125, 129, 297
Co^{III}–Fe^{II} 104, 125, 126, 127, 128, 129, 131, 135, 138, 139, 140, 155, 269, 296
Co^{III}–I^- 268
Co^{III}–Ir^{III} 140
Co^{III}–(organic radical) 238
Co^{III}–O_2^{2-} 306
Co^{III}–Ru^{II} 135, 137, 297
Co^{III}–Ti^{III} 116
Co^{III}–U^{III} 116
Co^{III}–V^{II} 116, 121, 126
Co^{III}–(various reductants) 306
Cr^{VI}–Cr^{III} 269, 286
Cr^V–(various reductants) 84
Cr^{III}–Co^{II} 106
Cr^{III}–Cr^{III} 119, 141
Cr^{III}–Cr^{II} 125, 129, 244, 269, 270, 272
Cr^{III}–Fe^{II} 64
Cr^{III}–Ir^{III} 135, 140
Cr^{III}–Mo^V 140
Cr^{III}–Pu^V 76, 93, 95, 140
Cr^{III}–Ru^{II} 131, 140, 155
Cr^{III}–U^V 93, 140
Cr^{III}–V^{III} 69, 70
Cr^{III}–V^{II} 116, 118–119
Cu^{II}–Cu^I 309
Cu^{II}–Fe^{II} 289
Eu^{III}–Eu^{II} 306
Fe^{IV}–Fe^{III} 306
Fe^{III}–Co^{II} 104
Fe^{III}–Fe^{II} 98, 108, 141, 268, 269, 270, 278, 286, 287, 290, 291, 305, 306, 309, 310
Fe^{III}–Ir^{III} 306
Fe^{III}–Mo^{IV} 286
Fe^{III}–Nb^{IV} 315
Fe^{III}–Pu^V 116, 119
Fe^{III}–Tl^I 269, 278, 306
H^+–Cl^- 269
H^+–H 5, 134
H^+–H^- 269
H–Co^{II} 94
H^+–Ru^{II} 170
H_2^+–He 41
H_2^+–Ar 42
H_2^+–Ne 42
H_2–$[Co^{II}]_2$ 94
He^I–H_2 41
Hg^{II}–Hg^I 60, 61
Hg^{II}–Hg^0 134, 135, 269
Li^I–Na 292
Mn^{VII}–Ag^I 269, 286

Mn^{IV}–Mn^{III} 269
Mo^{VI}–Mo^{V} 63
Mo^{VI}–V^{IV} 277, 278
Mo^{V}–Fe^{II} 286
Nb^{V}–Fe^{II} 314
Ne^{L}–H$_2$ 42
Ni^{III}–Ni^{II} 306, 309, 311
Np^{VII}–Co^{II} 116
Np^{VI}–Np^{V} 125
Os^{III}–Os^{II} 289
P^{V}–P^{III} 269
Pb^{IV}–Pb^{II} 269, 309
Pd^{IV}–Pd^{II} 307
Pr^{IV}–Pr^{III} 306
Pt^{IV}–Pd^{II} 307
Pt^{IV}–Pt^{II} 269, 305, 307
Pu^{V}–U^{V} 116
Rh^{III}–Rh^{II} 269
Ru^{III}–Br^{-} 285, 296
Ru^{III}–Cl^{-} 285, 296
Ru^{III}–I^{-} 269, 285, 296
Ru^{III}–H 170

Ru^{III}–NCS^{-} 285
Ru^{III}–Ru^{II} 269, 278, 288, 289
Ru^{IV}–S$_2$O$_3^{2-}$ 285
Sb^{V}–Sb^{III} 98, 269, 270, 276, 277, 278, 304, 309
Sn^{IV}–Sn^{II} 90
Ti^{IV}–Fe^{II} 277, 287
U^{V}–Pu^{V} 116
U^{V}–U^{V} 80, 98
U^{V}–U^{IV} 116
V^{V}–Mn^{V} 71
V^{V}–U^{V} 125
V^{V}–V^{IV} 269
V^{IV}–Fe^{III} 140
V^{IV}–Fe^{II} 140
V^{IV}–Ru^{II} 287
V^{IV}–V^{III} 306
V^{III}–V^{III} 140, 141
W^{VI}–V^{IV} 277, 278
W^{VI}–W^{V} 268
(Organic cation)–I^{-} 282
Organic charge-transfer complexes 292, 293
Organic radical-ions 291